Uncertainty Modeling in Vibration, Control and Fuzzy Analysis of Structural Systems

With a Foreword by Lotfi Zadeh

SERIES ON STABILITY, VIBRATION AND CONTROL OF SYSTEMS

Series Editors: Ardéshir Guran & Daniel J. Inman

About the Series

Rapid developments in system dynamics and control, areas related to many other topics in applied mathematics, call for comprehensive presentations of current topics. This series contains textbooks, monographs, treatises, conference proceedings and a collection of thematically organized research or pedagogical articles addressing dynamical systems and control.

The material is ideal for a general scientific and engineering readership, and is also mathematically precise enough to be a useful reference for research specialists in mechanics and control, nonlinear dynamics, and in applied mathematics and physics.

Selected Forthcoming Volumes

Series A. Textbooks, Monographs and Treatises

Dynamics of Gyroscopic Systems: Flow Induced Vibration of Structures, Gyroelasticity, Oscillation of Rotors and Mechanics of High Speed Axially Moving Material Systems
Authors: A. Guran, A. Bajaj, G. D'Eleuterio, N. Perkins and Y. Ishida

Adaptive Control of Nonlinear Systems
Authors: R. Ghanadan, A. Guran and G. Blankenship

Series B. Conference Proceedings and Special Theme Issues

Acoustic Interactions with Submerged Elastic Structures
Part II: Propagation, Ocean Acoustics and Scattering
Editors: A. Guran, G. Maugin and J. Engelbrecht

Part III: Acoustic Propagation and Scattering, Wavelets and Time-Frequency Analysis
Editors: A. Guran, A. de Hoop and D. Guicking

Part IV: Non-destructive Testing, Acoustic Wave Propagation and Scattering
Editors: A. Guran, A. Boström, A. Gerard and G. Maze

Structronic Systems: Smart Structures, Devices, and Systems (Part II)
Editors: H.-S. Tzou, A. Guran, G. L. Anderson and M. Natori

Proceedings of the First European Conference on Structural Control
Editors: A. Baratta and J. Rodellar

Nonlinear Dynamics: The Richard Rand 50th Anniversary Volume
Editor: A. Guran

SERIES ON STABILITY, VIBRATION AND CONTROL OF SYSTEMS

Series B Volume 10

Series Editors: Ardéshir Guran & Daniel J Inman

Uncertainty Modeling in Vibration, Control and Fuzzy Analysis of Structural Systems

With a Foreword by Lotfi Zadeh

Editors

Bilal M. Ayyub
Dept. of Civil Engineering
University of Maryland
College Park, Maryland

Ardéshir Guran
Electrical Engineering
University of Southern California
Los Angeles, California

Achintya Haldar
Dept. of Civil Engineering
University of Arizona
Tucson, Arizona

World Scientific
Singapore • New Jersey • London • Hong Kong

Published by

World Scientific Publishing Co. Pte. Ltd.
P O Box 128, Farrer Road, Singapore 912805
USA office: Suite 1B, 1060 Main Street, River Edge, NJ 07661
UK office: 57 Shelton Street, Covent Garden, London WC2H 9HE

Library of Congress Cataloging-in-Publication Data
Uncertainty modeling in vibration, control and fuzzy analysis of
structural systems / editors, Bilal M. Ayyub, Ardéshir Guran,
Achintya Haldar.
p. cm. -- (Series on stability, vibration, and control systems, Series B. ; v. 10)
Includes index.
ISBN 9810231342
1. Structural dynamics -- Mathematical models. 2. Uncertainty -
- Mathematical models. 3. Structural control (Engineering) 4. Fuzzy
systems. I. Ayyub, Bilal M. II. Guran, A. (Ardéshir)
III. Haldar, Achintya. IV. Series.
TA654.U53 1997
624.1'71'015192 -- dc21 97-26414
CIP

British Library Cataloguing-in-Publication Data
A catalogue record for this book is available from the British Library.

This book is printed on acid-free paper.

Printed in Singapore by Uto-Print

STABILITY, VIBRATION AND CONTROL OF SYSTEMS

Editor-in-chief: Ardéshir Guran
Co-editor: Daniel J. Inman

Advisory Board

Henry Abarbanel
University of California
San Diego
USA

Gary Anderson
Army Research Office
Research Triangle Park
USA

Jorge Angeles
McGill University
Montréal
CANADA

Leon Bahar
Drexel University
Philadelphia
USA

Anil Bajaj
Purdue University
Lafayette
USA

Anders Boström
Chalmers Technical Univ.
Goteborg
SWEDEN

Rafael Carbó-Fite
C.S.I.C.
Madrid
SPAIN

Fabio Casciati
Universitá di Pavia
Pavia
ITALY

Juri Engelbrecht
Estonian Academy of Sci.
Tallin
ESTONIA

Lucia Faravelli
Università di Pavia
Pavia
ITALY

Toshio Fukuda
Nagoya University
Nagoya
JAPAN

Alain Gérard
Laboratoire de Mécanique
Bordeaux
FRANCE

Hans Irschik
Johannes Kepler Universität
Linz
AUSTRIA

Heikki Isomäki
Helsinki Univ. of Tech.
Helsinki
FINLAND

Jer-nan Juang
Langley Research Center
Hampton
USA

John Junkins
Texas A&M University
College Station
USA

Youdan Kim
Seoul National University
Seoul
KOREA

James Knowles
California Inst. of Tech.
Pasadena
USA

Edwin Kreuzer
Technische Universität
Hamburg-Harburg
GERMANY

Oswald Leroy
Catholic University of
Louvain
BELGIUM

Jerrold Marsden
California Inst. of Tech.
Pasadena
USA

Hans Natke
Universität Hannover
Hannover
GERMANY

Paul Newton
University of S. California
Los Angeles
USA

Michihiro Natori
Inst. of Space & Astro.
Kanagwa
JAPAN

Friedrich Pfeiffer
Technische Universität
München
GERMANY

Raymond Plaut
Virginia Poly. Inst.
Blacksburg
USA

Karl Popp
Universität Hannover
Hannover
GERMANY

Richard Rand
Cornell University
Ithaca
USA

Jean Ripoche
University of Le Havre
Le Havre
FRANCE

Kazimirez Sobczyk
Polish Academy of Sci.
Warsaw
POLAND

Jon Juel Thomsen
Tech. Univ. of Denmark
Lyngby
DENMARK

Horn-Sen Tzou
University of Kentucky
Lexington
USA

Herbert Überall
Catholic Univ. of America
Washington
USA

Firdaus Udwadia
University of S. California
Los Angeles
USA

Dick van Campen
University of Technology
Eindhoven
NETHERLANDS

Jörg Wauer
Technische Universität
Karlsruhe
GERMANY

Joanne Wegner
University of Victoria
Victoria
CANADA

James Yao
Texas A&M University
College Station
USA

Lotfi Zadeh
University of California
Berkeley
USA

Franz Ziegler
Technische Universität
Wien
AUSTRIA

Foreword

There is hardly a domain of human activity in which uncertainty -- in one form or another -- does not play a significant role. This is especially true in the realm of design -- a realm in which most decisions are made in an environment of partial knowledge, partial uncertainty and partial truth.

And yet, as a scientific/engineering methodology, modeling of uncertainty does not have a long history. In part, this is due to the fact that the underlying issues are quite complex and that substantial computational resources are needed to translate theory into reality. Such resources have not been available in the past.

In recent years, substantial progress has been made in our ability to model and deal with uncertainty through computationally-oriented methods and techniques. It is this progress that is reflected in the highly informative and authoritative volume "Uncertainty Modeling in Vibration, Control and Fuzzy Analysis of Structural Systems," edited by Professors B. Ayyub, A. Guran and A. Haldar. For simplicity I will be referring to this volume as AGH.

In science and engineering, there is a long-standing tradition of assuming that probability theory is the only available tool for dealing with uncertainty. In recent years, however, the validity of this tradition has been called into question by the development of other tools -- centered largely on possibility theory and fuzzy logic -- for dealing with forms of uncertainty which are non-statistical in nature. Nevertheless, there are still some who claim that anything that can be done with alternative methodologies -- and especially possibility theory and fuzzy logic -- can be done equally well or better with probability-based methods.

The position of those who advocate the use of alternative methodologies -- and I am one of them -- is not that the alternative methodologies should replace probability-based methods. Rather, what is suggested is addition of the tools provided by the alternative methods to the tool chest of probability and statistics. In particular, in the case of possibility theory and fuzzy logic, the argument is that probability theory, on one side, and possibility theory and fuzzy logic on the other, are for the most part complementary and synergistic rather than competitive.

The collection of methods described in AGH reflect this philosophy. Of the ten chapters, four are given over to the exposition of probability-based methods; five relate to fuzzy - logic-based methods and genetic computing; and the last chapter provides a very insightful comparison between probabilistic and fuzzy-logic-based methods. In this way, the editors present a complete spectrum of methods that are available at this juncture for modeling and managing uncertainty in the realms of vibration, control and the analysis of structural systems.

There is an important and frequently misunderstood point that I should like to comment on -- a point which relates to the connection between probability theory and fuzzy logic.

In my view, there are two basic types of uncertainty: probabilistic uncertainty and possibilistic uncertainty. In this perspective, probability theory deals with probabilistic uncertainty while possibilistic theory deals with possibilistic uncertainty. An example of the latter is : John's income is between fifty and sixty thousand dollars. In this case, the interval [50 K, 60 K] is the possibility distribution of John's income, that is , the set of its possible values. There is no randomness in this example and there is no randomness in a fuzzified version of it: John's income is approximately between fifty and sixty thousand dollars. In this version of the example, the possibility distribution is a fuzzy set and possibility is a number in the interval [0, 1].

A point that is not widely recognized is that what fuzzy logic provides -- and no other methodology does so -- is a machinery for fuzzy information granulation. This machinery can be used in any theory, including probability theory.

There are two basic concepts which underlie fuzzy information granulation: they are the concepts of linguistic variable and fuzzy rule set. These concepts provide a basis for what might be called the calculi of fuzzy rules (CFR) and fuzzy graphs (CFG). Although they are not referred to as such in AGH, the calculi of fuzzy rules and fuzzy graphs play a prominent role in the chapters which deal with fuzzy-logic-based methods. In AGH, the exposition of these methods is clear, succinct and profusely illustrated by examples.

As was alluded to earlier, an interesting direction which falls beyond the scope of AGH involves application of the machinery of fuzzy information granulation to probability theory and probability-based methods. In this direction, linguistic variables and fuzzy rule sets are employed to describe fuzzy events and fuzzy probability distributions. This is one of the ways in which infusion into probability theory of concepts drawn from fuzzy logic can enrich probability theory and enhance its effectiveness in dealing with real-world problems. I do not expect that this vision will appeal to those who believe that standard probability theory is both necessary and sufficient for modeling uncertainty in all of its varied manifestations.

What is not debatable is the "Uncertainty Modeling in Vibration, Control and Fuzzy Analysis of Structural Systems," makes an important contribution to our understanding of how both probability-based and fuzzy-logic-based methods for modeling of uncertainty can be applied in the realms of vibration, control, structural analysis and related problem areas. The editors and the contributors to information volume deserve our thanks and congratulations for producing a work that is highly informative, authoritative, up-to-date and reader-friendly.

Lotfi A. Zadeh
Berkeley, California
February 2, 1997

Preface

The term uncertainty in the context of vibration, control, and fuzzy analysis of structural and mechanical systems can be defined as the lack of regularity in a system configuration, its output, and our knowledge about it. A strong link between information, its nature, and uncertainty exists. This uncertainty dimension can be observed in physical realization of most objects that are defined in a space-time context including structural and mechanical systems. The functionality of many modern structural and mechanical systems operating in real-world environment depends to a large extent on their ability to perform adequately and with a high level of reliability under uncertain conditions. Uncertainties in these systems result from such diverse phenomena as engine noise, turbulent flow, wind-induced irregular oscillations, soil variability, earthquake induced ground motion, rough pavement, random ocean waves, thermal and acoustic loading, and fatigue brought on by random vibration stresses.

This book gives an overview of the current state of uncertainty modeling in vibration, control, and fuzzy analysis of structural and mechanical systems. It is a coherent compendium written by leading experts and offers the reader a sampling of exciting research areas in several fast-growing branches in this field. Uncertainty modeling and analysis are becoming an integral part of system definition and modeling in many fields. The book consists of ten chapters that report the work of researchers, scientists and engineers on theoretical developments and diversified applications in engineering systems They deal with modeling for vibration, control, and fuzzy analysis of structural and mechanical systems under uncertain conditions. The book was designed for readers who are familiar with the fundamentals and wish to study a particular topic or use the book as an authoritative reference. It gives readers a sophisticated toolbox for tackling modeling problems in mechanical and structural systems in real-world situations. The book is a part of a series on Stability, Vibration and Control of Structures, and provides vital information in these areas.

The ten chapters of the book were contributed by twenty-two researchers from six countries worldwide. The first four chapters deal with stochastic vibration and dynamics of engineering systems. Schuëller and Pradlwarter provide, in chapter one, a primer for practical modeling and methods for the analysis of structural systems under stochastic dynamic loading. Procedures for simulating stochastic processes are presented for dealing with nonlinear stochastic dynamic problems. The second chapter, by Naess, deals with clumping effect approximations for narrow-band random vibration. The suggested approximate method account for the clumping effect of large peaks on first passage times and extreme values for both Gaussian and non- Gaussian processes by utilizing the joint crossing rates of the processes. Therefore, a correlation structure can be introduced. The third chapter, by Frey, deals with the dynamics of nonlinear systems by modeling the state space transport across separatrices. He shows that the separatrices might be breached in cases that involve the presence of weak external forces; therefore allowing the system to change dynamical modes. He models the complex dynamics of the system using the flux factor which is a leading order expression of the state space flux accross the separatrices. The models presented in this chapter can be used in dynamical control and analysis of relative stability. Kareem, Zhao, Tognarelli, and Gurley suggests the method of equivalent statistical linearization for dealing with nonlinear stochastic systems in chapter four of the book. According to this method, the nonlinearities in the excitation of practical systems are cast in an equivalent quadratic or cubic form rather than in an equivalent linear form. The authors close this chapter by a comparative examination of their techniques with other methods in the literature.

Chapters five to eight deal with modeling uncertainties in the control of structural and mechanical systems. In chapter five, the elements of fuzzy structural control are provided by Faravelli and Yao. The authors described and used fuzzy logic controllers in civil engineering applications that included both linear and nonlinear systems. Then, uncertainties in structural control are broadly discussed and described by Yao and Yao in chapter six. The sources of uncertainty in control are described in the context of civil engineering structures. Also, Yao and Yao outline some open questions in this area. Hassan and Ayyub provide a detailed methodology on structural fuzzy control in chapter seven. A general framework is described for the fuzzy control of structural systems. In addition to a suggested fuzzy control, Hassan and Ayyub describe a self-learning unit that can be used to define new rules in the fuzzy controller. The suggested method can be used to control structural systems to meet any stated objective of interest. In chapter eight, an application of fuzzy logic to structural vibration control is suggested by Furuta, Okanan, Kaneyoshi, and Tanaka. The method is suggested for the structural vibration control of earthquake-excited and wind- excited oscillations. Numerical examples are used to illustrate the usefulness of the suggested method in reducing the amplitude and acceleration of structural oscillations. The authors used genetic algorithms for self-tuning of the inference rules and their parameters that are used in the fuzzy controller.

Maglaras, Nikolaidis, Haftka, and Cudney provide an experimental comparison of probabilistic methods and fuzzy-set methods for designing under conditions of uncertainty in chapter 9. A cantilever truss structure is used as a test case. The location and tuning of viscoelastic dampers for structural vibration control were optimized using probabilistic and fuzzy-set methods for the purpose of a comparative evaluation. The optimized truss was then evaluated in terms of its reliability, and the methods were evaluated.

In chapter ten, Guran presents basic concepts and review of recent developments in autonomous intelligent control techniques, including neural network control, fuzzy logic control, neural fuzzy control, and learning control. The detailed design schemes, algorithms and procedures are presented in this chapter.

In conclusion, this book provides the fruits of a team effort by experts in the fields of uncertainty modeling of structural and mechanical systems. Put together, it provides the reader with a wealth of insight and information, and a unique global perspective in these fields. This volume will be updated periodically to include new developments in the areas presented herein, and also will be extended in scope to new areas using handbook-style chapters. The editors invite comments from the book readers to ensure suitable coverage of these fields and topics in future volumes.

Bilal M. Ayyub
College Park, Maryland

Ardéshir Guran
Encino, California

Achintya Haldar
Tucson, Arizona

Contributors

Bilal M. Ayyub
Department of Civil Engineering
University of Maryland
College Park, MD 20742
USA

Dan Boghiu
Department of Mechanical Engineering
Auburn University
Auburn, AL 36849
USA

Harley H. Cudney
Department of Aerospace Engineering
Virginia Polytechnic Institute
Balcksburg, VA 24061
USA

Lucia Faravelli
Department of Structural Mechanics
University of Pavia
via Abbiategrasso 211, 27100 Pavia
Italy

Michael Frey
Department of Mathematics
Bucknell University
Lewisburg, PA 17837
USA

Hitoshi Furuta
Department of Informatics
Kansai University
2-1-1, Ryozenji-Cho, Takatsuki
Osaka 569
Japan

Ardéshir Guran
American Structronics and Avionics
16661 Ventura Blvd
Encino, CA 91436
USA

Kurtis R. Gurley
Department of Civil Engineering
University of Notre Dame
Notre Dame, IN 46556-0767
USA

Rafael T. Haftka
Department of Aerospace Engineering
Virginia Polytechnic Institute
Balcksburg, VA 24061
USA

Maguid H.M. Hassan
Department of Civil Engineering
Higher Technological Institute, Box 228
10th of Ramadan City
Egypt

Masayoshi Kaneyoshi
Bridge Engineering Department
Hitachi Zosen Corporation
Sakai City
Japan

Ahsan Kareem
Department of Civil Engineering
University of Notre Dame
Notre Dame, IN 46556-0767
USA

George Maglaras
Department of Aerospace Engineering
Virginia Polytechnic Institute
Balcksburg, VA 24061
USA

Dan B. Marghitu
Department of Mechanical Engineering
Auburn University
Auburn, AL 36849
USA

Arvid Naess
Faculty of Civil Engineering
The Norwegian Institute of Technology
Trondheim N-7034
Norway

E. Nikolaidis
Department of Aerospace Engineering
Virginia Polytechnic Institute
Balcksburg, VA 24061
USA

Hiroo Okanan
Department of Construction Engineering
Osaka Prefectural Technical College
Japan

H.J. Pradlwarter
Institute of Engineering Mechanics
University of Innsbruck
Technikerst. 13, Innsbruck A-6020
Austria

Gerhart I. Schuëller
Institute of Engineering Mechanics
University of Innsbruck
Technikerst. 13, Innsbruck A-6020
Austria

Hiroshi Tanaka
Bridge Engineering Department
Hitachi Zosen Corporation
Sakai City
Japan

Michael A. Tognarelli
Department of Civil Engineering
University of Notre Dame
Notre Dame, IN 46556-0767
USA

James T-P. Yao
Department of Civil Engineering
Texas A&M University
College Station, TX 77843-3136
USA

Timothy Yao
Impact Forecasting, L.L.C.
230 West Monroe Street, 19th Floor
Chicago, IL 60606
USA

Jun Zhao
Barnett & Casbarian, Inc.
Houston, TX 77024-1596
USA

Contents

Uncertainty Modeling in Vibration and Fuzzy Analysis of Structural Systems, pp. 1–27
edited by B. M. Ayyub, A. Guran and A. Haldar
Series on Stability, Vibration and Control of Systems Series B: Vol. 10

STOCHASTIC STRUCTURAL DYNAMICS - A PRIMER FOR PRACTICAL APPLICATIONS

H.J. PRADLWARTER and G.I. SCHUËLLER
Institute of Engineering Mechanics
University of Innsbruck, A-6020 Innsbruck, Austria

ABSTRACT

Methods for the analysis of structural systems under stochastic dynamic loading are presented. First a brief review of the theoretical background along with procedures to simulate stochastic processes is given. In a next step the methodology for calculating the response of MDOF-systems under stochastic loading is discussed. This section closes with a treatment of procedures of non-linear stochastic dynamics where it is focussed on larger systems as generally encountered in engineering practice. The efficiency of the various computational procedures is highlighted.

1. Introduction

In structural mechanics time variant properties of the loading histories, in connection with the properties of the structures to be analyzed, may play an important role in view of the characteristics of their response. In other words, structural systems affected by time variant properties such as inertia effects fall into the class of structural dynamical problems. This classification, of course, applies as well when considering the stochastic characteristics of structural properties. In this section some basic principles of linear and nonlinear stochastic dynamics are presented. For more extensive information the reader is referred to e.g. [1–6].

In static analysis, for example, statistical fluctuations of systems loading and/or properties can be characterized by a set of random variables $\mathbf{X}^T = (X_1, X_2, .., X_n)^T$ whereas in stochastic structural dynamics the loading as well as the statistically varying environmental effects are considered as randomly time varying functions. Typical examples are traffic loading, earthquake excitation, wind and sea wave loading respectively. In fact, most phenomena observed in nature are to some extent unpredictable in a deterministic sense and also a function of time. Hence, the components X_i of the set describing random effects are not simple variables, but random functions $X_i(t)$ of time denoted as stochastic processes.

2. Stochastic Processes

2.1. General Considerations

A stochastic process is formally described by its joint distribution for an arbitrary number of instances t_i, i=1, 2, ..., n, defined by the probability

$$F_{\mathbf{X}(t_1)\mathbf{X}(t_2)\ldots\mathbf{X}(t_n)}(x_1, x_2, .., x_n) = P(\mathbf{X}(t_1) \le \mathbf{x}_1, \mathbf{X}(t_2) \le \mathbf{x}_2, \ldots.., \mathbf{X}(t_n) \le \mathbf{x}_n) \quad (1)$$

where $F_{\mathbf{X}}$ denotes the multidimensional cumulative distribution function and P denotes the probability. This, of course, is a formal definition of a stochastic process in the sense that any finite amount of data would not suffice for a *complete* description of the above multidimensional joint distribution. Some simplifying assumptions are generally required to make the mathematical theory of stochastic processes accessible to observed data and practical applications. In other words, eq. (1) implies an infinite amount of information which can be never realistic particularly for practical applications. Hence the task of the structural engineer is to make reasonable justifiable assumptions concerning the stochastic process in order to develop a workable model, representing all the available information in a consistent way.

As typical examples of loadings modeled as stochastic processes, measured wind velocity and an observed earthquake accelerations are shown in Fig. 1.

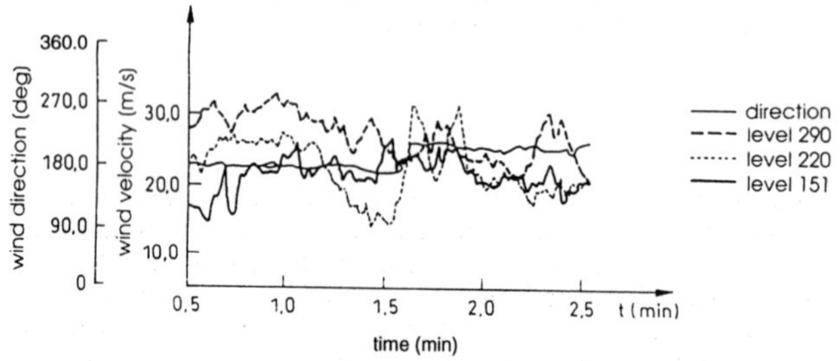

(a) Wind Velocity

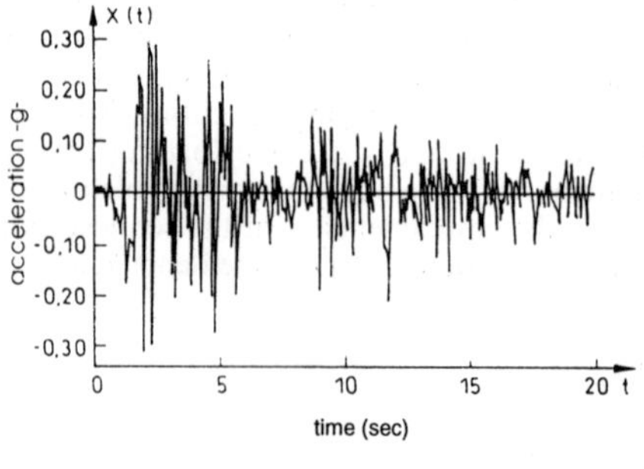

(b) Earthquake Excitation

Fig. 1: Measured Wind Velocity and Earthquake Excitation.

To model such a random behavior, the measured time series is regarded as a realization of an underlying stochastic process. A stochastic process itself might be specified by a (infinite) ensemble of such realizations, as sketched in Fig. 2.

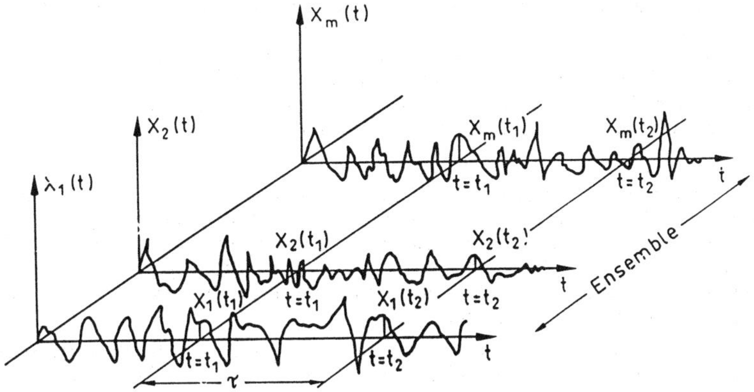

Fig. 2: Realizations of a Stochastic Process

For any fixed time t, a probability distribution or probability density function is defined as shown in Fig. 3 for the fixed instance t_1. Thus, for a fixed time the stochastic process is reduced to a random variable, characterized by its probability density function $f_{\mathbf{X}(t_1)}(\mathbf{x})$.

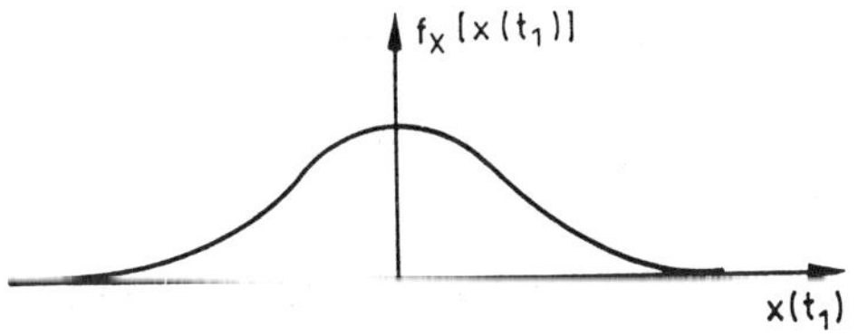

Fig. 3: Probability Density of a Stochastic Process for a Fixed Time t_1.

When observing relevant deterministically unpredictable processes in time, in most cases, only single realizations of the underlying stochastic process are obtained (e.g. earthquake excitation). The information content contained in such a realization is in general not sufficient to construct uniquely the underlying stochastic process, since, according to Fig. 2, the stochastic process is defined by the whole ensemble of realizations and not just by a single realization. Only in few cases, several realizations of a stochastic process might be available, which is surely helpful for a better description, but will not circumvent the general difficulty in defining uniquely the stochastic process in a mathematical sense. Thus, in engineering practice, the mathematical theory of stochastic processes should be regarded as a tool of information processing, where the theory is adjusted such, that for all observable information can be accounted for. In view of the necessity to relate the mathematical theory with observable data, simplifying assumptions for the stochastic process are indispensable. Such important simplifications are for example stationarity, ergodicity or the assumption of a normally distributed stochastic process, as shown in Fig. 3.

The most prominent characteristics of a random variable are its mean and variance respectively. This is certainly only a very approximate description, saying nothing about its distribution. However, in many practical applications, the two parameters might be the only information available. Only in cases, where it is justified - from observations and the physics of the process - to assume a normal distribution, these two pieces of information are sufficient for describing uniquely the distribution.

Similar characterizations are utilized to characterize a stochastic process, the main properties of which are described by its mean $\mathbf{m_X}(t)$ and covariance matrix $\mathbf{C_{XX}}(t_1,t_2)$ respectively,

$$\mathbf{m_X}(t) = E\{\mathbf{X}(t)\} = \int_{-\infty}^{+\infty} \mathbf{x}(t)\, f_{\mathbf{X}(t)}(\mathbf{x})\, d\mathbf{x} \tag{2}$$

$$\approx \overline{m}_{\mathbf{X}}(t) = \frac{1}{m}\sum_{i=1}^{m} \mathbf{X}_i(t)$$

$$\mathbf{C_{XX}}(t_1,t_2) = E\{(\mathbf{X}(t_1)-\mathbf{m_X}(t_1))\cdot(\mathbf{X}(t_2)-\mathbf{m_X}(t_2))^T\}$$

$$= E\{\mathbf{X}(t_1)\cdot\mathbf{X}^T(t_2)\} - \mathbf{m_X}(t_1)\cdot\mathbf{m_X^T}(t_2)$$

$$= \int_{-\infty}^{+\infty} (\mathbf{x}_1 - \mathbf{m_X}(t_1))\cdot(\mathbf{x}_1 - \mathbf{m_X}(t_1))^T\, f_{\mathbf{X}(t_1)\mathbf{X}(t_2)}(\mathbf{x}_1,\mathbf{x}_2)\, dx$$

$$\approx \overline{\mathbf{C}}_{\mathbf{XX}}(t_1,t_2) = \frac{1}{m-1}\sum_{i=1}^{m}(\mathbf{X}_i(t_1)-\overline{\mathbf{m}}_{\mathbf{X}}(t_1))\cdot(\mathbf{X}_i(t_2)-\overline{\mathbf{m}}_{\mathbf{X}}(t_2))^T \tag{3}$$

In the above relations, $\mathbf{X}(t)^T = (X_1(t), X_2(t),\ldots,X_n(t))^T$, is a vector valued function and $E\{\cdot\}$ denotes its expected value. Since this definition requires only some expected values, the mean and the covariance matrix of the process can be easily related to actual data by taking the sample mean among *m* given realizations as an unbiased estimator. It should be stressed, however, that mean and covariance matrix of a stochastic process respectively are deterministic parameters of the process, whereas the statistical estimators, i.e. the vector $\overline{\mathbf{m}}_{\mathbf{X}}(t)$ and the matrix $\overline{\mathbf{C}}_{\mathbf{XX}}(t_1,t_2)$ are random quantities, with a variance inversely proportional to m and $E\{\overline{\mathbf{m}}_{\mathbf{X}}(t)\} = \mathbf{m}_{\mathbf{X}}(t)$,and $E\{\overline{\mathbf{C}}_{\mathbf{XX}}(t_1,t_2)\} = \mathbf{C}_{\mathbf{XX}}(t_1,t_2)$. Only in the limiting case, where $m \rightarrow \infty$, the estimate coincides with the exact, i.e. true value.

2.2. Stationary Processes

In structural engineering one encounters a wide class of problems where the type and intensity of the observed loading phenomena varies quite slowly in comparison to the actual random fluctuations. Such random phenomena are suitably modeled by so called stationary stochastic processes. Typical examples are wind and sea wave loading, engine vibrations or random excitation due to imperfections of the railway track or roughness of road surfaces. Mathematically, a stationary stochastic process is defined by a constant mean value and a correlation matrix which does not depend explicitly on the time t_1 and t_2, but only on the time difference or time lag $\tau = t_2 - t_1$, i.e.

$$E\left\{\mathbf{X}(t_1)\mathbf{X}^T(t_2)\right\} = E\left\{\mathbf{X}(t)\mathbf{X}^T(t+\tau)\right\} = \mathbf{R}_{\mathbf{XX}}(\tau) \tag{4}$$

According to the above definition, the correlation matrix is real and symmetric for real valued processes. Assuming ergodicity, an unbiased estimate for the autocorrelation matrix can be established from a *single* observation of duration 2T regarded as a realization of an ergodic stationary process:

$$\mathbf{R}^T_{\mathbf{XX}}(\tau) = \frac{1}{2T-|\tau|}\int_{-T+|\tau|/2}^{T-|\tau|/2} \mathbf{X}(t-\frac{|\tau|}{2})\mathbf{X}^T(t+\frac{|\tau|}{2})dt \tag{5}$$

Ergodicity means, that all required information on the stochastic process is contained in a single realization. This, for example implies that the ensemble average can be replaced by the time average of a sample function (see Fig. 2).

The above representation of a stationary process is particularly convenient when utilizing the frequency domain. According to the theorem by *Wiener-Khintchine* the autocorrelation matrix and the power spectral density (PSD) matrix $\mathbf{S}_{\mathbf{XX}}(\omega)$, are related by fourier transforms as shown in eq. (6) and (7) below

$$\mathbf{S}_{\mathbf{XX}}(\omega) = \frac{1}{2\pi}\int_{-\infty}^{+\infty} \mathbf{R}_{\mathbf{XX}}(\tau)\, e^{-i\omega\tau} d\tau \tag{6}$$

$$\mathbf{R}_{\mathbf{XX}}(\tau) = \int_{-\infty}^{+\infty} \mathbf{S}_{\mathbf{XX}}(\omega)\, e^{i\omega\tau} d\omega \tag{7}$$

The spectral density matrix for real processes $\mathbf{X}(t)$ is Hermitian, i.e. $S_{X_iX_j}(\omega) = S^*_{X_jX_i}(\omega)$ where the "*" denotes its complex conjugate value implying real valued diagonal terms. In this context the autocorrelation function of the components $X_k(t)$ of the vector valued random process $\mathbf{X}(t)$ are of special interest:

$$E\{X_k(t)X_k(t+\tau)\} = R_{X_kX_k}(\tau) = 2\int_0^{\infty} S_{X_kX_k}(\omega)\cos\tau\omega\, d\omega \tag{8}$$

For $\tau = 0$, the constant mean square value or the so-called "power" of the component is obtained,

$$E\{X_k^2(t)\} = \sigma^2_{X_k} + m^2_{X_k} = 2\int_0^{\infty} S_{X_kX_k}(\omega)\, d\omega \tag{9}$$

where σ_{X_k} denotes the standard deviation and m_{X_k} the mean value respectively of the k-th component. Due to the relation,

$$\mathbf{R}_{\mathbf{XX}}(0) = E\{\mathbf{X}(t)\mathbf{X}^T(t)\} = \int_{-\infty}^{+\infty} \mathbf{S}_{\mathbf{XX}}(\omega)\, d\omega \tag{10}$$

the notion of "power" spectrum becomes obvious, since $\mathbf{S}_{\mathbf{XX}}(\omega)$ specifies how the average "power" of the process is distributed over the frequency range. Analogous to the autocorrelation matrix, the spectral density matrix can be estimated from a given time series of duration 2T - according to the following relation - as well as:

$$\mathbf{S}^T_{\mathbf{XX}}(\omega) = \frac{\pi}{T}\mathbf{X}_T(\omega)\mathbf{X}^H_T(\omega)\ ;\quad \mathbf{X}_T(\omega) = \frac{1}{2\pi}\int_{-T}^{+T}\mathbf{X}(t)e^{-i\omega t}dt \tag{11}$$

where $\mathbf{X}^H$ denotes the transposed complex conjugate vector of X. Unfortunately, this estimate for the spectral density is not unbiased, irrespectively of the length of T. To reduce the variance, the sample spectrum needs to bee smoothed e.g. by using a spectral window, leading to acceptable estimate for the spectral density function $\mathbf{S}_{\mathbf{XX}}(\omega)$. The necessity for smoothing might be seen also from the following relation between spectral density $\mathbf{S}_{\mathbf{XX}}(\omega)$ and the sample spectrum $\mathbf{S}^T_{\mathbf{XX}}(\omega)$, requiring the mean of $\mathbf{S}^T_{\mathbf{XX}}(\omega)$.

$$\mathbf{S}_{\mathbf{XX}}(\omega) = \lim_{T\to\infty} E\{\mathbf{S}^T_{\mathbf{XX}}(\omega)\} \tag{12}$$

Since the mean value is not available, an estimate for the mean is established by weighted averaging (smoothing) over neigboring frequency ranges.

A special case, of considerable theoretical interest and practical value, is defined by a density which assumes a constant value S_0 over the entire real axis. In analogy to the light of the sun, which covers all frequencies, such a stochastic process is generally called "white noise" denoted as *W(t)*. Such a mathematical abstraction of a process is, of course physically not realizable, since the power of the process is unbounded, i.e. approaches infinity (see eq.(10)). The process is, however, well defined either in terms of the constant spectral density or its autocorrelation function. Since the constant spectral density is related to its autocorrelation by

$$\mathbf{S}_0 = \mathbf{S}(\omega) = \frac{1}{2\pi} \int_{-\infty}^{+\infty} \mathbf{R}(\tau)\, e^{-i\omega\tau} d\tau \tag{13}$$

the autocorrelation must assume the function,

$$\mathbf{R}(\tau) = E\left\{\mathbf{W}(t)\mathbf{W}^T(t+\tau)\right\} = 2\pi\, \mathbf{S}_0\, \delta(\tau) \tag{14}$$

where $\delta(t)$ denotes Dirac's impulse function ($\delta(t \neq 0) = 0$, $\delta(0) \to \infty$, $\int_{-\varepsilon}^{+\varepsilon} \delta(t) dt = 1$, $\varepsilon > 0$). Thus, white noise is a non physical process where the amplitudes of the process are uncorrelated for any (arbitrary short) time difference $\tau > 0$. Although in nature such process is not observable, the use of white noise is justified and utilized quite frequently for a wide class of problems in engineering practice. This will be shown in section 4 discussing the stationary stochastic response of linear structural systems.

3. Monte Carlo Simulation of Stochastic Processes

Probabilities of the occurrence of events of engineering interest can be estimated from a sufficiently large sample of realizations (outcomes) of some random variables and stochastic processes. These realizations can be established either by measurements or artificially by Monte Carlo simulation (MCS). Since measurements are only in a limited number of cases feasible and moreover, quite expensive, the required large number of realizations are, whenever possible, generated by the computer. Since the outcome of the realizations is influenced by random effects, this type of simulation is denoted as Monte Carlo simulation. Randomness in MCS is established by utilizing a random number generator. In other words, MCS can be characterized as simulation of processes where its result depends on the (pseudo random) outcome of the random generator. Without considering further details, it can be assumed that each time the random generator is called a relization of a random variable ζ_i [$0 < \zeta_i < 1$] is generated. For an acceptable random generator it is required that the set $\{\zeta_i$

,i=1,2,,,N} tends towards an uniform distribution as N tends to infinity, and that there is no statistically observable correlation between any subset of random variables. Hence, each outcome of ζ_i the random generator can be regarded as uniformly distributed,

$$F_\zeta(\zeta) = \zeta \; ; \; f_\zeta(\zeta) = 1 \text{ for } 0 \le \zeta \le 1 \tag{15}$$

and statistically independent from other outcomes ζ_k ,k≠i, where $F_X(x)$ denotes the cumulative distribution function (CDF) of a random variable X and $f_X(x) = dF_X(x)/dx$ its probability density function (PDF). Other than uniform distributions $F_X(x)$ can be determined quite simply by setting $F_X(x) = F_\zeta(\zeta) = \zeta$ and utilizing the inverse of $F_X(x)$:

$$X_i = F_X^{-1}(\zeta_i) \tag{16}$$

The above relation corresponds to a map as shown in Fig. 4 and can be utilized for any distribution defined by its cumulative distribution function $F_X(x)$.

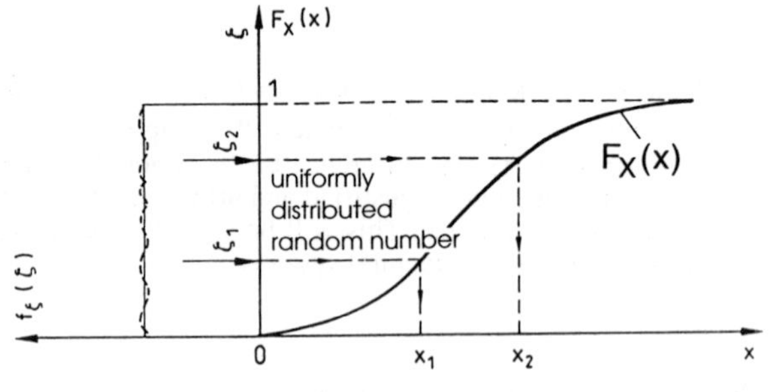

Fig. 4: Mapping of a Uniformly Distributed Random Variable ζ_k to a Random Variable X Defined by its Cumulative Distribution $F_X(x)$.

Due to the special role which the normal distribution plays in statistics and stochastics, the generation of normally distributed random variables is required most frequently. A normally distributed random variable (RV) is determined by its mean value m_X and standard deviation σ_X, and any realization X_i might be obtained from U_i following a standard normal distribution $U_i \in N(0,1)$ with zero mean and unit standard deviation.

$$X_i = m_X + U_i \sigma_X \tag{17}$$

Since the standard normal distribution $\Phi(u) = 1/\sqrt{2\pi} \cdot \int_{-\infty}^{u} \exp(-x^2/2)dx$ and its required inverse must be evaluated numerically in a computationally involved manner, other, more efficient procedures, have been developed for generating standard normal random variables. For example, the *Box-Muller* method transforms a pair of independent U(0,1) uniformly distributed RV's into a pair of independent standard normal distributed RV's:

$$U_i = \sqrt{-2\ln\zeta_i} \cdot \cos\, 2\pi\zeta_{i+1}\ ;\ U_{i+1} = \sqrt{-2\ln\zeta_i} \cdot \sin 2\pi\zeta_{i+1} \tag{18}$$

Furthermore, the *Polar Marsaglia* method, for example, which is a variation of the Box-Muller method, avoids the time consuming evaluation of the trigonometric functions.

MCS is most frequently utilized in stochastic dynamics to generate realizations of a stochastic process. One of the most important application of MCS is the generation of a stationary stochastic process, which is for a normally distributed process uniquely determined by the power spectrum ($\mathbf{G_{XX}}(\omega)$, $\omega \geq 0$) or the symmetric spectral density function $\mathbf{S_{XX}}(-\omega) = \mathbf{S_{XX}}(\omega)$. For simplicity, the one dimensional case, where $S_{XX}(\omega)$ is specified by a single symmetric function is considered first. It can be shown, that a stationary process can be represented exactly by an infinite sum of sinusoidal functions. Utilizing just a finite sum leads to a sufficiently accurate approximation,

$$\begin{aligned} X(t) &= \sum_{j=1}^{J} a_j \cos\omega_j t + b_j \sin\omega_j t \\ &= \sum_{j=1}^{J} c_j \cos(\omega_j t - \varphi_j) \end{aligned} \tag{19}$$

provided the constants a_j, b_j or c_j are determined such that the power of X(t) within some small frequency bands corresponds to the power of the theoretical spectrum. In the above representation, the variables a_j and b_j are both identically normally distributed RV's, c_j follows a Rayleigh distribution because $c_j^2 = a_j^2 + b_j^2$ and $\varphi_j = \mathrm{arctg}(b_j / a_j)$ is uniformly distributed within the range $[-\pi,+\pi]$. The RV's a_j and b_j, and consequently the RV c_j, are related to the power within *J* disjunct intervals as indicated in Fig. 5 where the number *J* of intervals is of the order 100÷300. The best representation of the stationary normally distributed process is obtained when the frequency interval $\Delta\omega_i$ is selected such that the power (area within the interval) is nearly equal for all *J* intervals. However, the generated stochastic process X(t) is often utilized as random excitation for determining the stochastic structural response. For such cases, smaller intervals should be selected within frequency ranges where the response is sensitive, since the aim of the MCS then is an accurate representation of the stochastic *response* and less of the excitation. Computationally most efficient is certainly the utilization of frequency intervals of equal length employing the Fast Fourier Transform (FFT). The efficiency of FFT allows for a fairly large number of J, so that both the excitation as well as the response are represented sufficiently accurate.

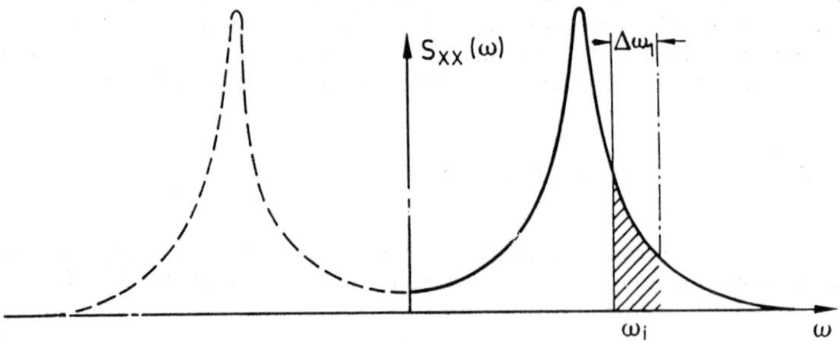

Fig. 5: Subdivision of the Power Spectrum into Discrete Intervals $\Delta\omega_j$.

Equating the power of the sinusoidal terms in the above equation with the power within the frequency band, leads to the relation

$$E\{a_j^2\} = E\{b_j^2\} = \frac{1}{2}E\{c_j^2\} = S_{XX}(\omega_j)\cdot\Delta\omega_j \tag{20}$$

where ω_j represents the center of the frequency interval. Selecting the phase angles φ_j in eq.(19) at random, the amplitudes c_j might be determined deterministically as

$$c_j = \sqrt{2S_{XX}(\omega_j)\Delta\omega_j}\,. \tag{21}$$

due to the central limit theorem - which states that the sum of a large number of independent random variables leads to a normal distribution irrespectively of the original distribution- the process will be normally distributed. In case FFT is employed, discrete complex amplitudes $c_j\exp(i\varphi_j)$ are generated in the frequency domain. FFT into the time domain leads to a realization of a stationary process with the specified spectral density function.

Consider next the general case where $\mathbf{S_{XX}}(\omega)$ is a n-dimensional matrix. For the special case where $\mathbf{S_{XX}}(\omega)$ is a (real) diagonal matrix, each component of the n-dimensional vector $\mathbf{X}(t)$ is uncorrelated and independent from other components, so that the above procedures immediately applies for each component using the spectral density $S_{X_kX_k}(\omega)$, $k=1,2,..,n$. In general, however, $\mathbf{S_{XX}}(\omega)$ is a n-dimensional *Hermitian* matrix. Solving the eigenvalue problem $\mathbf{S}\cdot\mathbf{T}=\mathbf{T}\cdot\tilde{\mathbf{S}}$, where $\tilde{\mathbf{S}}$ is a real valued diagonal matrix and $\mathbf{T}$ an orthogonal complex matrix, the transformation $\tilde{\mathbf{S}}=\mathbf{T}^{H}\cdot\mathbf{S}\cdot\mathbf{T}$ can be established. For each component of the real diagonal matrix $\tilde{\mathbf{S}}$, the amplitudes can be determined according to eq.(21) leading to a time domain representation $\tilde{X}(t,\omega_j)=\{c_j\}\cos(\omega_j+\varphi_j)$ for the transformed diagonal matrix $\tilde{S}(\omega_j)$. Utilizing further the transformation $X(t,\omega_j)=T(\omega_j)\cdot\tilde{X}(t,\omega_j)$, all components of the n-dimensional process can be established. Obviously, the orthogonal transformation defines the correlation between the components of the n-dimensional stationary process.

Another most important type of process generated by MCS is *band limited white noise.* Since, as already mentioned above, white noise itself is non-physical and hence a mathematical abstraction, white noise can not be generated nor is it required. In all practical applications it is just a convenient substitute for a flat (nearly constant) spectral density within a required frequency range. Thus, all what is really needed is a nearly constant spectral density within a specific frequency range $\omega\in[0,\Omega]$. A time series with this property can be simulated quite efficiently by a continuous process $\mathbf{X}(t)$ where the amplitudes $\mathbf{X}_k=\mathbf{X}(t_k)=\mathbf{X}(k\cdot\Delta t)$ at discrete times $t_k=k\cdot\Delta t$ are statistically independent and normally distributed. The covariance matrix of the vector process $\mathbf{X}(t_k)$ is related to a constant spectral density matrix $\mathbf{S}_0$ by:

$$E\{\mathbf{X}_k\mathbf{X}_k^H\} = \frac{2\pi}{\Delta t}\mathbf{S}_0 = \frac{\mathbf{I}}{\Delta t} \tag{22}$$

It should be noted that white noise is approximated by letting $\Delta t\to 0$, leading to an infinite variance and a discontinuous process $\mathbf{X}(t)$. The above continuous process has a nearly constant spectral density within half of the Nyquist frequency $\omega_N = 2\pi/\Delta t$. Hence Δt should be selected as $\Delta t < \pi/\Omega$. The generation of the vector $\mathbf{X}_k$ is straight forward in case $\mathbf{S}_0$ is a diagonal matrix, where each component of the vector $\mathbf{X}(t_k)$ can be considered separately and independently. Otherwise, similarly as for the generation of a stationary process with spectrum $\mathbf{S}(\omega)$, the Hermitian matrix $\mathbf{S}_0$ needs to be transformed to a diagonal matrix $\tilde{\mathbf{S}}_0=\mathbf{T}^{H}\cdot\mathbf{S}_0\cdot\mathbf{T}$ of which the associated vector $\tilde{X}(t_k)$ can be determined by evaluating the variance of each component. Then, the backward transformation $X(t_k)=T\cdot\tilde{X}(t_k)$ leads to desired amplitudes of the n-dimensional vector $\mathbf{X}(t_k)$. Note, that $\mathbf{X}(t)$ is purely real only for the case where $\mathbf{S}_0$ is real and positive definite.

Both representations as shown in eq.(19) and eq.(22) have different specific advantages a disadvantages. While eq.(19) is capable to represent all types of normally distributed stationary processes applicable for any conceivable spectral density function, eq.(22) is restricted to the special case of a constant spectral density within the frequency range $[0,\Omega]$. The sinusoidal form in (19) is especially suitable for strongly colored stationary excitation, i.e. where most of the contribution to the power $E\{X^2(t)\}$ stems from the power in small frequency ranges. For such cases, only few terms are required for an accurate representation, whereas for a wide band process a fairly large number J of terms are necessary to achieve a sufficient accuracy. The representation in (22) has the advantage that it can be easily extended to nonstationary band-limited white noise by utilizing a time modulating function f(t), i.e.:

$$I = I(t) = I_0 \cdot f(t) = 2\pi S_0 \cdot f(t) \tag{23}$$

In most technical applications as in wind-, ocean- and earthquake engineering, white noise is not a satisfactory representation of the loading process. There, a realistic spectral density has often one peak at a certain frequency. For such cases, *filtered white noise* might be used, i.e. the actual excitation is regarded a function of the output of a filter (usually a linear system) subjected to white noise. Consider for example the *Kanai-Tajimi* spectrum widely used in earthquake engineering to represent stationary ground motion acceleration a(t),

$$S_{aa}(\omega) = \frac{\omega_g^4 + 4\zeta_g^2\omega_g^2}{(\omega_g^2 - \omega^2)^2 + 4\zeta_g^2\omega_g^2\omega^2} S_0 \tag{24}$$

which has its maximum spectral density close to $\omega = \omega_g$, where ω_g can be interpreted as the natural frequency of the ground and ζ_g as the associated damping value. Then, it is possible to represent such a stationary stochastic process as a linear function of the filter output $(x_g(t), \dot{x}_g(t))$

$$a(t) = \omega_g^2 x_g(t) + 2\zeta_g\omega_g\dot{x}_g(t) \tag{25}$$

where the output is determined by the linear (2-nd order) system subjected to a constant white noise intensity $I = E\{W(t)W(t+\tau)\} = 2\pi S_0 \cdot \delta(\tau)$,

$$\ddot{x}_g(t) + 2\zeta_g\omega_g\dot{x}_g(t) + \omega_g^2 x_g(t) = W(t) \tag{26}$$

To generate a realization of the stochastic process a(t), the representation in eq.(22) is utilized for computing W(t). Integrating the above differential equation by traditional integration schemes (e.g. the *Newmark* step by step procedure) leads with eq.(25) to a realization of the ground acceleration a(t) which in turn is used further to obtain a realization of the stochastic response of a (linear or nonlinear) structural system. Since earthquake excitation is inherently nonstationary, white noise with a time varying intensity is generally employed.

Another general, and quite useful representation of a stochastic process, easy to generate by MCS, is the following,

$$\mathbf{X}(t) = \mathbf{a}_0(t) + \sum_{j=1}^{n} \xi_j \mathbf{a}_j(t) \tag{27}$$

where $\mathbf{a}_j(t)$ are deterministic functions of time t and ξ_j are independent zero mean standard normally distributed random variables. The above representation is known as *Karhunen Loeve* expansion of a stochastic process. Note, that the above representation comprizes the sinusoidal form used in eq.(19) as well as the white noise approximation of eq.(22), choosing $\mathbf{a}_j(t)$ of the form $\sin(\omega_j t)$ and $\cos(\omega_j t)$ or by selecting $\mathbf{a}_j(t) = 0$ for $|t-j\Delta t|>\Delta t$ and $\mathbf{a}_j(j\Delta t-\eta\Delta t) = |1-|\eta||$ for $|\eta|<1$. The above formulae might also be employed to represent a stochastic process in space where the time t is replaced by a spatial coordinate $\mathbf{z}$.

$$\mathbf{X}(\mathbf{z}) = a_0(\mathbf{z}) + \sum_{j=1}^{n} \xi_j \mathbf{a}_j(\mathbf{z}) \tag{28}$$

It should be noted that the above equation proved to be quite useful for random field representations.

4. Stochastic Response of Linear MDOF Systems

This section focuses on the stochastic response of linear MDOF systems subjected to loading represented as normally distributed stationary and nonstationary processes. For these special cases, the response is also normally distributed and can be conveniently represented by its first two moment properties, i.e. the mean vector $\mathbf{m}_\mathbf{Y}(t)$ and the autocorrelation matrix $\mathbf{R}_{\mathbf{YY}}(t_1,t_2)$. This simple representation is not valid, of course, for non-normal (non-Gaussian) loading, which complicates the analysis considerably.

The response of a linear MDOF system is governed by the equation of motion,

$$\mathbf{M}\cdot\ddot{\mathbf{y}}(t) + \mathbf{C}\cdot\dot{\mathbf{y}}(t) + \mathbf{K}\cdot\mathbf{y}(t) = \mathbf{x}(t) \tag{29}$$

where $\mathbf{y}(t)$ represents the generalized displacement vector of the structural response, $\mathbf{x}(t)$ the vector of external loading, and the quadratic matrices $\mathbf{M},\mathbf{C},\mathbf{K}$ denote the mass-, damping- and stiffness matrix, respectively. These matrices are usually constant, symmetric and positive definite. Linear systems, by their definition, possess the specific property that its response can be regarded as a superposition of a sum of individual effects as utilized in the following representation,

$$\mathbf{y}(t) = \int_{-\infty}^{t} [\mathbf{h}(t,\tau)]\cdot\mathbf{x}(\tau)\, d\tau \tag{30}$$

where $[\mathbf{h}(t,\tau)]$ is the dynamic response of the linear system at time t due to an impulse $\mathbf{x}(\tau)d\tau$ at instant τ. The response $\mathbf{h}(t,\tau)$ is commonly denoted as impulse response function. It defines a n x n matrix of time function $h_{jk}(t,\tau)$, $1 \le k, j \le n$, where n denotes the order of the system. As the name suggests, is the matrix $[\mathbf{h}(t,\tau)]$ determined as the response due to an impulse at time τ, defined by the equation of motion

$$\mathbf{M}\cdot\left[\ddot{\mathbf{h}}(t,\tau)\right]+\mathbf{C}\cdot\left[\dot{\mathbf{h}}(t,\tau)\right]+\mathbf{K}\cdot[\mathbf{h}(t,\tau)] \;=\; [\mathbf{E}]\cdot\delta(t-\tau) \tag{31}$$

where $\delta(t)$ denotes *Dirac*'s impulse function ($\delta(t\neq0) = 0$, $\delta(0) \to \infty$, $\int_{-\varepsilon}^{+\varepsilon}\delta(t)dt = 1,\ \varepsilon > 0$) and $[\mathbf{E}]$ represents the unity matrix. The impulse response function $[\mathbf{h}(t,\tau)]$ follows for $t > \tau$ the homogeneous solution,

$$\mathbf{M}\cdot\left[\ddot{\mathbf{h}}(t,\tau)\right]+\mathbf{C}\cdot\left[\dot{\mathbf{h}}(t,\tau)\right]+\mathbf{K}\cdot[\mathbf{h}(t,\tau)] \;=\; [\mathbf{0}] \tag{32}$$

with the initial condition at time $t = \tau$,

$$[\mathbf{h}(t,\tau)] = [\mathbf{0}] \text{ for } t \le \tau;\ \left[\dot{\mathbf{h}}(\tau,\tau)\right] = \mathbf{M}^{-1} \tag{33}$$

where $[\mathbf{0}]$ stands for the zero matrix.
For a SDOF system, the impulse response function reads with ($m = M$, $c = C$, $k = K$),

$$h(t,\tau) = \frac{e^{-\xi\omega_0(t-\tau)}}{m\omega_d}\sin\,\omega_d(t-\tau),\quad t \ge \tau \tag{34}$$

using the abbreviations:

$$\omega_0^2 = \frac{k}{m};\ \xi = \frac{c}{2\sqrt{mk}};\ \omega_d^2 = \omega_0^2(1-\xi^2) \tag{35}$$

As an alternative, the loading $\mathbf{x}(t)$ and the response $\mathbf{y}(t)$ can also be represented in the *frequency domain* utilizing Fourier transforms. For example, $\mathbf{x}(t)$ can be transformed to the frequency domain $\mathbf{x}(\omega)$ and vice versa:

$$\mathbf{x}(t) \;=\; \int_{-\infty}^{+\infty}\mathbf{x}(\omega)\, e^{i\omega t}d\omega \tag{36}$$

$$\mathbf{x}(\omega) \;=\; \frac{1}{2\pi}\int_{-\infty}^{+\infty}\mathbf{x}(t)\, e^{-i\omega t}dt \tag{37}$$

The introduction of eq. (36) into eq. (30) leads to the relation,

$$\mathbf{y}(t) = \int_{-\infty}^{t}[\mathbf{h}(t,\tau)]\cdot\int_{-\infty}^{+\infty}\mathbf{x}(\omega)\,e^{i\omega\tau}d\omega\,d\tau = \int_{-\infty}^{+\infty}\mathbf{H}(t,\omega)\mathbf{x}(\omega)d\omega \tag{38}$$

with

$$\mathbf{H}(t,\omega) = \int_{-\infty}^{t}[\mathbf{h}(t-\tau)]\,e^{i\omega\tau}d\tau = e^{i\omega t}\mathbf{H}(\omega) \tag{39}$$

$$\mathbf{H}(\omega) = \int_{0}^{+\infty}[\mathbf{h}(t)]\,e^{-i\omega t}dt = \int_{-\infty}^{+\infty}[\mathbf{h}(t)]\,e^{-i\omega t}dt \tag{40}$$

by making use of the fact that $[\mathbf{h}(t<0)] = 0$ and a variable substitution for $(t-\tau)$. Eq.(39) implies constant structural matrices M,C and K for which the equality $[\mathbf{h}(t,\tau)] = [\mathbf{h}(t-\tau)]$ holds. The matrix $\mathbf{H}(\omega)$, defined as transfer matrix, plays an important role in structural dynamics, since it relates in a simple way the loading $\mathbf{x}(\omega)$ to the response $\mathbf{y}(\omega)$ in the frequency domain:

$$\mathbf{y}(t) = \int_{-\infty}^{+\infty}\mathbf{H}(\omega)\mathbf{x}(\omega)\,e^{i\omega t}d\omega = \int_{-\infty}^{+\infty}\mathbf{y}(\omega)\,e^{i\omega t}d\omega \tag{41}$$

$$\mathbf{y}(\omega) = \mathbf{H}(\omega)\cdot\mathbf{x}(\omega) \tag{42}$$

A comparison of the above equation with eq. (30) shows that in time domain a convolution of the loading with the impulse response functions is required, whereas in the frequency domain a simple multiplication suffices. Moreover, the transfer matrix $\mathbf{H}(\omega)$ can be determined with relative ease. Consider the special loading $\mathbf{x}(t;\omega) = [\mathbf{E}]\cdot e^{\,i\omega t}$ which leads, according to the last relation, to the response of the form $\mathbf{y}(t,\omega) = \mathbf{H}(\omega)\cdot e^{\,i\omega t}$. Introducing $\mathbf{y}(t,\omega) = \mathbf{H}(\omega)\cdot e^{\,i\omega t}$ into the equation of motion (29) leads to the following relation for the transfer matrix $\mathbf{H}(\omega)$:

$$\mathbf{H}(\omega) = \left[-\omega^2\mathbf{M} + i\omega\mathbf{C} + \mathbf{K}\right]^{-1} \tag{43}$$

The above relations might be used for systems with only few DOFs. In general, however, employing eq.(43) is computationally not efficient, since the inversion might be required for several thousands of discrete frequencies ω_k. Computationally efficient alternatives to determine the transfer matrix are discussed in the following.

Consider the generalized eigenvalue solution,

$$\mathbf{K}\cdot\boldsymbol{\Phi} = \mathbf{M}\cdot\boldsymbol{\Phi}\cdot\mathrm{diag}\left[\omega_j^2\right] \tag{44}$$

and assume that the eigenvectors of $\mathbf{\Phi} = [\phi_1, \phi_2, \ldots, \phi_n]$ are normalized such that they satisfy $\mathbf{\Phi^T} \cdot \mathbf{M} \cdot \mathbf{\Phi} = [\mathbf{E}]$ unity matrix. Then,

$$\mathbf{\Phi^T} \cdot \mathbf{K} \cdot \mathbf{\Phi} = \operatorname{diag}\left[\omega_j^2\right] \quad \text{for} \quad \mathbf{\Phi^T} \cdot \mathbf{M} \cdot \mathbf{\Phi} = [\mathbf{E}] \tag{45}$$

are diagonalized by the above transformation. Using the classical eigenvalue decomposition of the mass- and stiffness matrix, the transformation of the damping matrix **C** will not be a diagonal matrix, i.e. $\mathbf{\Phi^T} \cdot \mathbf{C} \cdot \mathbf{\Phi}$ will not be diagonal. Since in structural analysis the damping matrix is quite often not known explicitly, but instead some estimates about the modal damping ratios ξ_j, it is nevertheless justifiable to assume

$$\mathbf{\Phi}^{\mathrm{T}} \cdot \mathbf{C} \cdot \mathbf{\Phi} = \operatorname{diag}\left[2\xi_j \omega_j\right] \tag{46}$$

to be diagonal. Introducing further the coordinate transformation $\mathbf{y}(t) = \mathbf{\Phi} \cdot \mathbf{z}(t)$ into eq. (29) and premultiplying by $\mathbf{\Phi}^{\mathrm{T}}$, leads to a decoupled equation of motion where each component z_j of the vector **z** can be determined independently:

$$\ddot{z}_j(t) + 2\xi_j \omega_j \dot{z}_j(t) + \omega_j^2 z_j(t) = \phi_j^{\mathrm{T}} \mathbf{x}(t) \tag{47}$$

Taking the Fourier transformation of the above equation, relates the complex valued amplitudes $z_j(\omega)$ to the Fourier transform of the loading $\mathbf{x}(\omega)$,

$$z_j(\omega) = \frac{\phi_j^{\mathrm{T}} \cdot \mathbf{x}(\omega)}{\omega_j^2 - \omega^2 + 2i\xi_j \omega_j \omega} \tag{48}$$

where $i = \sqrt{-1}$. Since $\mathbf{y}(\omega) = \mathbf{\Phi} \cdot \mathbf{z}(\omega) = \mathbf{H}(\omega) \cdot \mathbf{x}(\omega)$, the transfer matrix reads,

$$\mathbf{H}(\omega) = \sum_{j=1}^{n} \frac{\phi_j \cdot \phi_j^{\mathrm{T}}}{\omega_j^2 - \omega^2 + 2i\xi_j \omega_j \omega} \tag{49}$$

or the component $h_{kl}(\omega)$ of the matrix:

$$h_{kl}(\omega) = \sum_{j=1}^{n} \frac{\phi_{kj} \cdot \phi_{lj}}{\omega_j^2 - \omega^2 + 2i\xi_j \omega_j \omega} \tag{50}$$

In the general case, where **C** is explicitly given, the transformation $\mathbf{\Phi}^{\mathrm{T}} \cdot \mathbf{C} \cdot \mathbf{\Phi}$ will not be diagonal. Using *complex modal analysis* and solving the eigenvalue problem,

$$\left[\mathbf{K} + \lambda_j \mathbf{C} + \lambda_j^2 \mathbf{M}\right] \cdot \boldsymbol{\psi}_j = \mathbf{0} \quad \text{or} \quad \left[\mathbf{K} + \mathbf{\Lambda} \cdot \mathbf{C} + \mathbf{\Lambda}^2 \cdot \mathbf{M}\right] \cdot \mathbf{\Psi} = \mathbf{0} \tag{51}$$

leads to complex eigenvalues $\lambda_j = -\xi_j\omega_j + i\omega_j\sqrt{1-\xi_j^2}$ and (complex) eigenvectors $\psi_j \approx \phi_j$, in case the normalization coincides with the classical procedure for $\mathbf{C} = 0$, i.e. that

$$\left[\boldsymbol{\Psi}^T, \boldsymbol{\Lambda}^T\boldsymbol{\Psi}^T\right] \cdot \begin{bmatrix} \mathbf{C} & \mathbf{M} \\ \mathbf{M} & \mathbf{0} \end{bmatrix} \cdot \begin{bmatrix} \boldsymbol{\Psi} \\ \boldsymbol{\Psi\Lambda} \end{bmatrix} = 2\boldsymbol{\Lambda} \tag{52}$$

holds, leading to the relation:

$$\left[\boldsymbol{\Psi}^T, \boldsymbol{\Lambda}^T\boldsymbol{\Psi}^T\right] \cdot \begin{bmatrix} \mathbf{K} & \mathbf{0} \\ \mathbf{0} & -\mathbf{M} \end{bmatrix} \cdot \begin{bmatrix} \boldsymbol{\Psi} \\ \boldsymbol{\Psi\Lambda} \end{bmatrix} = -2\boldsymbol{\Lambda}^2 \tag{53}$$

The transformation $\mathbf{y}(t) = \boldsymbol{\Psi} \cdot \mathbf{z}(t)$ and the representation,

$$\begin{bmatrix} \mathbf{C} & \mathbf{M} \\ \mathbf{M} & \mathbf{0} \end{bmatrix} \cdot \begin{Bmatrix} \dot{\mathbf{y}} \\ \ddot{\mathbf{y}} \end{Bmatrix} + \begin{bmatrix} \mathbf{K} & \mathbf{0} \\ \mathbf{0} & -\mathbf{M} \end{bmatrix} \cdot \begin{Bmatrix} \mathbf{y} \\ \dot{\mathbf{y}} \end{Bmatrix} = \begin{Bmatrix} \mathbf{x}(t) \\ \mathbf{0} \end{Bmatrix} \tag{54}$$

premultiplyed by the matrix $\left[\boldsymbol{\Psi}^T, \boldsymbol{\Lambda}^T\boldsymbol{\Psi}^T\right]$ lead then to a decoupled equation of motion in terms of the vector $\mathbf{z}(t)$:

$$2\boldsymbol{\Lambda} \cdot \dot{\mathbf{z}}(t) - 2\boldsymbol{\Lambda}^2 \cdot \mathbf{z}(t) = \boldsymbol{\Psi}^T\mathbf{x}(t) \tag{55}$$

Taking the Fourier transformation of the above equation, relates the complex valued amplitudes $z_j(\omega)$ to the Fourier transform of the loading $\mathbf{x}(\omega)$,

$$z_j(\omega) = \frac{\psi_j^T \cdot \mathbf{x}(\omega)}{2\lambda_j(i\omega - \lambda_j)} \tag{56}$$

Since $\mathbf{y}(\omega) = \boldsymbol{\Phi} \cdot \mathbf{z}(\omega) = \mathbf{H}(\omega) \cdot \mathbf{x}(\omega)$, and to each eigenvalue λ_j with a positive imaginary part corresponds a complex conjugate eigenvalue λ_j^* with a negative imaginary part and associated complex conjugate eigenvectors , the transfer matrix reads:

$$\mathbf{H}(\omega) = \sum_{j=1}^{n} \left[\frac{\psi_j \cdot \psi_j^T}{2\lambda_j(i\omega - \lambda_j)}\right] + \left[\frac{\psi_j \cdot \psi_j^T}{2\lambda_j(-i\omega - \lambda_j)}\right]^* \tag{57}$$

All the above relations are generally valid for the deterministic as well stochastic excitation. Now, consider the case where the loading $\mathbf{x}(t)$ is represented as normally distributed stochastic process. To underline this assumption, capital letters are used to indicate in the following the stochastic nature of the excitation vector $\mathbf{X}(t)$ and the response vector $\mathbf{Y}(t)$. Since a normally distributed process may be a represented in form of eq.(27), and the stochastic response can be regarded as superposition of individual effects, the stochastic response must be normally distributed as well. For this special case, the mean $\mathbf{m_Y}(t)$ and the autocorrelation matrix $\mathbf{R_{YY}}(t_1,t_2) = E\{\mathbf{Y}(t_1)\mathbf{Y}^T(t_2)\}$ suffices to uniquely determine the response.

The mean $\mathbf{m_Y}(t) = E\{\mathbf{Y}(t)\}$ can be evaluated simply by taking the expectation of the equation of motion,

$$E\left\{\mathbf{M}\cdot\ddot{\mathbf{Y}}(t)+\mathbf{C}\cdot\dot{\mathbf{Y}}(t)+\mathbf{K}\cdot\mathbf{Y}(t)\right\} = E\{\mathbf{X}(t)\} \tag{58}$$

or

$$\mathbf{M}\cdot\ddot{\mathbf{m}}_\mathbf{Y}(t)+\mathbf{C}\cdot\dot{\mathbf{m}}_\mathbf{Y}(t)+\mathbf{K}\cdot\mathbf{m}_\mathbf{Y}(t) = \mathbf{m}_\mathbf{X}(t) \tag{59}$$

leading to the solution:

$$\mathbf{m}_\mathbf{Y}(t) = \int_{-\infty}^{t}[\mathbf{h}(t-\tau)]\cdot\mathbf{m}_\mathbf{X}(\tau)\,d\tau \tag{60}$$

For simplification of the subsequent expressions, it is advantageous to transform the stochastic process into zero mean processes $\tilde{\mathbf{X}}(t)$ and $\tilde{\mathbf{Y}}(t)$, i.e,

$$\tilde{\mathbf{X}}(t) = \mathbf{X}(t)-\mathbf{m}_\mathbf{X}(t)\ ; \qquad \tilde{\mathbf{Y}}(t) = \mathbf{Y}(t)-\mathbf{m}_\mathbf{Y}(t) \tag{61}$$

for which the relation (30) still holds, i.e.

$$\begin{aligned}\tilde{\mathbf{Y}}(t) &= \int_{-\infty}^{t}[\mathbf{h}(t-\tau)]\cdot\tilde{\mathbf{X}}(\tau)\,d\tau \\ &= \int_{-\infty}^{+\infty}\mathbf{H}(\omega)\tilde{\mathbf{X}}(\omega)\,e^{i\omega t}dt\end{aligned} \tag{62}$$

The above equation is then used to evaluate the covariance matrix:

$$\begin{aligned}E\left\{\tilde{\mathbf{Y}}(t_1)\tilde{\mathbf{Y}}^T(t_2)\right\} &= \\ &= \int_{-\infty}^{t_1}\int_{-\infty}^{t_2}[\mathbf{h}(t_1-\tau_1)]\cdot E\left\{\tilde{\mathbf{X}}(\tau_1)\tilde{\mathbf{X}}^T(\tau_2)\right\}\cdot[\mathbf{h}(t_2-\tau_2)]^T d\tau_1 d\tau_2 \\ &= \int_{-\infty}^{+\infty}\int_{-\infty}^{+\infty}\mathbf{H}(\omega_1)\cdot E\left\{\tilde{\mathbf{X}}(\omega_1)\tilde{\mathbf{X}}^H(\omega_2)\right\}\cdot\mathbf{H}^H(\omega_2)\,e^{i\omega_1 t_1-i\omega_2 t_2}d\omega_1 d\omega_2\end{aligned} \tag{63}$$

The above relation describes the general case of a nonstationary excitation $\mathbf{X}(t)$, the solution of which is somewhat burdensome due to the double integrals. This relation can not be further simplified, without a specific assumption on the stochastic process $\mathbf{X}(t)$, such as stationarity. As the approximation in eq.(19) already indicates, the amplitudes $\mathbf{X}(\omega_1)$ and $\mathbf{X}(\omega_2)$ are independent for $\omega_1 \neq \omega_2$ for a stationary process, leading to a delta correlation:

$$E\left\{\tilde{\mathbf{X}}(\omega_1)\tilde{\mathbf{X}}^*(\omega_2)\right\} = \mathbf{S}_{\tilde{\mathbf{X}}\tilde{\mathbf{X}}}(\omega_1)\,\delta(\omega_1-\omega_2) \tag{64}$$

Using the last expression in the above equation gives,

$$E\left\{\tilde{\mathbf{Y}}(t_1)\tilde{\mathbf{Y}}^T(t_2)\right\} = \mathbf{R}(t_1-t_2) = \int_{-\infty}^{+\infty} \mathbf{H}(\omega)\cdot\mathbf{S}_{\tilde{\mathbf{X}}\tilde{\mathbf{X}}}(\omega)\cdot\mathbf{H}^H(\omega)\, e^{i\omega(t_1-t_2)}d\omega \tag{65}$$

and with $\tau = t_1 - t_2$ to the stationary solution in the frequency domain:

$$\mathbf{R}_{\tilde{\mathbf{Y}}\tilde{\mathbf{Y}}}(\tau) = \int_{-\infty}^{+\infty} \mathbf{S}_{\tilde{\mathbf{Y}}\tilde{\mathbf{Y}}}(\omega)\, e^{i\omega\tau}d\omega \tag{66}$$

$$\mathbf{S}_{\tilde{\mathbf{Y}}\tilde{\mathbf{Y}}}(\tau) = \mathbf{H}(\omega)\cdot\mathbf{S}_{\tilde{\mathbf{X}}\tilde{\mathbf{X}}}(\omega)\cdot\mathbf{H}^H(\omega) \tag{67}$$

The representation of a nonstationary Gaussian excitation by eq.(27) leads to very simple relation for the normally distributed response. Solving for each deterministic loading function $\mathbf{a}_k(t)$, $k=0,1,2,..,m$, the associated deterministic response $\mathbf{r}_k(t)$,

$$\mathbf{M}\cdot\ddot{\mathbf{r}}_k(t)+\mathbf{C}\cdot\dot{\mathbf{r}}_k(t)+\mathbf{K}\cdot\mathbf{r}_k(t)=\mathbf{a}_k(t) \tag{68}$$

the stochastic response can be obviously represented similarly as the excitation in the form,

$$\mathbf{Y}(t) = \mathbf{r}_0(t)+\sum_{j=1}^{m}\zeta_j\cdot\mathbf{r}_j(t) \tag{69}$$

where ζ_j, $j=1,2,..,m$, are independent zero mean standard normally distributed RV's (unit standard deviation). Then, the second moment properties of the stochastic linear response read:

$$\mathbf{m}_{\mathbf{Y}}(t) = E\{\mathbf{Y}(t)\} = \mathbf{r}_0(t) \tag{70}$$

$$\mathbf{R}_{\mathbf{YY}}(t_1,t_2) = E\left\{\mathbf{Y}(t_1)\mathbf{Y}^T(t_2)\right\} = \sum_{k=0}^{m}\mathbf{r}_k(t_1)\mathbf{r}_k^T(t_2) \tag{71}$$

and the covariance matrix

$$\mathbf{C}_{\mathbf{YY}}(t_1,t_2) = \mathbf{E}\left\{\tilde{\mathbf{Y}}(t_1)\tilde{\mathbf{Y}}^T(t_2)\right\} = \sum_{k=1}^{m}\mathbf{r}_k(t_1)\mathbf{r}_k^T(t_2) \tag{72}$$

5. Non-Linear Stochastic Dynamic Analysis

Stochastic analyses of non-linear systems subjected to loading modeled as a stochastic process is considerably more involved when compared with the analysis of linear systems. This is due to the fact that the response can not be regarded as a superposition of responses due to single effects anymore and, moreover, the response is generally not normally distributed. Despite intensive research in this area, there are no generally applicable exact solutions available. Analytical procedures to analyse non-linear systems focus mainly on special types of SDOF systems subjected to white noise and an extension of these approaches to practically applicable MDOF systems is either not feasible or even impossible. Because of this lack of availability of exact solutions, approximate solutions must be utilized.

One of the few approaches applicable for non-linear structural MDOF systems is the method of statistical "equivalent linearization" (EQL). This method allows to estimate the second moment properties of the response in terms of the mean an the covariance matrix, where the standard deviations of the response might deviate from the correct value in the order 15% for strongly non-linear systems. Hence EQL is only capable to provide a very approximate estimate of the stochastic response of non-linear systems. It should therefore not be used as a basis for a further reliability analysis, since it does not provide sufficiently accurate information on the tails of the distribution which determine the estimates of probabilities of failure of a structure.

The method of statistical EQL is based on the assumption that the nonlinear equation of motion can be replaced by a linear one, for which analytical solutions are available. As an introduction to the problem, consider the following simple hysteretic SDOF system shown in Fig. 6a and its model in Fig. 6b.

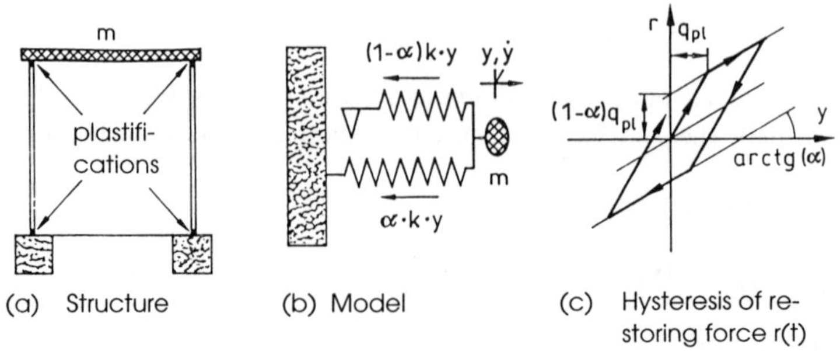

Fig. 6 Hysteretic elasto-plastic SDOF system

The equations of motion is given in the following equation, where the non-linear part *q(t)* of the restoring force undergoes hysteretic loops which are described by an additional non-linear differential equation (see Fig. 6c),

$$m\ddot{y}(t)+c\dot{y}(t)+\alpha k y(t)+(1-\alpha)kq(t) = x(t) \tag{73}$$

$$\dot{q}(t) = g(y,\dot{y},q;t) = \begin{cases} 0 & \text{for } |q| = q_{pl} \text{ and } \dot{y}q > 0 \\ \dot{y} & \text{for } |q| < q_{pl} \text{ or } \dot{y}q \le 0 \end{cases} \tag{74}$$

where $0 < \alpha < 1$ defines the linear contribution of the restoring force. Note that the first equation, i.e. eq.(73) is linear whereas the second is non-linear. The hysteresis is in this case uniquely determined by the displacement y(t), its velocity, and the so called auxiliary variable q(t) with the physical interpretation of displacements. In order to simplify the notation, all the components defining uniquely the non-linear response are comprized to a so called state vector **z**(t):

$$\mathbf{z}(t) = \begin{Bmatrix} z_1(t) \\ z_2(t) \\ z_3(t) \end{Bmatrix} = \begin{Bmatrix} y(t) \\ \dot{y}(t) \\ q(t) \end{Bmatrix} ; \quad \mathbf{x}(t) = \begin{Bmatrix} x_1(t) \\ x_2(t) \\ x_3(t) \end{Bmatrix} = \begin{Bmatrix} 0 \\ x(t) \\ 0 \end{Bmatrix} \tag{75}$$

In case of a stochastic excitation x(t), the response vector **z**(t) becomes also a stochastic process. Similarly as for the linear system, the stochastic response process can be described by the mean vector $\mathbf{m}_{\mathbf{z}}(t)$ and the correlation matrix $\mathbf{C}_{\mathbf{zz}}(t_1,t_2)$ (see eq.(2) and (3)). However, these quantities do not determine uniquely the distribution of the non-linear response, because, contrary to the linear response, naturally the non-linear response will be non-Gaussian. For a zero mean normally distributed loading process **x**(t), the stochastic response will also have zero mean, since the response is symmetric with respect to the state **z**(t). The non-linearity of the response is a function of the excitation **x**(t). For a very low or for a very high excitation (and $\alpha > 0$) the response behaves close to linear, resulting in a response close to a normal distribution, while the response will have considerable non-linear characteristics for an excitation in the middle range. In case of low excitation, the standard deviation $\sigma_y(t)$ will be considerably smaller than the yield limit q_{pl}, $\sigma_y(t) \ll q_{pl}$, leading to $q(t) \approx y(t)$ with a nearly constant stiffness k, whereas for a very high excitation the standard deviation $\sigma_y(t) \gg q_{pl}$ will exceed by far the yield limit, with a restoring force $\alpha k y(t) \pm q_{pl}$, where the linear contribution dominates. A considerable deviation from a linear behavior will be observed for excitation ranges where $\sigma_y(t) \approx q_{pl}$. This simple example demonstrates, that on one hand the non-linearity is function of the stochastic response and on the other hand, the stochastic response is obviously a function of the non-linearity. Thus, there is a mutual interaction between non-linearity and the response. This observation can be generalized for all non-linear systems.

For a description of the non-linear behavior, the notation

$$\dot{\mathbf{z}}(t) = \mathbf{g}(\mathbf{z}(t)) + \mathbf{G}\cdot\mathbf{x}(t) \tag{76}$$

might be used, where **z**(t) denotes the state vector of the system and **G** is a constant matrix of dimension (n x m) relating the excitation vector **x**(t) of dimension M to the components of the system of order n. The state vector **z**(t) must comprise all response quantities necessary for a complete description of the state of the structural system at any time t. These are in general the generalized displacement vector **y**(t), its velocity $\dot{\mathbf{y}}(t)$ and other quantities **q**(t) specifying time variant parameters of the non-linearities:

$$\mathbf{z}^T(t) = \left\{\mathbf{y}^T(t), \dot{\mathbf{y}}^T(t), \mathbf{q}^T(t)\right\}^T \tag{77}$$

For example, the component of the vector **q**(t) might contain parameters defining the non-linear restoring force, plasticity, degradation or damage. The vector **g**(**z**(t)) stands for a vector with possible non-linear function of the state vector **z**(t). To exemplify the notation, consider the above simple non-linear SDOF system specified in eq.(73) and (74). The state vector is defined by eq.(75), **g**(**z**(t)) and the matrix **G** read,

$$\mathbf{g}(\mathbf{z}(t)) = \begin{Bmatrix} g_1(\mathbf{z}) \\ g_2(\mathbf{z}) \\ g_3(\mathbf{z}) \end{Bmatrix} \quad \text{und} \quad \mathbf{G} = \begin{bmatrix} 0 \\ 1 \\ 0 \end{bmatrix} \tag{78}$$

with the interpretation:

$$\dot{z}_1(t) = \dot{y}(t) = g_1(\mathbf{z}(t)) = z_2(t) \tag{79a}$$

$$\dot{z}_2(t) = \ddot{y}(t) = g_2(\mathbf{z}(t)) = -\alpha\frac{k}{m}z_1(t) - \frac{c}{m}z_2(t) - (1-\alpha)\frac{k}{m}z_3(t) \tag{79b}$$

$$\dot{z}_3(t) = \dot{q}(t) = g_3(\mathbf{z}(t)) = \begin{cases} 0 & \text{for } |z_3(t)| = q_{pl} \text{ and } z_2(t)z_3(t) > 0 \\ z_2(t) & \text{for } |z_3(t)| < q_{pl} \text{ or } \quad z_2(t)z_3(t) < 0 \end{cases} \tag{79c}$$

The aim of equivalent linearization is the replacement of the non-linear differential equation (76) (e.g. (79c)) by an appropriate linear differential equation,

$$\mathbf{g}(\mathbf{z}(t)) \cong E\{\mathbf{g}(\mathbf{z}(t))\} + \mathbf{A}(t)\cdot(\mathbf{z}(t) - E\{\mathbf{z}(t)\}) \tag{80}$$

for which well developed procedures to determine the stochastic response are available. There are several procedures to determine the linearization coefficients of matrix **A**(t). The commonly utilized traditional approach minimizes the variance of the difference vector between the left and right hand side of eq.(80),

$$E\left\{\boldsymbol{\varepsilon}^T(t)\boldsymbol{\varepsilon}(t)\right\} \rightarrow \text{Min !} \tag{81}$$

$$\boldsymbol{\varepsilon}(t) = \tilde{\mathbf{g}}(\mathbf{z}(t)) - \mathbf{A}(t)\cdot\tilde{\mathbf{z}}(t) \tag{82}$$

using the abbreviations:

$$\tilde{\mathbf{g}}(\mathbf{z}(t)) = \mathbf{g}(\mathbf{z}(t)) - E\{\mathbf{g}(\mathbf{z}(t))\} \;;\; \tilde{\mathbf{z}}(t) = \mathbf{z}(t) - E\{\mathbf{z}(t)\} \tag{83}$$

Eq.(81) has the solution:

$$\mathbf{A}(t) \cdot E\{\tilde{\mathbf{z}}(t)\tilde{\mathbf{z}}^T(t)\} = E\{\tilde{\mathbf{g}}(\mathbf{z}(t))\tilde{\mathbf{z}}^T(t)\} \tag{84a}$$

or

$$\mathbf{A}(t) = E\{\tilde{\mathbf{g}}(\mathbf{z}(t))\tilde{\mathbf{z}}^T(t)\} \cdot E\{\tilde{\mathbf{z}}(t)\tilde{\mathbf{z}}^T(t)\}^{-1} \tag{84b}$$

The general validity of the above expression for the linearization coefficients in $\mathbf{A}$ is supported by the theorem that for any non-linear system for which the above expectations can be evaluated exactly, the associated linear system defined by the matrix $\mathbf{A}(t)$ will lead to the exact mean $\mathbf{m_Z}(t)$ and covariance matrix $\mathbf{C_{ZZ}}(t,t)$. However, the exact expectation in eq.(84) can only be determined in case the exact joint distribution $\mathbf{f_Y}(t)$ is known. Of course, linearization would not be required to determine the first two moments for such a case. Hence the linearization coefficient must be estimated without knowing the exact distribution of the response. This is generally done iteratively by assuming the non-linear response to be also normally distributed which are uniquely defined by their first two moments. The iteration is needed because of the interaction of the linearized system with the associated stochastic response. Under the assumption of a normally distributed response, the evaluation of the matrix $\mathbf{A}$ can be simplified considerably by utilizing the simple relation

$$A_{ij}(t) = E\left\{\frac{d\tilde{g}_i(\mathbf{z}(t))}{dz_j(t)}\right\} \tag{85}$$

for all required linearization coefficients. For certain types of non-linearities, closed form expressions for the linearization coefficients can be derived or found in the literature [6]. In general, however, one must resort to a numerical procedure for the evaluation of the expectations in eq.(85). For this purpose, it is suggested to generate a sufficiently large number of state vectors $\mathbf{z}(t)$ by Monte Carlo simulation procedure, and simply to estimate the expectation as sample mean.

For a fairly complete description of statistical equivalent linearization, it remains to be shown how the stochastic response of a time varying system with a non-symmetric matrix $\mathbf{A}(t)$ can be determined.

$$\dot{\mathbf{z}}(t) = \mathbf{A}(t) \cdot \mathbf{z}(t) + \mathbf{G} \cdot \mathbf{x}(t) \tag{86}$$

Analogous to eq.(30) the response of a general time varying linear system can be represented as a convolution of transition matrix $\mathbf{\Theta}(t,\tau)$ with the excitation,

$$\mathbf{z}(t) = \int_{-\infty}^{t} \mathbf{\Theta}(t,\tau) \cdot \mathbf{G} \cdot \mathbf{x}(\tau)\, d\tau \tag{87}$$

where the transition matrix satisfies the ordinary differential equation

$$\dot{\mathbf{z}}(t) = \mathbf{A}(t)\cdot\mathbf{z}(t),\ t > \tau \tag{88}$$

with

$$\mathbf{z}(t) = \boldsymbol{\Theta}(t,\tau)\cdot\mathbf{z}(\tau) \tag{89}$$

The time varying linearized system can be approximated by a step function, where $A(\tau_k) = A_k$ is assumed constant within short time intervals $\tau_k \in [t_k, t_{k+1}]$, for which the general solution reads:

$$\boldsymbol{\Theta}_k(\tau,t_k) = \boldsymbol{\Theta}_k(\tau - t_k) = e^{\mathbf{A}_k\cdot(\tau - t_k)}, \quad t_k < \tau \le t_{k+1} \tag{90}$$

The above fundamental solution can be utilized to obtain the stationary as well as the non-stationary response respectively.

The linear stochastic response has, according to eq.(67), the representation,

$$\mathbf{S}_{\mathbf{zz}}(\omega) = \mathbf{H}(\omega)\cdot\mathbf{S}_{\mathbf{xx}}(\omega)\cdot\mathbf{H}^{H}(\omega) \tag{91}$$

with

$$\mathbf{z}(\omega) = \mathbf{H}(\omega)\cdot\mathbf{x}(\omega) \tag{92}$$

where $\mathbf{H}(\omega)$ is determined by the transition matrix $\boldsymbol{\Theta}_k(\tau,t_k) = \boldsymbol{\Theta}_k(\tau - t_k)$ and the matrix $\mathbf{G}$:

$$\mathbf{H}(\omega) = \int_{-\infty}^{+\infty} \boldsymbol{\Theta}(t)\cdot\mathbf{G}\,e^{-i\omega t}dt \quad \text{with } \boldsymbol{\Theta}(t<0) = \mathbf{0} \text{ and } \boldsymbol{\Theta}(0) = \mathbf{E} \tag{93}$$

where $\mathbf{E}$ is the unity matrix. To derive the transfer matrix $\mathbf{H}(\omega)$ in a simple alternative manner, consider the special loading $\mathbf{x}(t;\omega) = [\mathbf{E}]\cdot e^{\,i\omega t}$ which leads to the response of the form $\mathbf{z}(t,\omega) = \mathbf{H}(\omega)\cdot e^{\,i\omega t}$ according to eq.(92). Introducing $\mathbf{y}(t,\omega) = \mathbf{H}(\omega)\cdot e^{\,i\omega t}$ into the equation of motion (86) leads to the following relation for the transfer matrix $\mathbf{H}(\omega)$:

$$\mathbf{H}(\omega) = [-\mathbf{A} + i\omega\mathbf{E}]^{-1}\cdot\mathbf{G} \tag{94}$$

For larger systems, the transfer matrix could be determined in a computationally efficient way by using the eigenvalue decomposition of the matrix $\mathbf{A}$. Solving the eigenvalue problem,

$$\mathbf{A}\cdot\boldsymbol{\Phi} = \boldsymbol{\Phi}\cdot\boldsymbol{\Lambda} \tag{95}$$

where the matrix $\boldsymbol{\Phi}$ contains all (complex valued) eigenvectors and $\boldsymbol{\Lambda}$ is a complex diagonal matrix, the transfer matrix reads:

$$\mathbf{H}(\omega) = \left[\sum_{j=1}^{m \le n} \frac{\phi_j \cdot \psi_j^T}{i\omega - \lambda_j} \right] \cdot \mathbf{G} \tag{96}$$

with

$$\mathbf{\Psi}^T = \mathbf{\Phi}^{-1} \tag{97}$$

and $\boldsymbol{\psi}_j$ is a left eigenvector of the eigenvalue problem,

$$\mathbf{A}^T \cdot \boldsymbol{\psi}_j = \lambda_j \cdot \boldsymbol{\psi}_j \tag{98}$$

where $m \le n$ is selected such that is covers with acceptable accuracy the frequency range of the excitation. All eigenvalues λ_j must have a negative real part. This assures the system to be stable and that a stationary solution exists.

The computation of the *non-stationary stochastic response*, as often required in earthquake engineering, is somewhat more involved. A closed form solution of the linear stochastic response can be utilized for the special case where the stochastic excitation is represented as a sum of a deterministic part $\mathbf{f}(t)$ and a purely stochastic part consisting of white noise $\mathbf{w}(t)$:

$$\mathbf{x}(t) = \mathbf{f}(t) + \mathbf{w}(t) \tag{99}$$

$$E\left\{\mathbf{w}(t) \cdot \mathbf{w}^T(t+\tau)\right\} = \mathbf{I}(t) \cdot \delta(\tau) \tag{100}$$

Then, representing the response vector $\mathbf{z}(t)$ in modal coordinates $\mathbf{u}(t)$,

$$\mathbf{z}(t) = \mathbf{\Phi} \cdot \mathbf{u}(t) \tag{101}$$

where $\mathbf{\Phi}$ comprises the right complex eigenvector of eq.(95), the ordinary differential equation (86) can be decoupled and reads,

$$\dot{\mathbf{u}}(t) = \mathbf{\Lambda} \cdot \mathbf{u}(t) + \mathbf{c}(t) + \mathbf{v}(t) \tag{102}$$

with the abbreviations,

$$\mathbf{c}(t) = \mathbf{\Psi}^T \cdot \mathbf{G} \cdot \mathbf{f}(t) \tag{103}$$

$$\mathbf{v}(t) = \mathbf{\Psi}^T \cdot \mathbf{G} \cdot \mathbf{w}(t) \tag{104}$$

following eq.(97). Given the mean vector $\mathbf{m_Z}(t_i)$ and the covariance matrix $\mathbf{C_{ZZ}}(t_i)$ at time t_i, both quantities defining a Gaussian distribution can be determined at a subsequent instant t_{i+1} as follows. Having determined the linearization coefficients in A_i, valid within the time interval $t \in [t_i, t_{i+1}]$, the eigenvalue problem in eqn.(95) and (98) is solved leading to the right and left eigenvectors $[\mathbf{\Phi}_i]$ and $[\mathbf{\Psi}_i]$, respectively. Then, the mean and the covariance matrix are transformed into the space of modal coordinates,

$$\{\boldsymbol{\mu}(t_i)\} = [\boldsymbol{\Psi}_i^T]\cdot\{\boldsymbol{\mu}_Z(t_i)\} \tag{105}$$

$$[\mathbf{S}(t_i)] = [\boldsymbol{\Psi}_i^T]\cdot\mathbf{X}_{ZZ}(t_i)\cdot[\boldsymbol{\Psi}_i^*] \tag{106}$$

where the "*" in the last equation denotes the conjugate complex matrix. The above quantities can be integrated to give for time $t_{i+1} = t_i + \tau_i$ for each component of the vector $\boldsymbol{\mu}$ and matrix $\mathbf{S}$:

$$\mu_k(t_{i+1}) = e^{\lambda_k \tau_i}\cdot\left[\mu_k(t_i) + \int_0^{\tau_i} c_k(t_i + s)\cdot e^{-\lambda_k s} ds\right] \tag{107}$$

$$S_{mn}(t_{i+1}) = e^{(\lambda_m + \lambda_n^*)\tau_i}\cdot\left[S_{mn}(t_i) + \int_0^{\tau_i} B_{mn}(t_i + s)\cdot e^{-(\lambda_m + \lambda_n^*)s}\, ds\right] \tag{108}$$

with

$$[\mathbf{B}(t_i + s)] = [\boldsymbol{\Psi}_i^T]\cdot\mathbf{G}\cdot\mathbf{I}(t_i + s)\cdot\mathbf{G}^T\cdot[\boldsymbol{\Psi}_i^*] \tag{109}$$

Transforming the above solutions back into modal coordinates, gives the desired result at time t_{i+1}:

$$\{\mathbf{m}_Z(t_{i+1})\} = [\boldsymbol{\Phi}_i]\cdot\{\boldsymbol{\mu}(t_{i+1})\} \tag{110}$$

$$[\mathbf{C}_{ZZ}(t_{i+1})] = [\boldsymbol{\Phi}_i]\cdot[\mathbf{S}(t_{i+1})]\cdot[\boldsymbol{\Phi}_i^H] \tag{111}$$

The above relation completes the evaluation of the non-stationary stochastic response within a time step. Since the linearization coefficients in matrix $\mathbf{A}_i$ depend on the stochastic response, an iteration is required which is continued until no significant change in the response (within one time interval) occurs. Because the response is changing over time, a time average of the linearization coefficients would be required. As an acceptable approximation, the linearization coefficients can be evaluated for the average response within the time interval:

$$\mathbf{m}_{Zi} = \frac{1}{2}\{\mathbf{m}_Z(t_i) + \mathbf{m}_Z(t_{i+1})\} \tag{112}$$

$$[\mathbf{C}_{ZZi}] = \frac{1}{2}[\mathbf{C}_{ZZ}(t_i) + \mathbf{C}_{ZZ}(t_{i+1})] \tag{113}$$

The most generally applicable and robust approach to estimate the stochastic response of non-linear structural systems is certainly the Monte Carlo simulation (MCS). Employing MCS, a larger number N of independent stochastic excitations is generated $\mathbf{X}_k(t)$, $0 < k \le N$ (see section 3). To each of these realizations $\mathbf{X}_k(t)$, $0 < k \le N$ the non-linear response $\mathbf{Y}_k(t)$, $0 < k \le N$ is determined by a suitable deterministic procedure integrating the nonlinear equation of motion. In engineering generally, non-linear FE codes are usually utilized for this purpose. However, any available procedure or code capable to determine the non-linear response due to prescribed deterministic loading $\mathbf{X}_k(t)$ might be used. Random properties enter the problem only at the very beginning, when the excitation or loading $\mathbf{X}_k(t)$ is generated. Hence, MCS provides a bridge between deterministic and stochastic analysis. Assuming that the non-linear equation of motion can be integrated with arbitrary accuracy, the estimates for the stochastic response depends only the sample size N of independent realizations, where the mean and covariance matrix can be estimated according to equation (2) and (3) for the stochastic process $\mathbf{Y}(t)$ as unbiased estimators:

$$\mathbf{m}_{\mathbf{Y}}(t) \approx \overline{\mathbf{m}}_{\mathbf{Y}}(t) = \frac{1}{N}\sum_{i=1}^{N}\mathbf{Y}_i(t) \tag{114}$$

$$\begin{aligned}\mathbf{C}_{\mathbf{YY}}(t_1,t_2) &\cong \overline{\mathbf{C}}_{\mathbf{YY}}(t_1,t_2) \\ &= \frac{1}{N-1}\sum_{i=1}^{N}(\mathbf{Y}_i(t_1)-\overline{\mathbf{m}}_{\mathbf{Y}}(t_1))\cdot(\mathbf{Y}_i(t_2)-\overline{\mathbf{m}}_{\mathbf{Y}}(t_2))^T\end{aligned} \tag{115}$$

The variance of the above estimators ($\overline{\mathbf{m}}_{\mathbf{Y}}(t)$ and $\overline{\mathbf{C}}_{\mathbf{YY}}(t_1,t_2)$ are inversely proportional to N, i.e.:

$$\mathrm{Var}\{\overline{m}_{Y_k}\} \approx \frac{1}{N(N-1)}\sum_{j=1}^{N}(Y_{jk}-m_{Y_k})^2 \tag{116}$$

$$\mathrm{Var}\{\overline{C}_{Y_k Y_i}\} \approx \frac{1}{N(N-1)}\sum_{j=1}^{N}[(Y_{jk}-m_{Y_k})(Y_{ji}-m_{Y_i})-\overline{C}_{Y_k Y_i}]^2 \tag{117}$$

Hence, in order to decrease the standard deviation of the estimates by a factor β, the sample size N needs to be increased by the factor β^2. This fact indicates drastically the limitation of the MCS approach. Due to the very slow convergence of MCS toward the exact values of the estimates, the number N is usually adapted to the need of the stochastic analysis.
It should be mentioned in this context, however, that the accuracy of the above estimates are independent from the dimension of the state vector $\mathbf{Y}(t)$. In this respect MCS has a clear advantage over all analytical procedures where the complexity grows quite fast with the dimension of the state.

6. References

1. Y.K. Lin, *Probabilistic Theory of Structural Dynamics* (McGraw-Hill, New York, 1967).
2. Y.K. Lin and C.Q. Cai, *Probabilistic Structural Dynamics: Advanced Theory and Applications* (McGraw-Hill, New York, 1995).
3. G.I. Schuëller and M. Shinozuka (Eds.), *Stochastic Methods in Stuctural Dynamics* (Martinus Nijhoff Publ., Dordrecht, The Netherlands, 1987).
4. G.I. Schuëller (Ed.), *Structural Dynamics - Recent Advances* (Springer Verlag, Berlin, Heidelberg, Germany, 1991).
5. T.T.Soong and M. Grigoriu, *Random Vibration of Mechanical and Structural Systems* (PTR Prentice Hall, Englewood Cliffs, New Jersey, 1995).
6. P.D. Spanos and J.B. Roberts, *Random Vibration and Statistical Linearization* (Wiley, New York 1990).

Uncertainty Modeling in Vibration and Fuzzy Analysis of Structural Systems, pp. 29–47
edited by B. M. Ayyub, A. Guran and A. Haldar
Series on Stability, Vibration and Control of Systems Series B: Vol. 10

CLUMPING EFFECT APPROXIMATIONS FOR NARROW-BAND RANDOM VIBRATION

ARVID NAESS
Faculty of Civil and Environmental Engineering
Norwegian University of Science and Technology
N-7034 Trondheim, Norway

ABSTRACT

An approximate method to account for the effect of clumping of large peaks on first passage times and extreme values of narrow-band random vibrations, both Gaussian and non-Gaussian, is proposed. The method is based on the concept of joint crossing rates of a stochastic process. This makes it possible to introduce a correlation structure to the sequence of peak values, allowing the introduction of an approximate estimate of the effect of clumping on large excursions of the underlying narrow-band process. The advantage of the proposed method is that explicit, closed-form expressons for the clumping effect on first passage times and extreme values are obtained. The method is illustrated by application to specific examples.

1. Introduction

A characteristic feature of narrow-band stationary (real) stochastic processes is the slowly varying size of consecutive peaks. A consequence of this behaviour is a tendency of large peaks to occur in clumps. It has been recognized for many years that this tendency leads to the effect that the extreme values of such processes are generally smaller than predicted by assuming the size of the individual peaks to be statistically independent variables.

The slowly varying size of the peaks of a narrow-band process clearly indicates a strong correlation between neighbouring peak values, and both theoretical work and Monte Carlo simulation studies have confirmed that the effect of such correlation is a reduction of observed extreme values and hence, increase of first passage times[1-3].

Even though the 'in- and exclusion' series of Rice[4] formally solves the problem of band-width effects on the first passage time and extreme value distribution, it is unfortunately a fact that its practical use is very difficult. This has lead to the development of a number of approximate solutions[5-11]. A survey of such solutions is given by Langley[11].

A typical feature of the available approximate solutions is that they apply only to Gaussian processes. Hence it would be of considerable practical interest if a simple approximate solution could be obtained that also applies to non-Gaussian processes. The derivation of such a method is the main purpose of this paper. The solution will be achieved by the introduction of a replacement process that has the same mean crossing rate as the given response process. The replacement process is obtained by a memoryless transformation of a stationary Gaussian process. The concept of mean joint crossing rate is subsequently applied to develop an approximate method for taking into account the effect of clumping of peaks on extreme values of the replacement process through the introduction of a correlation structure to the sequence of peak values. These estimates are then used to derive approximate expressions for the corresponding quantities of the original response process. The work presented in this chapter is an extension of previous work by the author[12].

To establish some notation and fix ideas, it is convenient to discuss briefly the simplest approximation to first passage and extreme values, in which the clumping effect is neglected.

2. The Poisson Approximation

Let $Z(t)$ denote a stationary, zero-mean process with continuous sample paths, and assume that an estimate of the mean upcrossing rate $\nu_Z^+(\zeta)$ for high levels ζ has been provided. In most cases of practical interest, $\nu_Z^+(\zeta)$ can be assumed given by a relation of the form

$$\nu_Z^+(\zeta) = \nu^+ \exp\{-g(\zeta)\} \tag{1}$$

where $g(\zeta)$ is some well-behaved, strictly increasing (for large ζ) function and ν^+ can be considered the mean crossing rate of some reference level, which is $\zeta = 0$ if $g(0) = 0$. Since this situation can always be achieved by a simple shift, it will be adopted throughout.

Let $\hat{Z}(T) = \max(Z(t), 0 \leq t \leq T)$ denote the largest value of $Z(t)$ during an interval of length T. Also, let $\Theta_B(\zeta)$ denote the random time from $t = 0$ to the first passage of $Z(t)$ through the level ζ. $\Theta_B(\zeta)$ is called the first passage time related to a one-sided, or B-type, barrier. Then $Prob\{\hat{Z}(T) \leq \zeta\} = Prob\{\Theta_B(\zeta) > T\}$, which clearly displays the complementary character of the extreme value and the first passage time problems. Similarly, let $\Theta_D(\zeta)$ denote the random time from $t = 0$ to the first passage of $|Z(t)|$ through the level ζ. $\Theta_D(\zeta)$ is then the first passage time for a two-sided, or D-type, barrier.

Let $N^+(\zeta; t)$ denote the random number of ζ-upcrossings of $Z(t)$ during a time interval of length t. Then clearly the following relation holds true

$$Prob\{\hat{Z}(T) \le \zeta\} = Prob\{Z(0) \le \zeta \ \text{ and } \ N^+(\zeta;T) = 0\} \tag{2}$$

Here, the main concern is with extreme values and first passage of high levels. Invoking the law of marginal probability, $Prob\{Z(0) \le \zeta$ and $N^+(\zeta;T) = 0\} \rightarrow Prob\{N^+(\zeta;T) = 0\}$ as $\zeta \rightarrow \infty$. In practical terms this means that for values of ζ pertinent to the distribution of extremes, one may adopt the approximation $Prob\{\hat{Z}(T) \le \zeta\} = Prob\{N^+(\zeta;T) = 0\}$. If it is assumed that the upcrossings of high levels are statistically independent events (the Poisson assumption), then the CDF of $\hat{Z}(T)$ is given by

$$F_{\hat{Z}(T)}(\zeta) = Prob\{\hat{Z}(T) \le \zeta\} = \exp\{-\nu_Z^+(\zeta)T\} \tag{3}$$

where $\nu_Z^+(\zeta) = E[N^+(\zeta;1)]$ denotes the mean rate of ζ-upcrossings of $Z(t)$. With the Poisson approximation, the survival probability $P_B(T;\zeta) = Prob\{\Theta_B(\zeta) > T\}$ is therefore also given as

$$P_B(T;\zeta) = \exp\{-\nu_Z^+(\zeta)T\}, \quad \zeta \rightarrow \infty \tag{4}$$

It has been shown that the stream of upcrossing events are asymptotically Poisson-distributed for stationary Gaussian processes under rather mild conditions[13]. In fact, one would expect this to be true for a wide class of processes met in engineering practice. Unfortunately, it turns out that the asymptotic convergence is usually exceedingly slow. Therefore, the practical consequences of this result is in some cases not very significant. Specifically, for narrow-band processes the inaccuracy implied by adopting the Poisson assumption may be quite considerable for practical values of level height and time interval. This makes it important to devise methods that make it possible to reduce the error incurred by the Poisson assumption.

3. Clumping Effect Corrections

A conspicuous feature of narrow-band processes is the fact that the magnitude of the peaks of the process changes relatively slowly with time. This implies that a large peak has large neighbouring peaks, that is, large peaks tend to occur in clumps. Hence the assumption about independent upcrossings of high levels is expectedly not accurate for narrow-band processes. In fact, it turns out that e.g. the actual first passage times of a narrow-band process are considerably longer than predicted by assuming independent upcrossing events. A large number of authors have studied this effect for stationary Gaussian processes, of which a few are listed as refs. [1-11].

From this discussion it follows that the bandwidth of the spectral density of a random process may have a significant influence on first-passage times and extreme values. More generally, this effect can be expected for any random process characterized by slowly varying peak values and up-crossing periods. In the present chapter the goal is to develop an approximate method to correct for the effect of bandwidth, or more precisely clumping of peaks, on first passage times and extreme values also for non-Gaussian processes. The basic idea is to exploit Eq. (1) to replace $Z(t)$ by a transformed Gaussian process in the following way. Let $X(t)$ denote a stationary $N(0,1)$ process with zero-upcrossing frequency ν^+. A function $h(\cdot)$ is chosen such that the process $Y(t) = h[X(t)]$ assumes the level crossing rate, cf. Eq. (1),

$$\nu_Y^+(\zeta) = \nu^+ \exp\{-g(\zeta)\} \tag{5}$$

To see how the function $h(\cdot)$ may be chosen, the following result connecting the level crossing rate ν_Y^+ of $Y(t)$ to that of $X(t)$ is useful[14,15]. Provided the necessary regularity conditions are satisfied, then

$$\nu_Y^+(\zeta) = \sum_{j=1}^{n} \nu_X^+(\xi_j) \tag{6}$$

where $\{\xi_1, \dots, \xi_n\} = h^{-1}(\zeta)$, that is $\xi_1, \dots, \xi_n$ are all possible x-solutions of the equation $\zeta = h(x)$, observing the domain of definition of h.

Assuming for simplicity that there is only one solution $\xi = h^{-1}(\zeta)$. Since $X(t)$ is Gaussian, $\nu_X^+(\xi) = \nu^+ \exp\{-\xi^2/2\}$, by the assumptions adopted. This result, together with Eqs. (5) and (6), then gives that $[h^{-1}(\zeta)]^2/2 = g(\zeta)$. Hence $g^{-1}(\xi^2/2)$ may serve as a guide for choosing $h(\xi)$.

On the basis of how $Y(t)$ is constructed, one would expect that when $Y(t)$ and $Z(t)$ are both narrow-band processes, then they must have very much the same period structure. This, together with the fact that $Y(t)$ and $Z(t)$ have the same crossing rates at high levels, leads one to assume that $Y(t)$ can be used to deduce information about the effect of bandwidth on extreme values and first passage times for $Z(t)$.

For this purpose the concept of joint mean crossing rate will be exploited in a manner similar to that adopted by Naess[16]. The assumption of a narrow-band process will be taken to imply an approximately one-to-one correspondence between peaks and zero-upcrossings. The mean number of peaks of $Y(t)$ during the time interval $(0,T)$ is therefore assumed to be $N = [\nu^+ T]$, where $[x]$ denotes the largest integer less than or equal to x.

Let A_j denote a peak of $Y(t)$ at $t = t_j$ where $0 \le t_1 \le t_2 < \dots < t_N \le T$. The PDF of $\hat{Y}(T)$ can then be written

$$\begin{aligned} F_{\hat{Y}(T)}(\zeta) &= Prob\,\{A_1 \le \zeta, \dots, A_N \le \zeta\} \\ &= Prob\,\{A_N \le \zeta \mid A_1 \le \zeta_1, \dots, A_{N-1} \le \zeta\} \\ &\quad \cdot Prob\,\{A_1 \le \zeta, \dots, A_{N-1} \le \zeta\} \end{aligned} \tag{7}$$

If the A_j's are assumed statistically independent, noting that they are identically distributed, then clearly $F_{\hat{Y}(T)}(\zeta) = F_{A_1}(\zeta)^N$, where $F_{A_1}(\cdot)$ denotes the CDF of A_1. This would lead to results similar to those obtained by assuming statistically independent upcrossings of high levels. The next approximation would quite naturally be to impose a Markov-like condition in the following way

$$Prob\{A_N \leq \zeta \mid A_1 \leq \zeta, \ldots, A_{N-1} \leq \zeta\} = Prob\{A_N \leq \zeta \mid A_{N-1} \leq \zeta\} \tag{8}$$

This equation combined with Eq. (7) leads to the relation

$$F_{\hat{Y}(T)}(\zeta) = \frac{F_{A_1A_2}(\zeta, \zeta)^{N-1}}{F_{A_1}(\zeta)^{N-2}} \tag{9}$$

where $F_{A_1A_2}(\cdot, \cdot)$ denotes the joint CDF of A_1 and A_2. In the Appendix, an estimate of the right hand side of Eq. (9) is derived. Specifically, the following approximation is obtained

$$F_{\hat{Y}(T)}(\zeta) = \exp\{-\nu^+ T p(\zeta)\} \tag{10}$$

It is shown that the function $p(\zeta)$ can be written as

$$p(\zeta) = \exp\{-g(\zeta)\}\, p_B(\zeta) \tag{11}$$

where

$$p_B(\zeta) = \frac{1 - \gamma_2 \exp\{-\frac{1 - \eta(r)}{1 + \eta(r)} \cdot g(\zeta)\}}{1 - \exp\{-g(\zeta)\}}, \tag{12}$$

$r = R_Y(T_0) / R_Y(0)$ and T_0 is the dominant period of the narrow-band vibrations. The function $\eta(r)$ is also defined in the Appendix. The parameter γ_2 accounts for difference in mean joint density of maxima and mean joint zero-upcrossing rate, as well as variability in period. As pointed out in the Appendix, a direct estimation of γ_2 from the defining relations is generally difficult. A more pragmatic approach is therefore adopted. From Eq. (12) it is seen that apart from γ_2, only the correlation coefficient r is needed to obtain the correction to $F_{\hat{Y}(T)}(\cdot)$. It is consequently expedient to assume that $\gamma_2 = q(r)$, that is, γ_2 is assumed to be a function only of r. What functional relation to choose will be studied subsequently by comparison with simulation results. By way of comment, it has been shown[17] that the clumping behaviour of Gaussian processes may show some sensitivity to the correlation structure of the process. Hence, one should not expect to find one function $q(r)$ that will work for all type of processes. However, here we shall be mainly concerned with damping-controlled random vibrations, and for a large class of such vibrations, one may expect the correlation structure to be similar.

Let $\tilde{\nu}_Y^+(\zeta) = \nu^+ p(\zeta)$. It is then seen that $\tilde{\nu}_Y^+(\zeta) = \nu_Y^+(\zeta)\, p_B(\zeta)$. Clearly, $0 \leq p_B(\zeta) \leq 1$, and $\tilde{\nu}_Y^+(\zeta)$ can be construed as a reduced level upcrossing rate that, grossly speaking, counts the effective number of independent upcrossings per unit time. Using this information concerning $Y(t)$ to correct the corresponding expression for $Z(t)$, leads to

$$\tilde{\nu}_Z^+(\zeta) = \nu_Z^+(\zeta)\, p_B(\zeta) \tag{13}$$

The extreme value distribution $F_{\hat{Z}(T)}(\zeta)$ and the survival probability $P_B(T;\zeta)$ for the single-sided barrier corrected for the bandwidth effect, would then assume the form, cf. Eqs. (3) and (4):

$$F_{\hat{Z}(T)}(\zeta) = P_B(T;\zeta) = \exp\{-\nu_Z^+(\zeta)\, p_B(\zeta) T\}, \quad \zeta \to \infty \tag{14}$$

To treat the case of the double-sided barrier, it is realized that the methodology used for the single-sided barrier is easily extended to deal also with this problem. It is reduced to the one-sided problem by introducing the process $S(t) = |Z(t)|$. By assuming that the function $g(\zeta)$ in equation (1) is symmetric, which is appropriate for the double-sided problem, it is obtained from equation (6) that $\nu_S^+(s) = 2\nu_Z^+(s)$. It is also realized that the dominant period of $S(t)$ is given by $T_0/2$. Therefore, by using the process $|Y(t)|$ to correct for the clumping behaviour, it is obtained that

$$P_D(T;\zeta) = \exp\{-2\nu_Z^+(\zeta)\, p_D(\zeta) T\}, \quad \zeta \to \infty \tag{15}$$

where

$$p_D(\zeta) = \frac{1 - \gamma_2 \exp\{-\dfrac{1-\eta(r')}{1+\eta(r')} \cdot g(\zeta)\}}{1 - \exp\{-g(\zeta)\}}, \tag{16}$$

$r' = R_Y(T_0/2)/R_Y(0)$, and γ_2 is to be replaced by $q(r')$ once the appropriate function $q(\cdot)$ has been determined.

4. Example Applications

4.1 White Noise Excited Linear Oscillator

For the linear oscillator with Gaussian white noise excitation, a lot of simulation results are available concerning the survival probability $P_s(T;\zeta)$. It is therefore possible in this case to test the assumption that the parameter γ_2 can be replaced by a function $q(r)$ of the correlation coefficient r only. The SDOF linear oscillator has the equation of motion

$$\ddot{Z} + 2\kappa\omega_0 \dot{Z} + \omega_0^2 Z = N(t) \tag{17}$$

where κ denotes relative damping, ω_0 is the undamped natural frequency and $N(t)$ is Gaussian white noise. The autocorrelation function of the response process $Z(t)$ of Eq. (17) is given by $(\tau > 0)$[18,19]

$$R_Z(\tau) = \frac{\pi S_0}{2\kappa\omega_0^3} e^{-\kappa\omega_0\tau}\{\cos\omega_d\tau + (\kappa\omega_0/\omega_d)\sin\omega_0\tau\} \tag{18}$$

where S_0 is the constant spectral density of $N(t)$ and $\omega_d = \omega_0\sqrt{1-\kappa^2}$. For κ reasonably small, it follows from Eq. (18) that $r = r_X \approx \exp\{-2\pi\kappa\}$. Note that the relation $r \approx \exp\{-2\pi\kappa\}$ would apply also to the case of a linear oscillator excited by a non-Gaussian load process.

Since $0 \le q(r) \le 1$, and $q(r) \to 1$ as $r \to 1$, it seems reasonable to try the relation $q(r) = r^n$ for some n for the range of values $0 << r \le 1$, which is the region of primary interest here. Comparison with available simulation results for the Gaussian case has shown that $n = 2$ does in fact provide quite good agreement. In this case, $g(\zeta) = \zeta^2/2$ in Eq. (11) and the modifying function $p_B(\zeta)$ is taken to be

$$p_B(\zeta) = \frac{1 - r^2 \exp\{-\frac{1-r}{1+r}\frac{\zeta^2}{2}\}}{1 - \exp\{-\frac{\zeta^2}{2}\}} \tag{19}$$

with $r = \exp\{-2\pi\kappa\}$. In Figure 1a, $p_B(\zeta)$ and corresponding simulation results[1] are plotted for various values of κ. Similarly, in Figure 1b, the function $p_D(\zeta)$ for the double-sided barrier is compared with simulation[1,8]. In this case $r' = \exp\{-\pi\kappa\}$. It is seen that the level of agreement is generally very good.

In several design applications one uses the expected largest value, or the level not exceeded by a prescribed probability during a given time, as a characteristic largest value in a reliability format. Let the response variable in question be described by a narrow-band stationary Gaussian process $Z(t)$ with a mean level upcrossing rate[18,19] $\nu_Z^+(\zeta) = \nu^+\exp\{-\zeta^2/(2\sigma^2)\}$ where $\sigma^2 = Var[Z(t)]$ = variance of $Z(t)$. We now proceed to calculate the expected largest value of $Z(t)$ during a given time interval T, that is $E[\hat{Z}(T)]$, using Eq. (14). It is convenient to write $F_{\hat{Z}(T)}(\zeta) = \exp\{-\exp\{-\psi(\zeta)\}\}$ where $\psi(\zeta) = \zeta^2/(2\sigma^2) - \ln p_B(\zeta) - \ln(\nu^+T)$. $E[\hat{Z}(T)]$ can then be calculated using the formula[16,20]

$$E[\hat{Z}(T)] \approx \zeta_0 + \frac{\gamma}{\psi'(\zeta_0)} - \frac{1}{2}\left(\frac{\pi}{6} + \gamma^2\right)\frac{\psi''(\zeta_0)}{(\psi'(\zeta_0))^3} \tag{20}$$

where $\gamma = 0.5772...$ denotes Euler's constant, ζ_0 is the appropriate solution of the equation $\psi(\zeta) = 0$, $\psi'(\zeta) = d\psi(\zeta)/d\zeta$, and $\psi''(\zeta) = d^2\psi(\zeta)/d\zeta^2$. It can be shown that for the special case of a Gaussian process

$$\zeta_0 = (2\sigma^2 Q)^{1/2} \tag{21}$$

where

$$Q = \ln[\nu^+ T(1 - r^2(\nu^+ T)^{-\delta})] \tag{22}$$

and

$$\delta = \frac{1 - r}{1 + r} \tag{23}$$

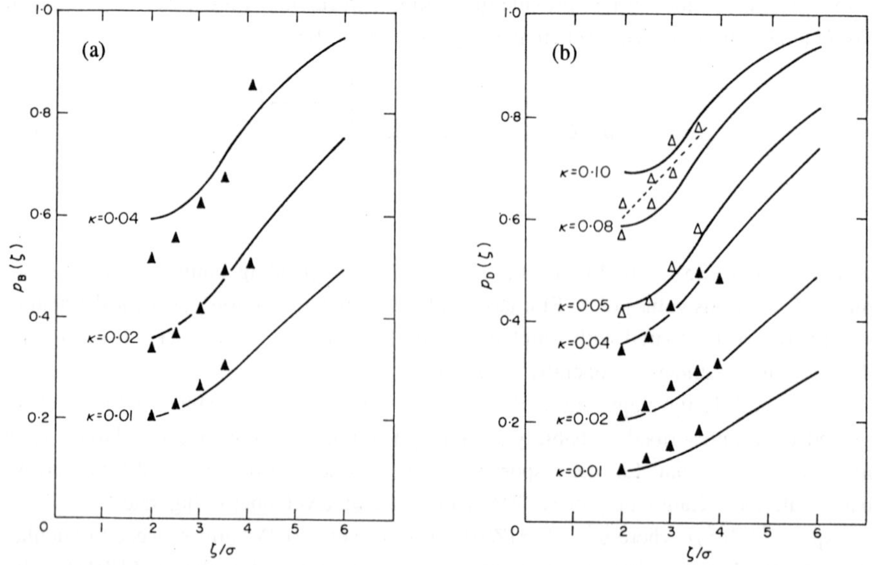

Figure 1a-b Variation of $p_B(\zeta)$ and $p_D(\zeta)$ for different values of the damping ratio κ. ----: Present method; ▲: Simulation results[1]; △: Simulation results[8].

This leads to the following expression

$$E[\hat{Z}(T)] \approx (2\sigma^2 Q)^{1/2}\{1 + \frac{\gamma}{2} Q^{-1} - \frac{1}{8}(\frac{\pi^2}{6} + \gamma^2) Q^{-2}\} \tag{24}$$

Preumont[17] has given simulation results for the so-called peak factor of a white noise excited linear oscillator. In our notation, the peak factor can be defined as $E[\hat{Z}(T)]/\sigma$. Table 1 compares Preumont's results with the corresponding estimates derived from equation (24). The maximum discrepancy in the table is about 4 %.

Table 1 Peak factor for a linear oscillator with white noise excitation

Damping ratio κ	Peak factor	
	Simulations	Present method
0·001	2·01	2·03
0·002	2·29	2·37
0·003	2·46	2·55
0·005	2·62	2·73
0·008	2·81	2·89
0·01	2·87	2·95
0·02	3·06	3·14
0·04	3·22	3·29
0·06	3·30	3·34
0·08	3·36	3·37
0·10	3·36	3·39
0·15	3·39	3·40
0·20	3·38	3·40
0·25	3·37	3·40

The choice $q(r) = r^2$ for the Gaussian case corresponds to

$$q(r) = \eta(r)^2 \tag{25}$$

in the general case, cf Appendix. This functional relation is tentatively adopted in the following, and γ_2 in Eq. (12) is to be replaced by $\eta(r)^2$. Likewise, γ_2' in Eq. (16) is to be replaced by $\eta(r')^2$.

4.2 The Duffing Oscillator

The Duffing oscillator may serve as a good example of the use of the general method outlined in this chapter. The case of the white noise excited oscillator with a stiffening spring will be considered. The following equation of motion applies ($\varepsilon > 0$)

$$\ddot{Z} + 2\kappa\omega_0\dot{Z} + \omega_0^2(1 + \varepsilon Z^2)Z = N(t) \tag{26}$$

The mean upcrossing rate of this oscillator is given by[18,19]

$$\nu_Z^+(\zeta) = \nu_Z^+(0)\exp\{-\frac{1}{2}b^2 - \frac{1}{4}\varepsilon^* b^4\} \tag{27}$$

where $b = \zeta/\sigma_o$, $\sigma_o^2 = Var[Z_o(t)]$, $Z_o(t)$ denotes the response process for $\varepsilon = 0$, and $\varepsilon^* = \varepsilon\sigma_o^2$.

From Eq. (27) it is seen that in this particular case

$$g(\zeta) = g(\zeta;\varepsilon^*) = \frac{1}{2}b^2 + \frac{1}{4}\varepsilon^* b^4 \tag{28}$$

This leads to the following choice of the function $h(\cdot)$

$$h(\xi) = sign(\xi)\,\{\frac{1}{\varepsilon^*}[\sqrt{2\varepsilon^*\xi^2 + 1} - 1]\}^{1/2} \tag{29}$$

Of particular interest now is the relation between r and r_X given by the equation $r = \psi(r_X)$, cf. Eq. (A13). This relationship has been plotted in Figure 2 for various values of the nonlinearity parameter ε^*. It is seen that this relationship is, in fact, to a large extent unaffected by the degree of nonlinearity of the oscillator. This leads us to assume that $r \approx r_X = \exp\{-\kappa T_o\}$, where $T_o = 1/\nu_Z^+(0)$. That is, instead of estimating r from the response time series, it is assumed that the Gaussian process $X(t)$ is the normalized response of the (linear) oscillator with $\varepsilon = 0$ and $\omega_o = 2\pi\nu_Z^+(0)$. This is supported by a study of the simulated time series.

As examples, we have presented in Figure 3 the results of 100 numerical simulations of response time series to estimate the expected extreme response for two values of κ and two values of ε^*. The corresponding predictions calculated from Eq. (20), which has general applicability, are also given, and it is seen that the agreement is quite satisfactory. For comparison, the predictions provided by neglecting the correlation effect (Poisson assumption) are also shown.

In the case of a nonlinear oscillator subjected to random excitation, like the Duffing oscillator, the correlation coefficient r may not be readily available. In such a case, it can be estimated, e.g., by a direct correlation analysis of a time series of the peak values of the response $Z(t)$, obtained either by simulation or measurement.

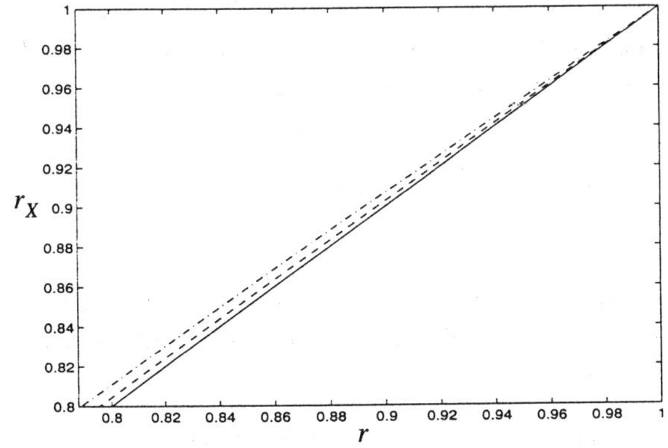

Figure 2 Plots of $r_X = \eta(r)$ for $\varepsilon^* = 0$ (——), $\varepsilon^* = 1.0$ (-·-·-), $\varepsilon^* = 5.0$ (----)

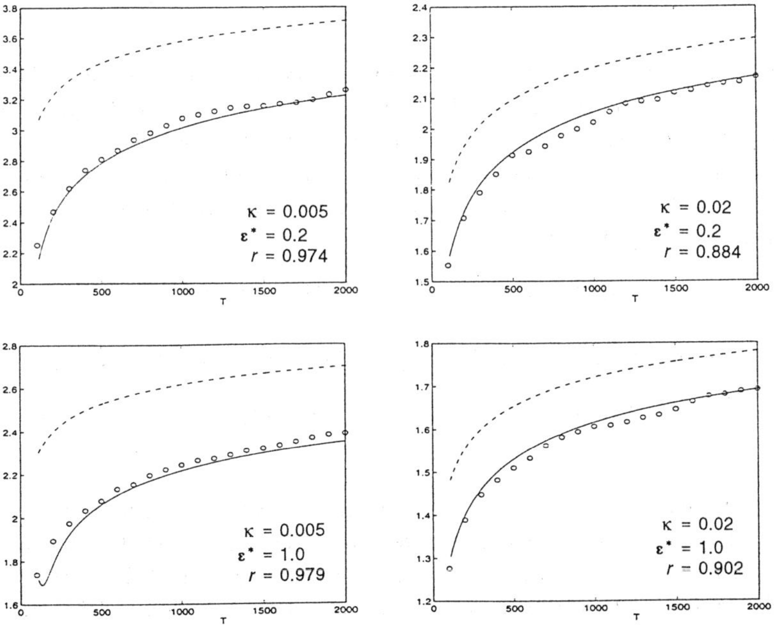

Figure 3 Plots of $E[\hat{Z}(T)]$. Equation (20) (——), Poisson assumption (---), Simulations (o)

4.3 Non-Gaussian Excited Linear Oscillator

The correction procedure developed will also be tested on the response of a SDOF linear oscillator described by the left side of Eq. (17), which is excited by a non-linear, second order transformation of a Gaussian process. Specifically, let $G(t)$ denote a stationary, zero-mean Gaussian process; then the excitation force is given by

$$F(t) = \int_0^\infty \int_0^\infty k_2(\tau_1,\tau_2)\, G(t-\tau_1)\, G(t-\tau_2)\, d\tau_1\, d\tau_2 \tag{30}$$

where $k_2(\tau_1,\tau_2)$ is a function characterizing the force mechanism. If the linear oscillator in Eq. (17) is excited by $F(t)$, then the response process can be written as[21]

$$Z(t) = \int_0^\infty \int_0^\infty h_2(\tau_1,\tau_2)\, G(t-\tau_1)\, G(t-\tau_2)\, d\tau_1\, d\tau_2 \tag{31}$$

where

$$h_2(\tau_1,\tau_2) = \int_0^\infty l(t)\, k_2(\tau_1 - t, \tau_2 - t)\, dt \tag{32}$$

and $l(t)$ denotes the linear impulse response function of the oscillator. The response statistics of a process given as in Eqs. (30) and (31) has been studied, e.g., by Naess[21-23]. In particular, it has been shown that under certain conditions, the so-called slow-drift response of a moored offshore structure can be described by Eq. (31). It has been shown that in such cases the mean level upcrossing rate of the slow-drift response can be estimated by $(\zeta > 0, \mu_1 > 0)$

$$\nu_Z^+(\zeta) = \nu^+ \sqrt{\zeta/\mu}\, \exp\{-\zeta/2\mu\}[1 - a(\zeta/\mu)^{-1}]\,, \quad \zeta \to \infty \tag{33}$$

where the parameters μ, ν^+ and a are given in reference [21]. In this case

$$g(\zeta) = (\zeta/2\mu) - \frac{1}{2}\ln(\zeta/\mu) - \ln[1 - a(\zeta/\mu)^{-1})], \qquad \zeta \to \infty \tag{34}$$

It is seen that $g(\zeta) \approx \zeta/(2\mu)$ for large values of ζ. Since we are primarily interested in the effect of clumping on extreme values, it is appropriate to adopt this approximation for calculating $\eta(r)$ in the present case. It is then obtained that $h(\xi) = \mu|\xi|\xi$, and the relationship between r and r_X can be calculated. According to Eq. (18) and the subsequent discussion, $r = R_Z(T_0)/R_Z(0) \approx \exp\{-2\pi\kappa\}$ where $T_0 = 2\pi/\omega_0$ is the slow-drift period if the spectral density of the excitation force is approximately constant over the resonance peak of the linear transfer function of the dynamic model, which is a reasonable assumption in the present case.

Zhao and Faltinsen[24] have carried out extensive Monte Carlo simulations of the slow-drift response of a moored infinitely long cylinder in deep-water beam sea waves: that is, unidirectional waves with the wave crests parallel to the cylinder axis. This results in a two-dimensional problem. The axis of the undisturbed cylinder is located in the mean free surface and the beam at the water line (diameter) is 30 m. The linear mooring system is assumed to give a natural sway period of 100 s. The wave elevation process $G(t)$ in Eq. (31), is synthesized from a (one-sided) spectral density $S_G(\omega)$ of the ISSC (International Ship Structures Congress) type

$$S_G(\omega) = (173\, H_S^2 / T_2^4\, \omega^5) \exp\{ -691 / T_2^4\, \omega^4 \} \tag{35}$$

where the significant wave height $H_S = 4\sqrt{m_0}$ and the average wave period $T_2 = 2\pi\sqrt{m_0/m_2}$. Here m_k denotes the kth spectral moment of $G(t)$.

The estimates for the expected largest sway excursions of the moored cylinder based on numerical simulations and the theory presented here, are given in Figures 4 - 6 for different values of relative damping and sea state parameters. Corresponding results obtained by assuming the response to be a Gaussian process are also presented. The reason for this is that one may sometimes see the following argument for assuming a response process to be Gaussian: In the limit of zero damping, the response of a linear dynamic system tend to become Gaussian for a wide range of excitation processes. Therefore, in the case of low damping, it may seem a reasonable approximation to assume that the response is Gaussian. From the following discussion it will be recognized that for the purpose of estimating extreme responses, such an approximation should be used with some care.

From Figure 4 it is seen that in the case of relatively high damping, $\kappa = 0.145$, the predicted effect of clumping is small. The agreement between theoretical prediction and simulation results is fairly good. Assuming the response to be Gaussian is seen to give significant underprediction.

In Figure 5 the results correspond to a damping ratio $\kappa = 0.043$. In this case it is seen that the predicted effect of clumping is quite large, and that the theoretical estimates corrected for this effect agrees well with the numerical simulation results.Also in this case it is seen that the assumption of a Gaussian response process would lead to significant underprediction.

Entirely similar comments apply to the results given in Figure 6, which concern a case where the damping ratio $\kappa = 0.036$.

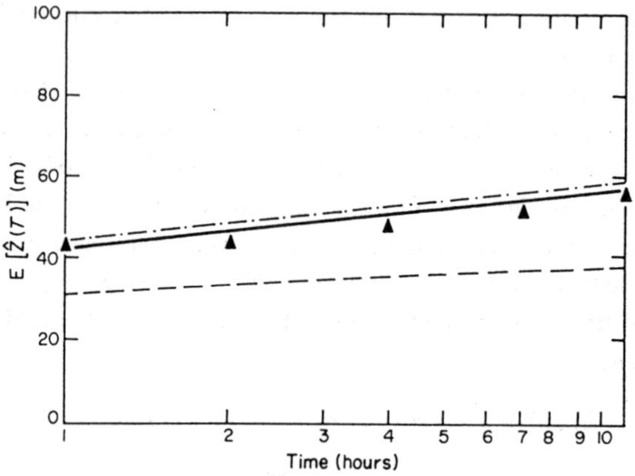

Figure 4 Expected largest slow-drift sway response of moored cylinder (mean value included) for H_s = 10 m and T_2 = 8 s. Damping ratio κ = 0.145. ———, estimates corrected for clumping effects; -·-·-·, estimates without correction; - - - -, estimates based on Gaussian assumption; ▲, simulation results[24].

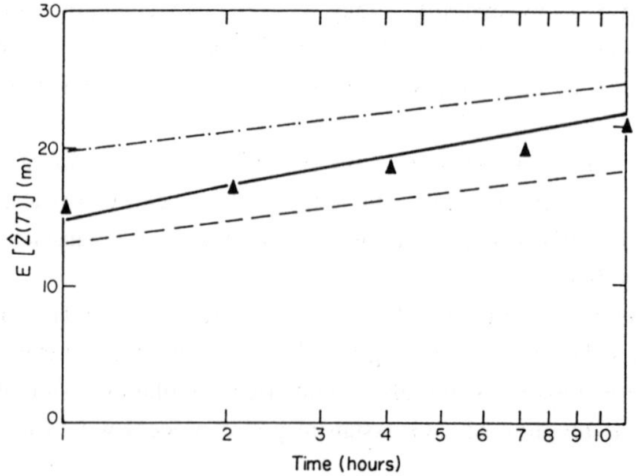

Figure 5 As Figure 4, but H_s = 11 m, T_2 = 12 s and κ = 0.043.

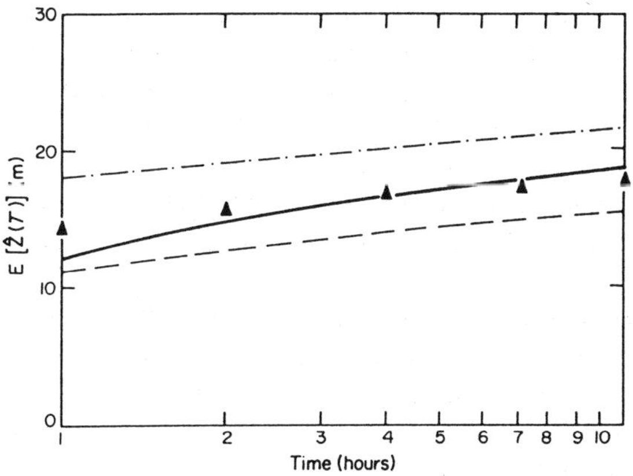

Figure 6 As Figure 4, but H_s = 5 m, T_2 = 8 s and κ = 0.036.

Acknowledgements The author is grateful to Mr. T.E. Hagen for carrying out some of the numerical work presented in this chapter.

5. References

1. S H Crandall, K L Chandiramani and R G Cook, *Journal of Applied Mechanics*, **33** (1966) 532-538.
2. S H Crandall, *Journal of Sound and Vibration*, **12** (1970) 285-299.
3. J N Yang and M Shinozuka, *Journal of Applied Mechanics*, **38** (1971) 1017-1022.
4. S O Rice, in *Selected Papers on Noise and Stochastic Processes*, ed. N. Wax, (Dover, New York, 1954 reprint).
5. J N Yang, *Journal of the Engineering Mechanics Division American Society of Civil Engineers*, **101** (1975) 361-372.
6. E H Vanmarcke, *Journal of Applied Mechanics*, **42** (1975) 215-220.
7. J B Roberts, *Journal of Sound and Vibration*, **46** (1976) 1-14.
8. L D Lutes, Y T-T Chen and S H Tzuang, *Journal of the Engineering Mechanics Division American Society of Civil Engineers*, **106** (1980) 1111-1124.
9. A B Mason and W D Iwan, *Journal of Applied Mechanics*, **50** (1983) 641-646.
10. A Naess, *Journal of Sound and Vibration*, **94**(1) (1984) 87-103.
11. R S Langley, *Journal of Sound and Vibration* , **122** (1988) 261-275.
12. A Naess, *Journal of Sound and Vibration*, **138**(3) (1990) 365-380.

13. M R Leadbetter, G Lindgren and H Rootzen, *Extremes and related properties of random sequences and processes*, (Springer-Verlag, New York, 1983).
14. M Grigoriu, *Journal of Engineering Mechanics*, ASCE, **110**(EM4) (1984) 610-620.
15. A Naess, *Ocean Engineering*, **10**(5) (1983) 313-324.
16. A Naess, *Applied Ocean Research*, **7**(1) (1985) 35-50.
17. A Preumont, *Journal of Sound and Vibration*, **100** (1985) 15-34.
18. Y K Lin, *Probabilistic Theory of Structural Dynamics*, (McGraw-Hill, New York, 1967).
19. N C Nigam, *Introduction to Random Vibrations*, (The MIT Press, Cambridge, 1983).
20. E J Gumbel, *Statistics of Extremes*, (Columbia University Press, New York, 1958).
21. A Naess, *Journal of Ship Research* **29**(4) (1985) 270-284.
22. A Naess, *Proceedings 5th International Conference on Offshore Mechanics and Arctic Engineering*, Tokyo, Japan, **1** (1986) 317-329.
23. A Naess, *Journal of Sound and Vibration*, **115** (1987) 103-129.
24. R Zhao and O M Faltinsen, *Proceedings 7th International Conference on Offshore Mechanics and Arctic Engineering*, Houston, Texas, (1988).

6. Appendix

Let $\mu_Y^+(\zeta)$ denote the mean member of peaks of $Y(t)$ above the level ζ per unit time. Then

$$Prob\{A_1 > \zeta\} = \frac{\mu_Y^+(\zeta)}{\mu_Y^+(-\infty)} \tag{A.1}$$

For the processes considered here it will be assumed that $\mu_Y^+(\zeta)/\nu_Y^+(\zeta) \to 1$ as $\zeta \to \infty$. That is, between an upcrossing and the subsequent downcrossing of a high level there tend to be only one peak. One may therefore write

$$Prob\{A_1 > \zeta\} = \gamma_1 \frac{\nu_Y^+(\zeta)}{\nu_Y^+(0)}, \quad \zeta \to \infty \tag{A.2}$$

where $\gamma_1 = \nu_Y^+(0)/\mu_Y^+(-\infty)$, $0 < \gamma_1 \leq 1$. With the assumption that there is a one-to-one correspondence between peaks and upcrossings, one gets $\gamma_1 = 1$. With this approximation and Eq. (5) it follows that

$$Prob\{A_1 > \zeta\} = \exp\{-g(\zeta)\} \tag{A.3}$$

Similarly, let $\mu_Y^{++}(\zeta, \zeta; t)$ denote the mean joint number of maxima of $Y(t)$ above ζ

at separation t[16]. Then

$$Prob\{A_1 > \zeta,\, A_2 > \zeta \mid t\} = \frac{\mu_Y^{++}(\zeta,\zeta\,;t)}{\mu_Y^{++}(-\infty,\ -\infty;\ t)} \tag{A.4}$$

Denoting the PDF of the time between consecutive maxima, $\tilde{T}$, by $f_{\tilde{T}}(\cdot)$, it follows that

$$Prob\{A_1 > \zeta,\, A_2 > \zeta\} = \int_0^\infty Prob\{A_1 > \zeta,\, A_2 > \zeta \mid t\} f_{\tilde{T}}(t)\, dt \tag{A.5}$$

In the limit of a narrow-band process $f_{\tilde{T}}(t) \approx \delta(t - T_0)$ where T_0 is the dominant period of the process and δ denotes Dirac's delta symbol. We shall have more to say about the choice of the dominant period T_0 below. In the limit of high thresholds one also expects that the ratio $\mu_Y^{++}(\zeta,\zeta\,;t)\,/\,\nu_Y^{++}(\zeta,\zeta\,;t)$ approaches 1.0, where $\nu_Y^{++}(\zeta,\zeta\,;t)$ denotes the mean joint upcrossing frequency at separation t of $Y(t)$[16]. The following approximation is now adopted

$$Prob\{A_1 > \zeta,\, A_2 > \zeta\} = \gamma_2\, \frac{\nu_Y^{++}(\zeta,\zeta\,;T_0)}{\nu_Y^{++}(0,0\,;T_0)} \tag{A.6}$$

Here $\gamma_2 = \gamma_3 \cdot \gamma_4$, where $\gamma_3 = \nu_Y^{++}(0,0\,;T_0)\,/\,\mu_Y^{++}(-\infty,-\infty\,;T_0)$ and γ_4 is a factor compensating for the effect of putting $f_{\tilde{T}}(t) = \delta(t - T_0)$. It can be shown that $0 < \gamma_2 \leq 1$. In general it is very difficult to estimate γ_3 and γ_4 directly, and even if $\gamma_1 = 1$ is a good approximation, this may not be so for γ_2. Instead of entering a detailed discussion of possible estimation procedures for γ_2, a more pragmatic approach has been chosen. This is discussed in the main text of the chapter.

It can be shown[16] that a result very similar to Eq. (6) also holds for mean joint crossing rates. Specifically, it follows that ($\sqrt{2g(\zeta)} = h^{-1}(\zeta)$)

$$\nu_Y^{++}(\zeta,\zeta\,;T_0) = \nu_X^{++}(\sqrt{2g(\zeta)},\, \sqrt{2g(\zeta)};\ T_0) \tag{A.7}$$

According to Naess[16],

$$\nu_X^{++}(\xi,\xi\,;T_0) = \nu_X^{++}(0,0\,;T_0)\exp\{-\frac{\xi^2}{1 + r_X}\} \tag{A.8}$$

where

$$r_X = R_X(T_0)\,/\,R_X(0) = R_X(T_0) \tag{A.9}$$

and $R_X(\tau) = E[X(t)\,X(t + \tau)]$ denotes the autocorrelation function of $X(t)$. The equality in Eq. (A8) depends on choosing T_0 such that $R_X'(T_0) = dR_X(t=T_0)/dt = 0$. In general $T_m \le T_0 \le T_z$ where T_z (T_m) denotes the mean time interval between zero upcrossings (peaks) of $X(t)$. For a narrow-band process $T_0 \approx T_z$. A closer study of the choice of dominant period reveals that a slightly different value than the one chosen here would be appropriate if the derivations had been based on joint number of peaks instead of joint rate of crossings. This difficulty will here be absorbed in the problem of chosing the correction factor γ_2. A further discussion of this issue will therefore not be pursued here.

It is seen from Eqs. (A7) and (A8) that to determine the left side of Eq. (A7), an estimate of the correlation coefficient r_X of $X(t)$ is required. What can be assumed available is a correlation coefficient r derived from the original process $Z(t)$. Having adopted $Y(t)$ as an approximation to $Z(t)$, it is natural to require that r is chosen in such a manner that we may adopt the relation $r_Y = R_Y(T_0)/R_Y(0) = r$. Due to the fact that the upcrossing periods of nonlinear vibrations depend in a systematic manner on the response level, it is not appropriate to use the correlation coefficient $r = R_Z(T_0)/R_Z(0)$ for any specific choice of T_0. The reason being that the correlation information of relevance is directly related to the sequence of peak values, which would clearly not be properly accounted for by the correlation coefficient $r = R_Z(T_0)/R_Z(0)$ since the time separation of peak values of $Z(t)$ would in general depend on the response level. To find the relation between r_Y and r_X, the following formula is used

$$\begin{aligned} R_Y(T_0) &= E[Y(t)\,Y(t+T_0)] = E[h[X(t)]\,h[X(t+T_0)]] \\ &= \int_{-\infty}^{\infty}\int_{-\infty}^{\infty} h(u)\,h(v)\,\varphi_2(u,v;r_X)\,du\,dv \end{aligned} \tag{A.10}$$

where $\varphi_2(\cdot,\cdot;r)$ denotes the joint PDF of two correlated $N(0,1)$-variables with correlation coefficient r. That is,

$$\varphi_2(u,v;r) = \frac{1}{2\pi\sqrt{1-r^2}}\exp\{\frac{-1}{2(1-r^2)}[u^2 - 2ruv + v^2]\} \tag{A.11}$$

It is seen from Eq. (A10) that $R_Y(T_0)$ is only a function of the correlation coefficient r_X. We shall henceforth denote this function by $\psi_0(r_X)$. It is also seen that

$$R_Y(0) = \psi_0(1) \tag{A.12}$$

Hence it is obtained that

$$r = \psi(r_X) = \frac{\psi_0(r_X)}{\psi_0(1)} \tag{A.13}$$

Assuming that $\psi(\cdot)$ is invertible, we may write

$$r_X = \eta(r) \tag{A.14}$$

where $\eta(x) = \psi^{-1}(x)$.

From Eqs. (A6) - (A8) and (A14) the following approximation is obtained.

$$Prob\{A_1 > \zeta, A_2 > \zeta\} = \gamma_2 \exp\{-\frac{2g(\zeta)}{1+\eta(r)}\} \tag{A.15}$$

Eqs. (9), (A3) and (A15) are now combined to give

$$F_{\hat{Y}(T)}(\zeta) = [1 - m(\zeta)]\{1 - \frac{m(\zeta) - n(\zeta)}{1 - m(\zeta)}\}^{N-1} \tag{A.16}$$

where

$$m(\zeta) = \exp\{-g(\zeta)\} \tag{A.17}$$

and

$$n(\zeta) = \gamma_2 \exp\{-\frac{2g(\zeta)}{1+\eta(r)}\} \tag{A.18}$$

Let

$$p(\zeta) = \frac{m(\zeta) - n(\zeta)}{1 - m(\zeta)} \tag{A.19}$$

Then as N, or equivalently T, increases, the relevant range of ζ-values changes so that $Np(\zeta)$ remains effectively bounded. Hence as N increases, ζ also increases, and the following approximation can be adopted.

$$F_{\hat{Y}(T)}(\zeta) = \exp\{-Np(\zeta)\} = \exp\{-\nu^+ Tp(\zeta)\}, \quad \zeta \to \infty \tag{A.20}$$

By rewriting Eq. (A.19), it follows that $p(\zeta)$ can be expressed as

$$p(\zeta) = \exp\{-g(\zeta)\}\, p_B(\zeta) \tag{A.21}$$

where

$$p_B(\zeta) = \frac{1 - \gamma_2 \exp\{-\frac{1-\eta(r)}{1+\eta(r)} \cdot g(\zeta)\}}{1 - \exp\{-g(\zeta)\}}, \tag{A.22}$$

Uncertainty Modeling in Vibration and Fuzzy Analysis of Structural Systems, pp. 49–100
edited by B. M. Ayyub, A. Guran and A. Haldar
Series on Stability, Vibration and Control of Systems Series B: Vol. 10

STATE SPACE TRANSPORT ACROSS SEPARATRICES: DYNAMICAL CONTROL AND RELATIVE STABILITY

MICHAEL FREY
Department of Mathematics, Bucknell University
Lewisburg, PA 17837, USA

ABSTRACT

A separatrix in the state space of a dynamical system is a boundary between regions of qualitatively dissimilar modes of dynamics. In the presence of weak external forcing the separatrix can be breached, allowing the system to change dynamical modes. Such breaches are a root cause of chaos in externally forced, continuous-time systems. The flux factor, a leading order expression of the state space flux across the separatrix, measures the "propensity" for chaotic dynamics. Two new uses of the flux factor are given. In the first of these, the flux factor is an objective function for a control against crossing a separatrix to maintain a system within a potential well. In the second, the flux factor is the basis of a measure of the relative stability of competing potential wells. This measure of relative stability is used to formulate and prove a "blowtorch" theorem for potential wells.

1. Introduction

The separatrices of a dynamical system partition the system state space into regions with qualtitatively different long-term dynamics. This idea is useful to understanding the different modes of dynamics of which a system is capable. In certain particular senses too—among them the possibility of chaos in externally forced systems—separatrices play a central role. This chapter's subject is two newly recognized roles for separatrices and the state space flux across them: dynamical control and relative stability. The chapter begins with a review of some standard terminology related to dynamical systems and then illustrates the idea of a separatrix with three examples of two-dimensional dynamical systems: the Lotka-Volterra model of species competition, the Duffing oscillator and the ocean vessel capsizing model.

1.1. Dynamical Systems and Separatrices

A continuous-time dynamical system is one whose state changes with time t according to an n-dimensional, first-order differential equation of the form

$$\frac{d\vec{z}}{dt} = \dot{\vec{z}} = \vec{\pi}(\vec{z}) \tag{1}$$

with velocity field $\vec{\pi}(\vec{z}) = (\pi_1(\vec{z}), ..., \pi_n(\vec{z}))$ where $\vec{z} = \vec{z}(t) = (z_1(t), ..., z_n(t))$ is the state of the system at time t and takes values in the system's state space $\mathcal{X}$. Solutions of system Eq. (1) are assumed to be given by a flow $\phi_t(\vec{z})$ defined such that

$$\frac{\partial}{\partial t}\phi_t(\vec{z}) = \vec{\pi}(\phi_t(\vec{z})) \ , \quad \phi_0(\vec{z}) = \vec{z} \tag{2}$$

for each $\vec{z} \in \mathcal{X}$. Flows have the property that $\phi_s(\phi_t(\vec{z})) = \phi_{s+t}(\vec{z})$ for each $\vec{z} \in \mathcal{X}$ and $s, t \in \Re$. Geometrically, the solution $\{\phi_t(\vec{z}_0), t \in \Re\}$ of system Eq. (1) describes a curve (orbit, trajectory, path) in state space which passes through $\vec{z}_0$ at $t = 0$. The set of all orbits of a system flow is called the state space portrait of the system.

A separatrix $\partial\mathcal{S}$ in the state space $\mathcal{X}$ of a dynamical system must meet two conditions. First, $\partial\mathcal{S}$ must be the boundary of an open set $\mathcal{S}$ where $\mathcal{S}$ and the complement $\mathcal{T} = (\mathcal{S} \cup \partial\mathcal{S})^c$ of its closure are nonempty and disjoint. Second, the sets $\mathcal{S}$ and $\mathcal{T}$ must each be invariant under the system flow $\phi_t(\vec{z})$; that is, $\phi_t(\mathcal{S}) \subseteq \mathcal{S}$ and $\phi_t(\mathcal{T}) \subseteq \mathcal{T}$ for all t. A dynamical system may have more than one separatrix. Some examples of system separatrices follow.

The Lotka-Volterra model is a two-dimensional dynamical system describing the populations of two species competing for a common food supply[7]. Making standard assumptions about the growth rate of each population and the conflict that arises between the populations we have

$$\begin{aligned} \dot{u} &= 4u - u^2 - 3uv \\ \dot{v} &= 2v - v^2 - uv \end{aligned} \tag{3}$$

where $u(t)$ and $v(t)$ are the populations of the two species at any given time t and the notation $\dot{u}$, for example, signifies the time derivative of $u(t)$. Here numerical values have been assigned arbitrarily to the model's various parameters. The state space portrait of the system is shown in Figure 1(a). As seen there, if the system is at any time in a state above and to the left of the bold solid line in the portrait then it remains so always, eventually reaching the point $(u, v) = (0, 3)$ of extinction of the first species. Starting out in any state below and to the right of the bold solid line, the two populations necessarily evolve toward a point $(u, v) = (4, 0)$ of extinction of the second species. The bold solid line in the state space portrait is therefore a separatrix. Absent any outside influence, the system state cannot cross the separatrix and the destiny of each species is determined by the initial state of the system. As we shall see, if an external force is included in the model, passage across the separatrix becomes possible.

The Duffing oscillator is a simplified mathematical model of a variety of physical systems, among them a buckled beam and a slender, suspended, magnetoelastic strip[16]. The Duffing oscillator takes the form

$$\begin{aligned} \dot{x} &= v \\ \dot{v} &= x - x^3 \end{aligned} \tag{4}$$

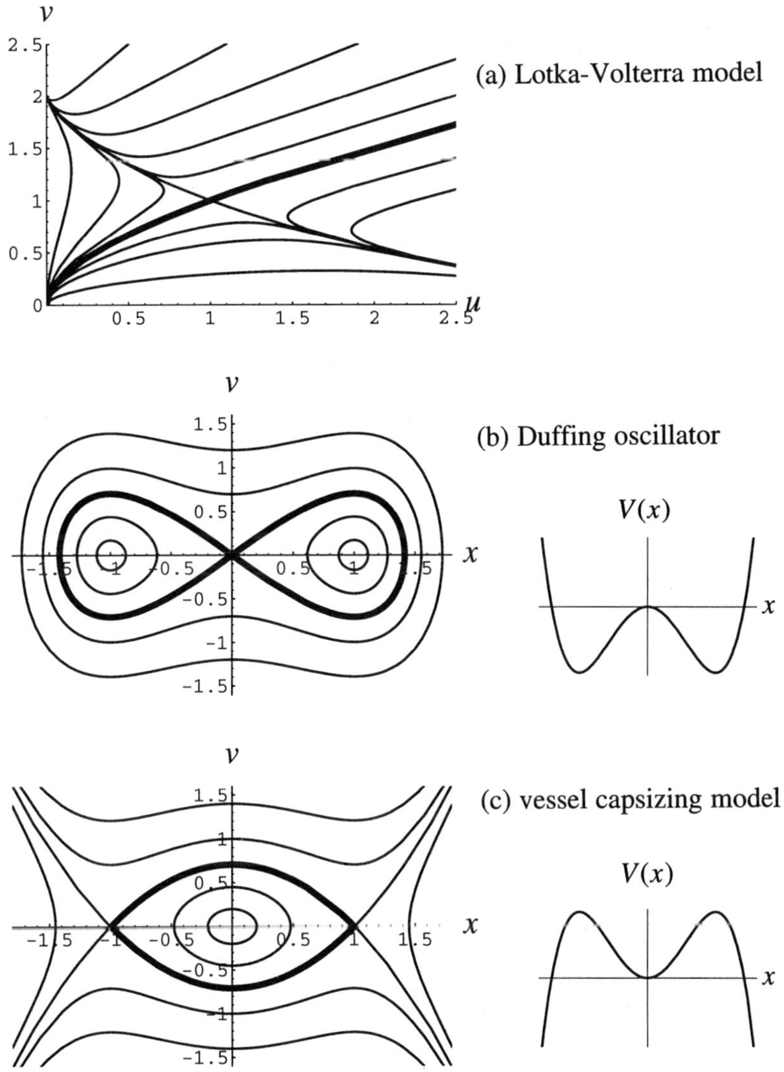

Figure 1. Separatrices in the state spaces of three dynamical systems.

with $x(t)$ interpreted as displacement at time t. This system can be rewritten as a single second-order equation: $\ddot{x} = -V'(x)$ where $V(x) = x^4/4 - x^2/2$ and $\ddot{x}$ is the second time derivative of $x(t)$. The Duffing oscillator is therefore a Newtonian system[29] and so can be interpreted as a particle with unit mass moving under the influence of the energy potential $V(x)$. Unlike the Lotka-Volterra model this system is conservative: the conserved quantity here is the Hamiltonian energy $H(x,v) = V(x) + v^2/2$, the sum of the particle's potential and kinetic energies. Referring to the plot of the energy potential $V(x)$ shown in Figure 1(b), one readily sees that if the particle is initially in a state near the bottom of one of the potential wells of $V(x)$ and has small or zero initial velocity, then the particle gently oscillates forever back and forth, never leaving the well. This is similarly true for the other well. These motions correspond to the roughly circular orbits within the bold figure-eight in the state space portrait (Figure 1(b)) of the Duffing oscillator. If the system is initially at a position of high potential energy or if its initial speed, and consequently its kinetic energy, is large, then the particle moves from one well to the other of $V(x)$ crossing the potential barrier at $x = 0$. After entering and passing through the bottom of the second well the particle is turned back at the far wall of the well and recrosses the potential barrier, returning to the first well. This process of entering a well and being turned back to the previous well repeats endlessly. The dynamics associated with this process corresponds to orbits outside the figure-eight in the state space portrait. From within the figure-eight the system never leaves its initial potential well while from outside the figure-eight the system state periodically changes wells. Thus the figure-eight is a separatrix for the system, partitioning the state space into regions of distinctly dissimilar dynamics. Each orbit in the portrait has a distinct total energy $H(x,v) = H(x_0, v_0)$ determined by the system's initial position and velocity, x_0 and v_0. The total energy is conserved so without an external withdrawal or addition of energy the system cannot stray from one orbit to another and, in particular, the system state cannot cross the separatrix.

The third example of a separatrix is based on the ocean vessel capsizing model

$$\begin{aligned} \dot{x} &= v \\ \dot{v} &= x^3 - x \end{aligned} \tag{5}$$

representing a vessel's roll in a calm sea[18]. Here $x(t)$ is the roll angle (from the vertical) and $v(t)$ is the roll velocity. Like the Duffing oscillator, this system is Newtonian with the potential energy $V(x) = x^2/2 - x^4/4$ shown in Figure 1(c). If the initial roll angle x_0 is near zero and if the initial roll velocity v_0 is also not too large, the vessel rolls back and forth but never capsizes. If the initial roll velocity is too large, the vessel rolls more and more to port or starboard until it capsizes, this being the physical analogue of mathematically crossing one or the other of the potential barriers of $V(x)$ and escaping the potential well. These two types of motion are shown in Figure 1(c) along with the separatrix which isolates them. As in the two previous examples, the state of this system cannot, without the intrusion of some external factor such as ocean beam waves, penetrate the system's separatrix to jump between safe and capsizing paths.

The examples given here of separatrices of dynamical systems are but three among many available from diverse fields. Among these, the four given by Wiggins[40] from fluid vortex dynamics, celestial mechanics, bubble dynamics and molecular dissociation are useful reading.

1.2. Safe Regions, System Failure and State Space Transport

Many systems must for proper operation remain within a potential well. In the ocean vessel capsizing model, for example, the vessel capsizes if the system escapes from its potential well. For systems like the Duffing oscillator and the ocean vessel capsizing model which admit an energy potential, associated with each well of the system potential is a region in state space bounded by a separatrix. To escape from a potential well the system's state must pass through the separatrix bounding the corresponding region in state space. This region in state space corresponding to the potential well is therefore called a safe region and the system is said to fail if or when the system exits the safe region. States within a safe region have insufficient energy to escape from the corresponding potential well. States outside the safe region correspond to states outside the potential well together with states within the potential well having too much energy to remain. The separatrix forms an impenetrable barrier between these two sets of states and if initially within the safe region the system state never leaves and no failure ever occurs. This global picture persists even if the system is weakly perturbed by, for example, external forcing, though now the separatrix is replaced by a pseudo-separatrix through which the system can possibly pass to exit the safe region and, in so doing, escape from the potential well and fail [4,8,40]. This is the only mechanism for escape from the potential well and the state space flux, defined as the rate of transport of state space across the pseudo-separatrix, is a measure of the possibility of and time required for escape from the potential well and hence system failure. Failure probabilities, mean failure rates and the distribution of times to failure in weakly forced dynamical systems are treated from essentially this viewpoint by Hsieh, Troesch and Shaw[18] and by Simiu and Frey[12,34].

State space transport across the separatrices associated with the potential wells of Newtonian systems is closely connected with Melnikov processes and chaos[8]. All separatrices in Newtonian systems are composed of superposed stable and unstable manifolds. These manifolds, in the presence of external forcing, separate to a definable distance provided the forcing is uniformly sufficiently weak. This distance, the Melnikov distance between stable and unstable invariant manifolds, is given to leading order by a stochastic process called the Melnikov process. The Melnikov process has its origins in the Melnikov function which measures the separation between stable and unstable invariant manifolds in the presence of weak nonrandom periodic forcing[2,25]. In a succession of generalizations, the generalized Melnikov function was introduced for quasiperiodic forcing[39], the Melnikov transform for uniformly bounded, uniformly continuous forcing[26], and the Melnikov process for uniformly bounded, ensemble uniformly continuous random forcing processes[8]. Many forcing processes, including Gaussian, dichotomous and shot noise processes, are limits of uniformly bounded, ensemble uniformly continuous processes and hence, in a limiting sense, they too fall within the purview of Melnikov processes[8,9,33]. The Melnikov process is determined by the forcing model and the relatively simple geometry of the unperturbed system and relates both to chaos and back to state space transport. First, the time-average of the Melnikov process provides a leading order expression for the average state space flux out of the safe region. This quantity, called the flux factor, has an ensemble average counterpart which admits formulae for a wide variety of random and deterministic, additive and multiplicative forcing models[8,9]. Second, the presence of zeros of the Melnikov function or process is a renowned criterion for chaos in weakly perturbed dynamical systems[2,8,26]. The state space flux for its part has been interpreted as the system's "propensity" for chaos[4]; in particular, a positive flux factor is a necessary condition for chaos.

As an asymptotic expression for the time-averaged state space flux, the flux factor measures a potential well's capability as a barrier to escape in the face of external forcing. A system is said to be stable if its state is maintained against external forcing within a potential well with high probability (suitably defined) for a long time. Therefore, in this sense, the flux factor measures the system's stability within the potential well. This is illustrated in Section 3 with a numerical example.

The flux factor has much to recommend it as a measure of stability. First, as already noted, the flux factor can be expressed in closed form for a broad range of perturbation models. In particular, the impact of external forcing (including that of any applied control) on the flux factor is mediated by one or more filters called orbit filters. In important cases the orbit filter is linear and time-invariant with impulse response derived directly from the separatrix. The orbit filter represents the system, in effect, often allowing an otherwise difficult nonlinear problem to be recast in linear form. Finally, although the flux factor is certain to satisfactorily approximate the flux only in the limiting case of weak perturbation, this case arises often in practice since only rarely is a strongly perturbed system highly stable—the design demands are too great. There is, moreover, some numerical evidence to suggest that the Melnikov theory to which the flux factor is related is "robust" in the limit of weak forcing[27,30].

A system is said to fail if or when its state exits the safe region associated with a potential well. From this viewpoint a system is perfectly stable if it never fails and, in general, its stability within a potential well is measured by the state space flux out of the associated safe region. If initially within the safe region, then, in the absence of external forcing, the system never escapes from the safe region and the state space flux is zero. In the presence of very weak amplitude external forcing, the system may only after a long time, if ever, exit the safe region. At this level of external forcing, the system within the potential well is still fairly stable and the flux out of the safe region is small. As the forcing level increases the state space flux increases, the walls of the potential well are less and less barriers to escape and the system becomes less stable. The relative stability of a system at, for example, two different levels or forms of external forcing can be assessed by a corresponding ratio of flux factors[11]. This ratio is called the flux ratio relative stability (FRRS) and its use can be extended even to comparing different potential wells with different or identical forcing. Another quantity, probability ratio relative stability (PRRS), is also available for comparing potential wells. A "blowtorch" theorem is known for PRRS which states that the PRRS of a well is necessarily decreased by any increase in forcing intensity anywhere in the well[21,38]. This result for PRRS is known only for white Gaussian forcing and addresses no other forcing models. Proceeding from the definition of FRRS, a blowtorch theorem is proved which is valid for all white, wide-sense stationary forcing models including white Gaussian forcing. It is also shown that a blowtorch theorem is not generally valid for colored Gaussian or deterministic forcing. This last result indicates that introducing the appropriate state-dependent excitation within a potential well can increase the system's stability within that well, in a sense extending the life of the system before escape and failure.

The flux factor has important properties relating to its use as an objective for controlling system stability. Given the flux factor as a control objective, closed-loop controls depending explicitly on the system state have, for a very broad class of forcing and control models, equivalent open-loop counterparts which depend only on the forcing. This significantly reduces the class of controls to be considered and greatly simplifies any ultimate implementation of the control. Also, minimizing the flux factor to increase system stability is equivalent to minimizing the referred mean-square cancellation error due, for example, to time lag in the control, insufficient control power or a mismatched control power spectrum. This referred error is the mean-square difference in the forcing and control processes as they appear at the output of the orbit filters. Equivalence of the flux factor and the referred cancellation error holds generally for wide-sense stationary forcing and allows us to show that for this class of forcing models the optimal control, given sufficient available control power, incorporates a Wiener filter. For available power below a certain threshold, the optimal filter is not Wiener. For this case a filter is proposed which, though not optimal, has some appealing properties.

A system's stability has been identified in this introduction with maintenance of the system state against perturbing forces within a potential well for a long time with high probability. The state space flux—specifically, the flux factor—out of the associated safe region was then tendered as a measure of this stability. In the following five sections this theme is developed to propose controls to increase stability and to make comparisons of relative stability. The first of these sections introduces the class of Newtonian systems to be considered. The second section sets out properties of the flux factor with a numerical example of the relationship of the flux factor to the mean rate of escape from a well. Controls for system stability based on the flux factor are treated in the third section. The fourth section uses the flux factor to assess relative stability for different levels and types of forcing, the aim there being a blowtorch theorem for FRRS. The last section contains final remarks.

2. Model

We consider the Newtonian system[29]

$$\ddot{x} = -V'(x) \tag{6}$$

with potential energy $V(x)$ where $x = x(t)$ is the position of the system at time t. This system is expressible as the two-dimensional dynamical system

$$\dot{x} = v\ , \quad \dot{v} = -V'(x) \tag{7}$$

with state $\vec{z} = (x, v) \in \Re^2$. In particular, system Eq. (6) is a one degree-of-freedom Hamiltonian system

$$\dot{x} = \frac{\partial}{\partial v} H(\vec{z})\ , \quad \dot{v} = -\frac{\partial}{\partial x} H(\vec{z}) \tag{8}$$

with Hamiltonian $H(\vec{z}) = H(x,v) = V(x) + v^2/2$. The potential $V(x)$ is assumed to be such that system Eq. (7) has at least one hyperbolic saddle. Note: a point $\vec{z}_0$ in the state space of the dynamical system $\dot{\vec{z}} = \vec{\pi}(\vec{z})$ is a hyperbolic saddle if $\vec{\pi}(\vec{z}_0) = 0$ and the derivative matrix $D\vec{\pi}(\vec{z}_0)$ has at least one eigenvalue with a positive real part, at least one eigenvalue with a negative real part and no eigenvalues with zero real part. Hyperbolic saddles are structurally stable[2]; that is, they persist as hyperbolic saddles under weak perturbations of the velocity field $\vec{\pi}$. For two-dimensional systems, there are four special trajectories associated with a hyperbolic saddle. Two of these, called stable trajectories, approach the hyperbolic saddle as $t \to \infty$ while the other two, called unstable trajectories, approach the hyperbolic saddle as $t \to -\infty$. The stable manifold of the hyperbolic saddle is the union of the hyperbolic saddle and its two stable trajectories; its unstable manifold is the union of the hyperbolic saddle and its two unstable trajectories. An orbit which is both a stable and an unstable trajectory for the same hyperbolic saddle is homoclinic; it is heteroclinic if it is stable for one hyperbolic saddle and unstable for another.

2.1. Energy Potentials

The energy potential $V(x)$ governs the dynamics of system Eq. (7). Two archetypal classes $\mathcal{V}_{\mathsf{M}}$ and $\mathcal{V}_{\mathsf{W}}$ of energy potential exist for this system, the first being represented by that of the ocean vessel capsizing model and the second by the Duffing potential. Potentials $V(x) \in \mathcal{V}_{\mathsf{M}}$ have global maxima at exactly two points, x_1 and $x_2 > x_1$ such that $V(x_1) = V(x_2)$, $V'(x_1) = V'(x_2) = 0$ and $V(x) < V(x_1)$ for all $x \notin \{x_1, x_2\}$. These maxima define a potential well on the interval $[x_1, x_2]$. For $V(x) \in \mathcal{V}_{\mathsf{M}}$, system Eq. (7) has hyperbolic saddles at $\vec{z}_1 = (x_1, 0)$ and $\vec{z}_2 = (x_2, 0)$ connected by a pair of heteroclinic orbits $\vec{z}_{h1}(t) = (x_{h1}(t), v_{h1}(t))$ and $\vec{z}_{h2}(t) = (x_{h2}(t), v_{h2}(t)) = -\vec{z}_{h1}(t)$. These orbits satisfy the conditions

$$\lim_{t\to\infty} \vec{z}_{h2}(t) = \lim_{t\to-\infty} \vec{z}_{h1}(t) = \vec{z}_1\ , \quad \lim_{t\to\infty} \vec{z}_{h1}(t) = \lim_{t\to-\infty} \vec{z}_{h2}(t) = \vec{z}_2 \tag{9}$$

and can be found as solutions of $H(x,v) = V(x_2)$. Together these orbits form a closed curve which is a separatrix for the system. The interior of the separatrix is the safe region corresponding to the potential well of $V(x)$.

In our second class $\mathcal{V}_W$ of potentials, $V(x)$ has relative minima at $x_1 < 0$ and $x_2 > 0$ such that $V(0) > V(x)$ for all $x \in [x_1, x_2]$. Furthermore, there exists an $x < x_1$ such that $V(x) = V(0)$ and $V'(x) < 0$ and an $x > x_2$ such that $V(x) - V(0)$ and $V'(x) > 0$. Under these conditions $V(x)$ has two potential wells separated by a potential barrier at $x = 0$. For $V(x)$ in this class, system Eq. (7) has a hyperbolic saddle at the origin $\vec{z} = (0,0)$ in state space connected to itself by a pair of homoclinic orbits $\vec{z}_1(t) = (x_1(t), v_1(t))$ and $\vec{z}_2(t) = (x_2(t), v_2(t)) = -\vec{z}_1(t)$. These orbits satisfy the conditions

$$\lim_{t\to\pm\infty} \vec{z}_{h1}(t) = \lim_{t\to\pm\infty} \vec{z}_{h2}(t) = (0,0) \tag{10}$$

and can be found as solutions of $H(x,v) = V(0)$. Each homoclinic orbit forms a closed curve which is a separatrix for the system. The interiors of the two separatrices are the safe regions corresponding to the two wells of $V(x)$.

Members of $\mathcal{V}_M$ and $\mathcal{V}_W$ are most easily distinguished by their number of wells and hyperbolic saddles: members of $\mathcal{V}_M$ have one well fixed between two hyperbolic saddles while members of $\mathcal{V}_W$ have two wells separated by a single hyperbolic saddle. Further, the safe region associated with the well of a $\mathcal{V}_M$ potential is bounded by a pair of heteroclinic orbits while those associated with the two wells of a $\mathcal{V}_W$ potential are each bounded by a single homoclinic orbit. These differences determine, as we shall see, the number of terms in the formulae for the flux factor for transport from a safe region. More subtly, important features of the safe region's orbit filters are affected[8]. The two classes $\mathcal{V}_M$ and $\mathcal{V}_W$ of energy potential detailed here are not exhaustive; one can easily concoct potentials with more than two wells or systems with both heteroclinic and homoclinic orbits whose potentials combine features from both classes. $\mathcal{V}_M$ and $\mathcal{V}_W$ do, however, include most of the commonly treated models in the literature.

As an illustration of foregoing concepts, consider the diamond potential

$$V(x) = -D(1 - |x|)^2 \tag{11}$$

shown in Figure 2. The diamond potential is a member of $\mathcal{V}_M$, has global maxima at $x_2 = 1$ and $x_1 = -1$ and a potential well with depth D on the interval $[-1, 1]$. A state space portrait of system Eq. (7) with potential Eq. (11) is shown in Figure 2. The system has saddles at $\vec{z}_1 = (-1, 0)$ and $\vec{z}_2 = (1, 0)$ connected by heteroclinic orbits $\vec{z}_{h1}(t)$ and $\vec{z}_{h2}(t) = -\vec{z}_{h1}(t)$ which form a separatrix. Potential Eq. (11) takes its name from the shape of the safe region enclosed by the separatrix. From the Hamiltonian equation $H(\vec{z}) = H(x,v) = V(x) + v^2/2 = 0$, the heteroclinic orbit $\vec{z}_{h1}(t)$ is

$$x_{h1}(t) = \begin{cases} e^{\sqrt{2D}t} - 1, & t \le 0 \\ 1 - e^{-\sqrt{2D}t}, & t > 0 \end{cases} \tag{12}$$

with

$$v_{h1}(t) = \sqrt{2D}e^{-\sqrt{2D}|t|}. \tag{13}$$

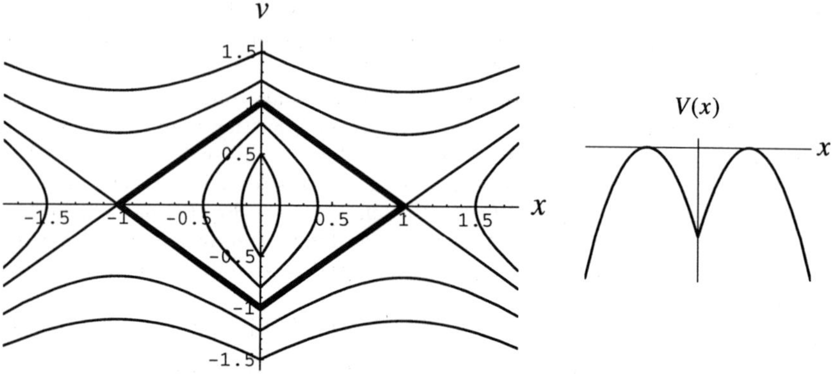

Figure 2. Newtonian system with diamond potential.

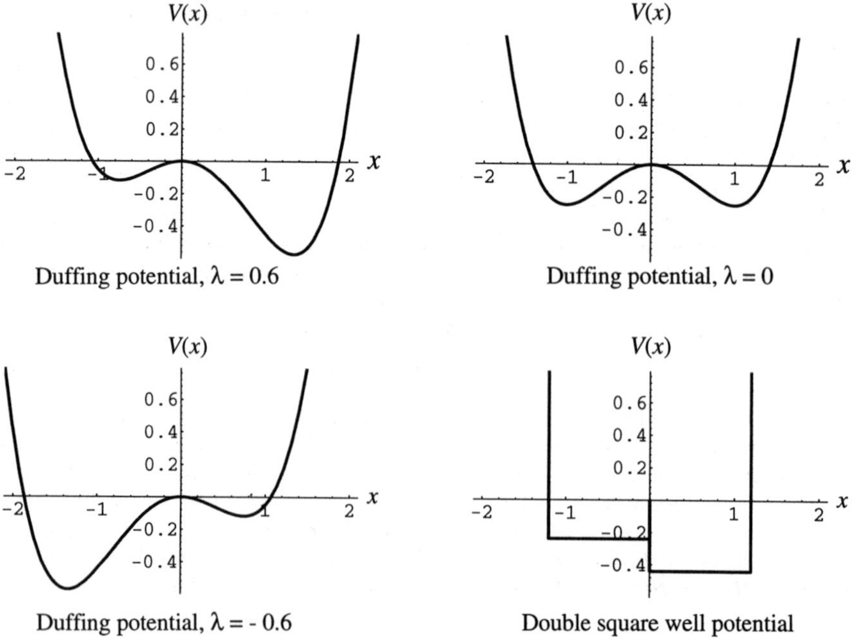

Figure 3. Examples of double square well and asymmetric Duffing potentials.

The orbit $\vec{z}_{h1}(t)$ parametrically traces out the upper half of the diamond-shaped path in the state space portrait in Figure 2 beginning at $t = -\infty$ at the hyperbolic saddle $(x, v) = (-1, 0)$, arriving at time $t = 0$ at state $(0, \sqrt{2D})$ and ending at time $t = \infty$ at the second hyperbolic saddle $(x, v) = (1, 0)$. Beginning at this latter saddle, the orbit $\vec{z}_{h2}(t) = -\vec{z}_{h1}(t)$ traces out a return path passing through the state $(0, -\sqrt{2D})$. The separatrix formed by $\vec{z}_{h1}(t)$ and $\vec{z}_{h2}(t)$ divides the orbits within the well into two groups, those with insufficient energy for escape $(H(x, v) < 0)$ and those which do escape.

The area A of a system's safe region is often a part of the flux factor for transport out of the region. In the case of the Newtonian system with the diamond potential Eq. (11), the safe region has a simple geometric shape and its area is immediately $A = \sqrt{2D}$. In general, the area A of the safe region of a potential $V(x) \in \mathcal{V}_{\mathsf{M}}$ is

$$A = 2\int_{x_1}^{x_2} v_{h1}(x)\, dx \tag{14}$$

$$= 2\int_{-\infty}^{\infty} v_{h1}^2(t)\, dt. \tag{15}$$

The first integral for A is obtained by noting that the safe region is symmetric with its upper half bounded by $(x_{h1}(t), v_{h1}(t))$. The second integral is obtained from the first using $v_{h1}(t) = \dot{x}_{h1}(t)$. Potentials of class $\mathcal{V}_{\mathsf{W}}$ have two safe regions, each with a similar pair of integral expressions for their areas. These areas A_1 and A_2 are

$$A_1 = 2\int_{x_o}^{0} v_{h1}(x)\, dx \tag{16}$$

$$= \int_{-\infty}^{\infty} v_{h1}^2(t)\, dt \tag{17}$$

and

$$A_2 = 2\int_{0}^{x_o} v_{h2}(x)\, dx \tag{18}$$

$$= \int_{-\infty}^{\infty} v_{h2}^2(t)\, dt \tag{19}$$

where x_o in Eq. (16) is the minimum of $x_{h1}(t)$ and x_o in Eq. (18) is the maximum of $x_{h2}(t)$.

As an example of $\mathcal{V}_{\mathsf{W}}$ potentials we consider the parametric family of asymmetric Duffing potentials[5]

$$V(x) = \frac{x^4}{4} - \frac{\lambda x^3}{3} - \frac{x^2}{2} \tag{20}$$

with asymmetry parameter λ. Potential Eq. (20) has two wells with relative minima at

$$x_{min1} = x_{min1}(\lambda) = \frac{\lambda}{2} - \sqrt{\frac{\lambda^2}{4} + 1} \tag{21}$$

and

$$x_{min2} = x_{min2}(\lambda) = \frac{\lambda}{2} + \sqrt{\frac{\lambda^2}{4} + 1} \tag{22}$$

separated by a potential energy barrier at $x = 0$ with height $V(0) = 0$. The depth of the well at x_{min2} is therefore $V(0) - V(x_{min2}) = -V(x_{min2})$. We have

$$\frac{d}{d\lambda}\{-V(x_{min2}(\lambda))\} = \frac{1}{3}\left(\frac{\lambda}{2} + \sqrt{\frac{\lambda^2}{4} + 1}\right)^3 \tag{23}$$

so the depth of the well at x_{min2} is a strictly increasing function of λ. Similarly, the depth of the well at x_{min1} is a strictly decreasing function of λ. Therefore, well 2 is deeper than well 1 for $\lambda > 0$, shallower for $\lambda < 0$ and of equal depth for $\lambda = 0$. Examples of the asymmetric Duffing potential are shown in Figure 3.

The homoclinic orbits $\vec{z}_1(t)$ and $\vec{z}_2(t)$ for the asymmetric Duffing potential can be obtained from the Hamiltonian equation $H(x, v) = V(x) + \dot{x}^2/2 = 0$. We have[5]

$$x_1(t) = \frac{-1}{B\cosh t + \lambda/3}, \quad x_2(t) = \frac{1}{B\cosh t - \lambda/3} \tag{24}$$

where $B^2 = \lambda^2/9 + 1/2$. The corresponding velocity components of the orbits are

$$v_1(t) = \frac{-B\sinh t}{(B\cosh t + \lambda/3)^2}, \quad v_2(t) = \frac{B\sinh t}{(B\cosh t - \lambda/3)^2}. \tag{25}$$

The areas A_1 and A_2 of the safe regions of the asymmetric Duffing potential Eq. (20) can be expressed in closed form using Eqs. (17) and (19). The result is

$$A_i = \frac{4\sqrt{2}}{3}\lambda B^2\left(\sin^{-1}\frac{\lambda}{3B} + (-1)^i\frac{\pi}{2}\right) + \frac{4}{3} + \frac{4\lambda^2}{9} \tag{26}$$

for $i = 1, 2$. Evidently, $A_2 > A_1$ for $\lambda > 0$.

The double square well potential is an idealized two well potential given by

$$V(x) = \begin{cases} 0, & x = 0 \\ -d_1, & -w < x < 0 \\ -d_2, & 0 < x < w \\ \infty, & x \geq w \\ \infty, & x \leq -w \end{cases} \tag{27}$$

where $d_1 > 0$ and $d_2 > 0$ are the depths of the two wells and w is the wells' common width. See Figure 3. System Eq. (7) with the double square well potential Eq. (27) requires careful interpretation since this potential is not a member of the class of continuously differentiable C^1 potentials usually treated[2,40]; system Eq. (7) is not even defined at the putative hyperbolic saddle $\vec{z} = (0, 0)$. This difficulty can be resolved by considering a sequence of C^1 potentials converging pointwise to the double square well potential—in this case, the sequence of C^1 potentials

$$V_n(x) = \begin{cases} d_1 \left[e^{-(nx/w)^2} + (x/w)^{2n} - 1 - (x/w)e^{-n^2} \right], & x \leq 0 \\ d_2 \left[e^{-(nx/w)^2} + (x/w)^{2n} - 1 - (x/w)e^{-n^2} \right], & x > 0 \end{cases} \tag{28}$$

for integers $n \geq 1$. Following this approach, system Eq. (7) with potential Eq. (27) is defined to inherit all its properties (hyperbolic saddles, homoclinic orbits, etc.) from the sequence of systems with potentials $V_n(x)$. Figure 4 shows system Eq. (7) with the double square well potential to have a hyperbolic saddle at $\vec{z} = (0,0)$ and a pair of rectangular homoclinic orbits. The areas A_1 and A_2 of the two safe regions are the width w of the well multiplied by twice the speed (unsigned velocity) along the homoclinic orbit. This speed can be found from the Hamiltonian equation $v_i^2(t)/2 + V(x_i(t)) = 0$. $V(x_i(t)) = -d_i$ so the speed is $\sqrt{2d_i}$ and $A_i = 2w\sqrt{2d_i}$ for $i = 1, 2$.

2.2. Perturbed Newtonian Systems

If system Eq. (7) is modified to include a time-dependent perturbative term, and if this perturbation is sufficiently weak, the system separatrices become "porous" as noted above, forming proximate pseudo-separatrices. For certain ranges of system parameters these pseudo-separatrices can be penetrated, allowing the system state to exit the safe region. This is the situation which occurs in the weakly ($0 \leq \varepsilon \ll 1$) perturbed Newtonian system

$$\ddot{x} = -V'(x) + \varepsilon[\gamma F(t) - k\dot{x} - \alpha C(t)] \tag{29}$$

in which the stochastic process $F(t)$ is an externally applied force. The forcing $F(t)$ introduces the possibility of escape from a potential well, reducing the stability of the system within the well. The process $C(t)$ in Eq. (29) is a control process designed to counteract $F(t)$, delaying or eliminating escapes and increasing the stability of the system state within the well. The perturbed system considered here is subject to weak damping as represented by the term $-k\dot{x}$ in Eq. (29). Intuitively, the damping is a source of passive resistance to escape internal to the system in contradistinction to the external control's active resistance. For simplicity only linear damping is considered. The parameters $\gamma, k, \alpha \geq 0$ in Eq. (29) determine the presence and relative amounts of forcing, damping and control in the system.

The forcing $F(t)$ in Eq. (29) will generally be assumed to be a wide-sense stationary (WSS) process[43] with zero mean, autocovariance $c_F(t) = \mathrm{Cov}[F(s), F(s+t)]$ and extant spectral density

$$\hat{c}_F(\nu) = \int_{-\infty}^{\infty} c_F(t) e^{-j\nu t}\, dt. \tag{30}$$

Members $F(t)$ of this class of forcing processes are expressible as a filtered white noise: $F(t) = \mathcal{F}[\dot{W}](t) = (f * \dot{W})(t)$ where $\mathcal{F}$ is a time-invariant linear filter with square-integrable impulse response $f(t)$, $\dot{W}(t)$ is a (generalized) white noise process[43]

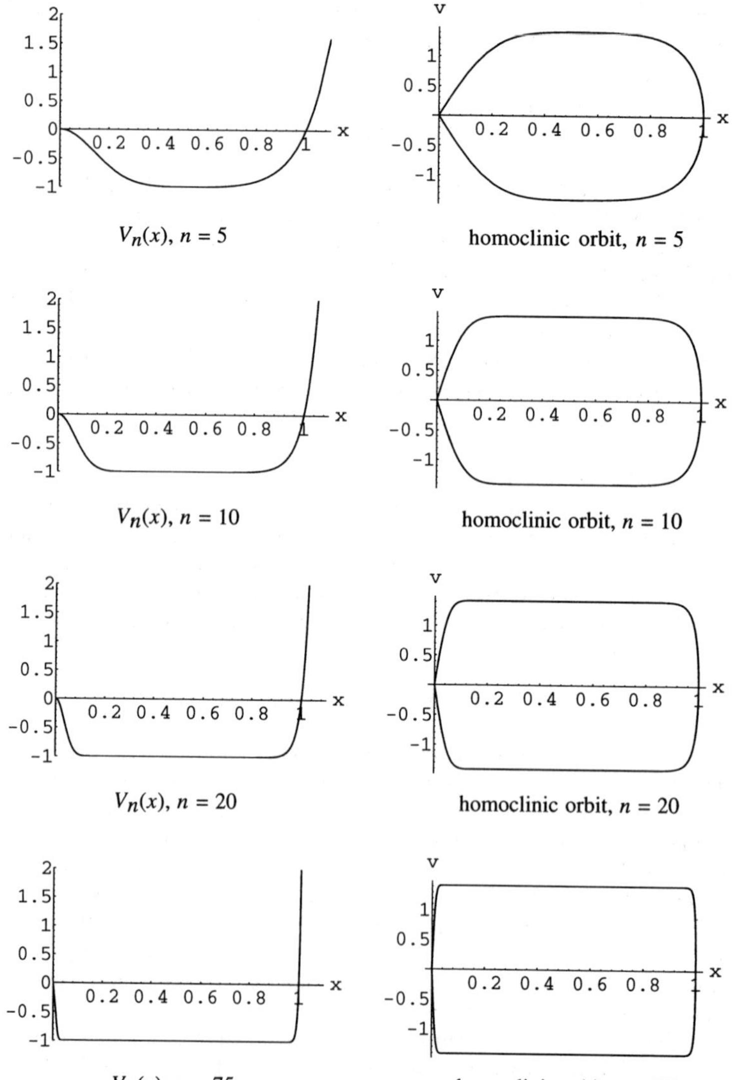

$V_n(x)$, $n = 5$ — homoclinic orbit, $n = 5$

$V_n(x)$, $n = 10$ — homoclinic orbit, $n = 10$

$V_n(x)$, $n = 20$ — homoclinic orbit, $n = 20$

$V_n(x)$, $n = 75$ — homoclinic orbit, $n = 75$

Figure 4. Energy potential $V_n(x)$ with corresponding homoclinic orbit for $w = 1$, $d_2 = 1$, and $n = 5, 10, 20, 75$. The limiting square well potential has a rectangular homoclinic orbit.

and $(f * \dot{W})(t)$ is the convolution of $f(t)$ and $\dot{W}(t)$. Equivalently, $F(t) = \mathcal{F}[\dot{W}](t)$ is the stochastic integral

$$F(t) = \int_{-\infty}^{\infty} f(t-s)\, dW(s) \tag{31}$$

where $W(t)$ is an orthogonal increment process satisfying, for all $t_1 \leq t_2$ and $t_3 \leq t_4$,

$$E[(W(t_1) - W(t_2))(W(t_3) - W(t_4))] = l([t_1, t_2] \cap [t_3, t_4]) \tag{32}$$

where $l([s,t]) = t - s$ denotes the length of the interval $[s,t]$. Model Eq. (31) includes Gaussian, shot noise and dichotomous noise forcing models, and is, for example, a colored Gaussian process if $W(t)$ is Brownian motion. Filtered white noise processes of type Eq. (31) have the property that

$$\mathrm{Cov}\left[\int_{-\infty}^{\infty} w_1(t)\, dW(t), \int_{-\infty}^{\infty} w_2(t)\, dW(t)\right] = \int_{-\infty}^{\infty} w_1(t) w_2(t)\, dt \tag{33}$$

provided $w_1(t)$ and $w_2(t)$ are square-integrable[6]. Given Eq. (31), the autocovariance of $F(t)$ is

$$c_F(t) = \int_{-\infty}^{\infty} f(s) f(s+t)\, ds \tag{34}$$

with $\hat{c}_F(\nu) = |\hat{f}(\nu)|^2$. Here and throughout we adhere to the convention that, for example, $\hat{f}(\nu)$ is the Fourier transform of $f(t)$.

Three classes of excitation can be identified for system Eq. (29) with increasing degrees of generality. An excitation $\chi X(t)$ such as $\gamma F(t)$ or $\alpha C(t)$ in Eq. (29) is called additive if χ is a constant independent of the system state. More generally, if $\chi = \chi(\vec{z})$ varies with system state, then the excitation $\chi(\vec{z}(t))X(t)$ is multiplicative. Still more general are excitations of the form $\chi(X(t), \vec{z}(t))$. The effect of an external excitation on the transport of state space across a system's pseudo-separatrix and, more specifically, on that transport's flux factor is mediated by a filter called the orbit filter[8,9]. This filter depends explicitly on the geometry of the homoclinic or heteroclinic orbit from which the pseudo-separatrix is derived. For the excitation $\chi(X(t), \vec{z}(t))$ the orbit filter Θ_X takes the form[26]

$$\Theta_X[\chi(\vec{z}, X)](t) = \int_{-\infty}^{\infty} v_h(\tau) \chi(X(\tau + t), \vec{z}_h(\tau))\, d\tau \tag{35}$$

where $v_h(t)$ is the velocity component of the heteroclinic or homoclinic orbit $\vec{z}_h(t)$. The filter Θ_X is here neither linear nor time-invariant. For multiplicative excitations, though, including the additive case, Θ_X is both linear and time-invariant,

$$\Theta_X[\chi(\vec{z})X](t) = \int_{-\infty}^{\infty} v_h(\tau) \chi(\vec{z}_h(\tau)) X(\tau + t)\, d\tau. \tag{36}$$

In fact, in this case the orbit filter response can be written as the convolution integral

$$\Theta_X[\chi(\vec{z})X](t) = \int_{-\infty}^{\infty} h(t - \tau) X(\tau)\, d\tau \tag{37}$$

where

$$h(t) = \chi(\vec{z}_h(-t))v_h(-t) \tag{38}$$

is the orbit filter impulse response[9] and $\hat{h}(\nu)$ is the corresponding filter transfer function. We assume throughout that all excitations in the perturbed system Eq. (29) are multiplicative so all orbit filters arising from system Eq. (29) will be linear and time-invariant.

3. Flux Factor

All separatrices in the Newtonian system $\ddot{x} = -V'(x)$ are homoclinic or heteroclinic orbits of the system. These orbits are defined by the coincidence of the stable and unstable manifolds emanating from the system's hyperbolic saddles. In the perturbed Newtonian system Eq. (29) these manifolds no longer coincide and, instead, separate to a distance given by the Melnikov distance. A pseudo-separatrix proximate to the original separatrix can be defined from segments of these manifolds. State space transport across the pseudo-separatrix is possible depending on the orientation of the separated manifolds and, indeed, this is the only avenue of escape from the associated potential well. These structures—the Melnikov distance between the separated manifolds, the pseudo-separatrix and the state space flux across the pseudo-separatrix—are defined provided the perturbation to the system is sufficiently weak.

3.1. Flux Factor Formulae

Both the Melnikov distance between separated manifolds and the state space flux across the pseudo-separatrix are difficult to compute and useful exact expressions are unavailable in even the simplest of cases. However, each admits convenient leading order approximations valid for $\varepsilon \to 0$. In particular, for a Newtonian system with a potential $V(x) \in \mathcal{V}_{\mathsf{W}}$ the time-averaged state space flux out of the safe region corresponding to a well of $V(x)$ is proportional[31] to $\varepsilon\Xi + O(\varepsilon^2)$ where the flux factor Ξ is[8]

$$\Xi = E[(X_F - X_C - kA)^+] \tag{39}$$

with A defined as

$$A = \int_{-\infty}^{\infty} v_h^2(t)\, dt. \tag{40}$$

From Eq. (15) it follows that A in Eq. (39) is the area of the safe region. The notation x^+ in Eq. (39) denotes the positive part of the real number x; $x^+ = x$ for $x > 0$ and $x^+ = 0$ otherwise. The distribution of the random pair (X_F, X_C) in Eq. (39) is equal to the stationary mean (SM) distribution[15] of the vector process $(\Theta_F[F](t), \Theta_C[C](t))$. If the forcing and control excitations in the perturbed system Eq. (29) are multiplicative so that Θ_F and Θ_C are linear and time-invariant and if $F(t)$ and $C(t)$ are jointly WSS processes, then the SM distribution of $(\Theta_F[F](t), \Theta_C[C](t))$ is identically its marginal distribution. For a Newtonian system with potential $V(x) \in \mathcal{V}_{\mathsf{M}}$ the flux factor Ξ for transport out of the safe region has two terms, one for each of the two heteroclinic orbits bounding the safe region:

$$\Xi = E[(X_{F1} - X_{C1} - kA/2)^+] + E[(X_{F2} - X_{C2} - kA/2)^+] \tag{41}$$

where here too A is the area of the safe region. The random pairs (X_{F1}, X_{C1}) and (X_{F2}, X_{C2}) in Eq. (41) are defined similarly to (X_F, X_C) in Eq. (39). The distribution of (X_{F1}, X_{C1}), for example, is the marginal distribution of the vector process $(\Theta_{F1}[F](t), \Theta_{C1}[C](t))$ provided the forcing and control excitations are multiplicative and $F(t)$ and $C(t)$ are jointly WSS.

Equations (39) and (41) for the flux factor require $F(t)$ and $C(t)$ to have certain technical properties, among them ergodicity, uniform boundedness and ensemble uniform continuity. Many processes including Gaussian, shot noise and dichotomous processes fail to meet one or more of these conditions. Yet by treating these processes as limits of processes which do have the required properties, such processes can be brought within the purview of Eqs. (39) and (41). The reader is referred to Frey and Simiu[8,9,34] and Hsieh, Troesch and Shaw[18] for discussion on this point.

For additive forcing and control excitations, $\gamma F(t)$ and $\alpha C(t)$, the impulse responses of the orbit filters Θ_F and Θ_C associated with the flux factor in Eq. (39) are

$$\gamma v_h(-t)\ , \quad \alpha v_h(-t) \tag{42}$$

respectively. Similarly, in the case of the flux factor in Eq. (41) the impulse responses of the orbit filters Θ_{F1}, Θ_{C1}, Θ_{F2} and Θ_{C2} are, respectively,

$$\gamma v_h(-t)\ , \quad \alpha v_h(-t)\ , \quad -\gamma v_h(-t)\ , \quad -\alpha v_h(-t). \tag{43}$$

Thus for additive excitation we identify a common orbit filter Θ with impulse response $h(t) = v_h(-t)$. The flux factor Eq. (39) for transport out of a safe region of a system with a $\mathcal{V}_{\mathsf{W}}$ potential, then, is

$$\Xi = E[(X - kA)^+] \tag{44}$$

while for transport out of the safe region of a system with a $\mathcal{V}_{\mathsf{M}}$ potential the flux factor Eq. (41) is

$$\Xi = E[(X - kA/2)^+] + E[(X + kA/2)^-] \tag{45}$$

where in both Eqs. (44) and (45) the distribution of the random variable X is identically the SM distribution of the process $\Theta[\gamma F - \alpha C](t)$. The notation $x^- = |x| - x^+$ in Eq. (45) denotes the negative part of x. The variance σ^2 of X is called the referred mean-square cancellation error of the control $C(t)$.

Proposition 1: Assume that the random variable X in Eqs. (44) or (45) for the flux factor Ξ has a finite second moment with variance $\sigma^2 = \mathrm{Var}[X]$. Then Ξ is a nondecreasing function of σ^2. If the support of X is the real line then Ξ is a strictly increasing function of σ^2.

Proof: Let $F_Z(z) = P(Z \le z)$ be the cumulative distribution function of the standardized random variable $Z = X/\sigma$. Then

$$\begin{aligned} E[(X - kA)^+] &= \sigma \int_{\frac{kA}{\sigma}}^{\infty} (s - \frac{kA}{\sigma})\, dF_Z(s) \\ &= \sigma \int_{\frac{kA}{\sigma}}^{\infty} \int_{\frac{kA}{\sigma}}^{s} dz\, dF_Z(s) \\ &= \sigma \int_{\frac{kA}{\sigma}}^{\infty} \int_{z}^{\infty} dF_Z(s)\, dz \\ &= \sigma \int_{\frac{kA}{\sigma}}^{\infty} [1 - F_Z(z)]\, dz. \end{aligned} \tag{46}$$

Thus the flux factor for a $\mathcal{V}_{\mathsf{W}}$ potential is

$$\Xi = \sigma \int_{\frac{kA}{\sigma}}^{\infty} [1 - F_Z(z)]\, dz. \tag{47}$$

Combining Eq. (46) with a similar calculation for $E[(X + kA/2)^-]$, we have for the case of a $\mathcal{V}_{\mathsf{M}}$ potential

$$\Xi = \sigma \int_{-\infty}^{-\frac{kA}{2\sigma}} F_Z(z)\, dz + \sigma \int_{\frac{kA}{2\sigma}}^{\infty} [1 - F_Z(z)]\, dz. \tag{48}$$

The result then follows from inspection of Eqs. (47) and (48) for Ξ.

Equations (47) and (48) suggest a concise notation for the flux factor. It follows from Eq. (47) that, for each of the wells of a $\mathcal{V}_{\mathsf{W}}$ potential,

$$\Xi = kA\, \eta(\frac{\sigma}{kA}) \tag{49}$$

where, with $F_Z(z)$ as in Eq. (47),

$$\eta(x) = x \int_{1/x}^{\infty} [1 - F_Z(z)]\, dz \tag{50}$$

for $x > 0$. For many classes of excitation—deterministic sinusoidal excitations and Gaussian excitations among them—the SM distribution of $\Theta[\gamma F - \alpha C](t)$ is symmetric. Then the random variable X in Eq. (45) is symmetric, $F_Z(-z) = 1 - F_Z(z)$ and, for the well of a $\mathcal{V}_{\mathsf{M}}$ potential,

$$\Xi = kA\, \eta(\frac{2\sigma}{kA}). \tag{51}$$

The function $\eta(x)$ has the following easily established properties: it is convex with $\eta(0) = \eta'(0) = 0$, $\eta'(x) \ge 0$, $\eta''(x) \ge 0$ and

$$\eta'_\infty = \lim_{x \to \infty} \eta'(x) = \int_0^{\infty} [1 - F_Z(z)]\, dz. \tag{52}$$

Furthermore, $\eta(x)$ has the piece-wise linear approximation

$$\eta(x) \doteq \begin{cases} 0, & 0 \leq x \leq x_o \\ \eta'_\infty (x - x_o), & x > x_o \end{cases} \tag{53}$$

where $x_o = (1 - F_Z(0))/\eta'_\infty$. The flux factor Ξ is a nondecreasing function of σ since $\eta'(x) \geq 0$. This reiterates the result in Proposition 1 and implies that a control $C(t)$, if present, reduces the flux factor Ξ and increases system stability only if the introduction of $C(t)$ reduces σ^2. Indeed, for large σ the relationship between Ξ and σ is nearly linear; for $\sigma \to \infty$, $\Xi \sim \sigma E[|Z|]$ for $\mathcal{V}_{\mathsf{W}}$ potentials and $\Xi \sim 2\sigma E[|Z|]$ for $\mathcal{V}_{\mathsf{M}}$ potentials.

3.2. Gaussian Excitations

In the special case where the forcing $F(t)$ and control $C(t)$ are each Gaussian with zero mean, the random variable $Z = X/\sigma$ is standard Gaussian and $\eta(x)$ in Eq. (50) is[8]

$$\eta(x) = x\phi(1/x) + \Phi(1/x) - 1 \tag{54}$$

where $\phi(z) = (2\pi)^{-1/2}\exp(-z^2/2)$ is the standard Gaussian density and $\Phi(z)$ is the corresponding cumulative distribution function. The function $\eta(x)$ in Eq. (54) is shown in Figure 5 together with its asymptotic approximation

$$\eta(x) \doteq \frac{x}{\sqrt{2\pi}} - \frac{1}{2}, \quad x > \sqrt{\frac{\pi}{2}} \tag{55}$$

derived from Eq. (53).

Example: The flux factor has been proposed as a measure system stability within a potential well where stability is defined as maintenance of the system state within the well in opposition to external forcing. We show here in an example provided by Simiu and Franaszek[35] that there is a direct relationship between the flux factor for a potential well and the mean rate of escape from the well. We consider the perturbed Newtonian system

$$\ddot{x} = -V'(x) + \varepsilon[\gamma F(t) - k\dot{x}] \tag{56}$$

with Duffing potential $V(x) = x^4/4 - x^2/2$ and colored Gaussian noise $F(t)$. This system has two safe regions each with area $A = 4/3$. Using data[35] for the case $\varepsilon = 0.1$, $k = 0.45$ and low-pass forcing spectral density $2\pi/5$, we obtain the result in Figure 6. For this particular example the relationship between the Monte Carlo-estimated mean rate of escape from the well and the well's calculated flux factor is roughly linear.

4. Dynamical Control

With properties of the flux factor established in the previous section we are now prepared to consider controls to reduce the flux factor for a potential well and thereby increase the stability of the system within the well. We assume throughout this section

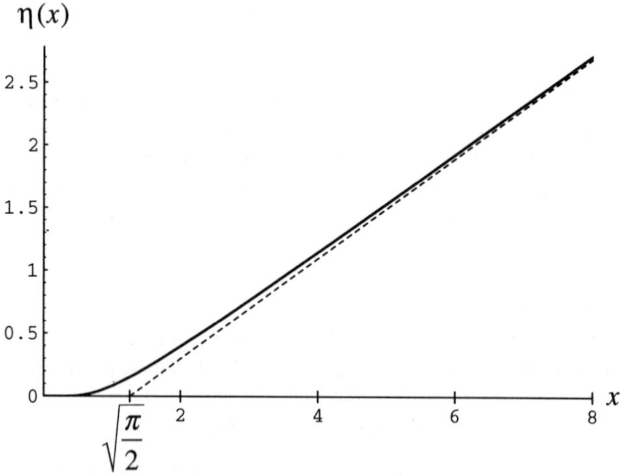

Figure 5. A plot of $\eta(x) = x\phi(1/x) + \Phi(1/x) - 1$ and its linear aymptotic approximation.

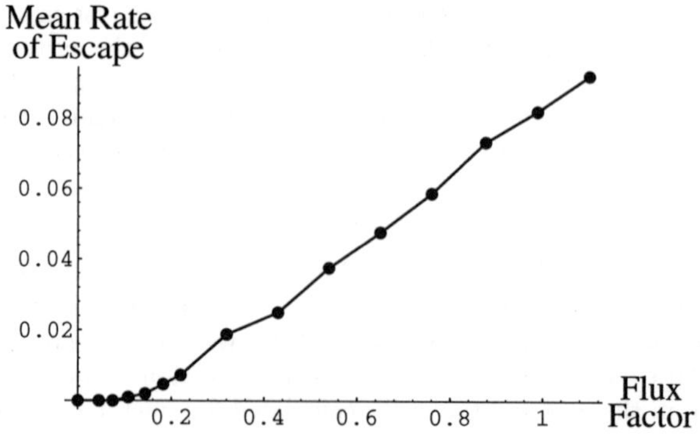

Figure 6. Monte Carlo-estimated mean rates of escape from a well of the Duffing potential.

that the forcing $F(t)$ in the perturbed Newtonian system Eq. (29) is a zero mean, WSS process as in Eq. (31).

4.1. Control Model

To give structure to the control process $C(t)$ in the perturbed Newtonian system Eq. (29) we observe that if the control and forcing processes $C(t)$ and $F(t)$ are uncorrelated then the presence of $C(t)$ cannot decrease the rate of escape from the potential well. This is intuitively obvious and is also clear from consideration of the flux factor. If, on the other hand, the path realized by $F(t)$ is completely known so that $C(t)$ can be made perfectly correlated with $F(t)$, then $C(t)$ completely cancels the effect of $F(t)$ for $\alpha = \gamma$. Neither situation is realistic; the control $C(t)$ is, in the first instance, ineffectual and, in the second, impractical. Reasonably, the control $C(t)$ should be able to at least track low frequency components of $F(t)$, perhaps with some positive time lag ℓ. We therefore model the control process by

$$C(t) = \beta[F](t - \ell) \tag{57}$$

where $\ell \geq 0$ is a constant representing the control lag and β is a causal, time-invariant, linear filter with impulse response $b(t)$ and transfer function $\hat{b}(\nu)$. Typically, the control filter β will be low-pass, reflecting the control's inability to track high frequency components of $F(t)$. Given Eq. (57), $C(t)$ can be written as the convolution $C(t) = (b * \delta_\ell * F)(t)$ where $\delta(t)$ is the Dirac-δ function and $\delta_\ell(t) = \delta(t - \ell)$. Then using Eq. (31) for $F(t)$, the control process $C(t)$ is the stochastic integral

$$C(t) = \int_{-\infty}^{\infty} (b * \delta_\ell * f)(t - s)\, dW(s). \tag{58}$$

4.2. Closed- vs. Open-Loop Control

A schematic diagram of the perturbed Newtonian system Eq. (29) with multiplicative forcing and control is shown in Figure 7. Multiplicative controls depend on the current system state $\vec{z}$ and are a form of closed-loop control. If the control is additive then there is no dependence on system state, there is no feedback path and the control is open-loop.

The effect of the excitation $\chi(X(t), \vec{z}(t))$ on the flux factor is determined by its orbit filter Θ_X. This filter is linear and time-invariant if the excitation is multiplicative. Thus with the flux factor as the control objective, the multiplicatively forced and controlled system in Figure 7 can be represented as in Figure 8 with linear, time-invariant filters Θ_F and Θ_C. Figure 8 depicts an open-loop control problem: the choice of control process $C(t)$ is determined by the forcing process $F(t)$ and not by the system state. In fact, we arrive at this same diagram even for the more general forcing and control excitations $\gamma_F(F(t), \vec{z}(t))$ and $\alpha_C(C(t), \vec{z}(t))$. However, for these more general excitations, the orbit filters Θ_F and Θ_C are nonlinear and no general analysis of the open-loop control problem posed by Figure 8 is available. Our analysis of Figure 8, being based on mean-square spectral properties of the forcing and

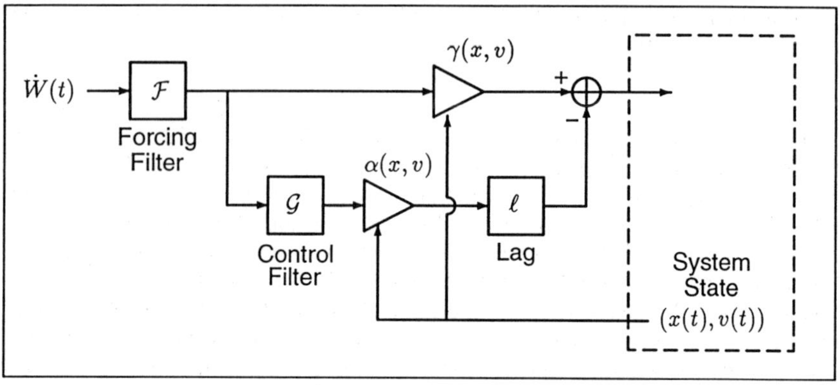

Figure 7. Dynamical system with multiplicative forcing and control.

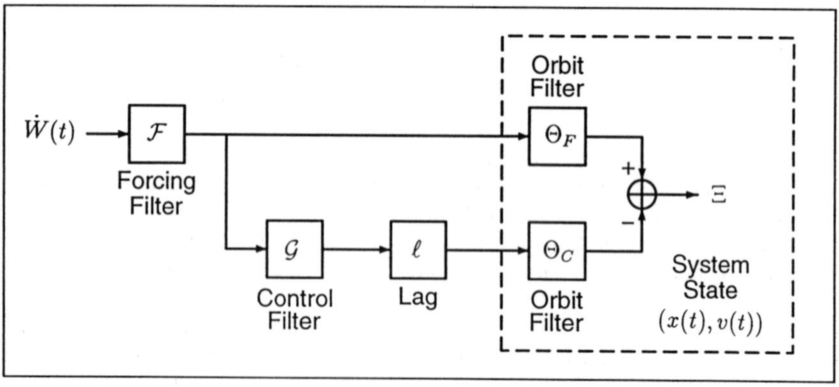

Figure 8. Equivalent dynamical system with multiplicative forcing and control given the flux factor as the control objective.

orbit filters, requires linear, time-invariant orbit filters and is therefore restricted to multiplicative excitations.

If the forcing and control are additive then, as discussed in Subsection 3.1, they have a common orbit filter Θ and the situation diagrammed in Figure 8 simplifies to that shown in Figure 9. If we set $\Theta = \Theta_F$ in the multiplicatively forced and controlled system in Figure 8 and a filter β' exists such that $\beta'\Theta_F = \beta\Theta_C$, then the system can be represented as shown in Figure 10. Remarkably, the representation in Figure 10 is the same as that in Figure 9 for the additively forced and controlled system. This shows that if the flux factor is the control objective and the filter β' exists, then multiplicative control and/or forcing has an additive representation and means, in particular, that multiplicative closed-loop control is equivalent to open-loop control.

The filter β' satisfying $\beta'\Theta_F = \beta\Theta_C$ does not exist for all combinations of orbit filters Θ_F and Θ_C and control filters β. While our results are not substantively changed in the absence of β', their derivation is significantly complicated by the need to account for two orbit filters—Θ_F and Θ_C—rather than just Θ. For simplicity, then, we assume that either the forcing and control are additive or, if multiplicative, that β' exists and restrict our attention to the open-loop control problem represented in Figure 9.

A variety of filters $\mathcal{F}$, β and Θ with corresponding impulse responses $f(t)$, $b(t)$ and $h(t)$ have now been introduced. The filter $\mathcal{F}$ models the spectrum of the forcing $F(t)$ and need not be causal. The filter Θ associated with the homo/heteroclinic orbits of the potential well is necessarily noncausal because its impulse response $h(t)$ is nonzero for $t < 0$. Neither case presents a logical problem since neither of these filters is itself directly attributable to any physical process or structure. The control filter β, however, represents some mechanical or electronic device and is therefore causal with real impulse response function $b(t) = 0$ for $t < 0$.

4.3. Control Strength

The forcing and control processes $F(t)$ and $C(t) = \beta[F](t-\ell)$ are jointly WSS given the forcing model Eq. (31). The orbit filter Θ is linear and time-invariant for the excitation models treated here so, for any fixed t, $\Theta[F](t)$ and $\Theta[C](t)$ are zero mean random variables with covariance matrix

$$\begin{pmatrix} \sigma_F^2 & \sigma_{FC} \\ \sigma_{FC} & \sigma_C^2 \end{pmatrix}. \tag{59}$$

The control $C(t)$ to some degree cancels $F(t)$ only if $C(t)$ and $F(t)$ are positively correlated; if negatively correlated, $C(t)$ is useless. Negatively correlated controls, therefore, need not be considered and, with no loss of generality, we assume $\sigma_{FC} > 0$ in Eq. (59). Using the notation in Eq. (59), the control's referred mean-square cancellation error σ^2 is the variance of the zero mean random variable $\gamma\Theta[F](t) - \alpha\Theta[C](t)$,

$$\sigma^2 = \mathrm{Var}[\gamma\Theta[F](t) - \alpha\Theta[C](t)] = \gamma^2\sigma_F^2 + \alpha^2\sigma_C^2 - 2\gamma\alpha\sigma_{FC}. \tag{60}$$

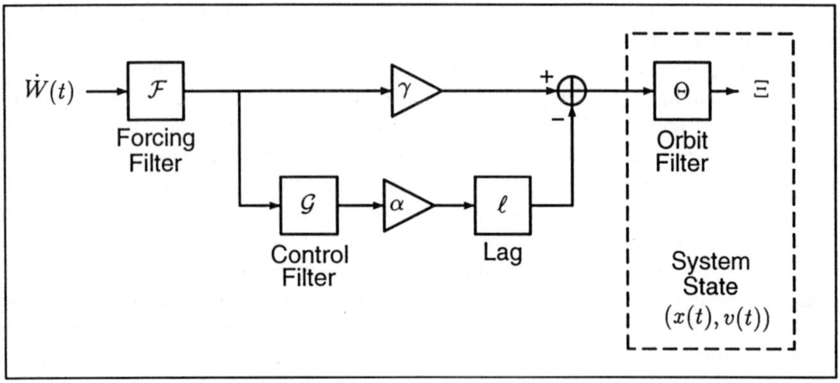

Figure 9. Dynamical system with additive forcing and control given the flux factor as the control objective.

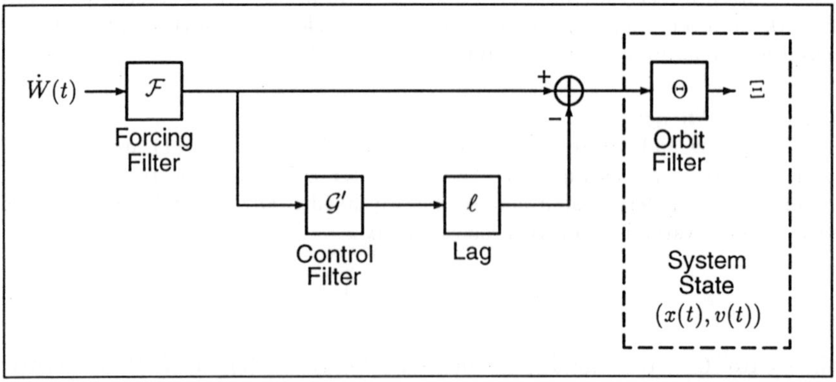

Figure 10. Equivalent additive (open-loop) representation of the multiplicatively forced and controlled dynamical system given the flux factor as the control objective.

According to Proposition 1, the control $C(t)$ reduces the flux factor Ξ and increases the stability of the system within the potential well only if the presence of $C(t)$ reduces σ^2 in Eq. (60) below its value $\gamma^2\sigma_F^2$ for zero control strength $\alpha = 0$. The control $C(t)$ therefore increases the stability of the system state within the potential well only if $\Delta < 0$ where

$$\Delta = \sigma^2 - \gamma^2\sigma_F^2 = \alpha^2\sigma_C^2 - 2\alpha\gamma\sigma_{FC}. \tag{61}$$

We have, in turn, that $\Delta < 0$ if and only if

$$\frac{\alpha}{\gamma} < 2\frac{\sigma_{FC}}{\sigma_C^2}. \tag{62}$$

Condition Eq. (62) on the relative control strength α/γ is necessary for the control to increase stability. The optimal relative control strength satisfying Eq. (62) is found by elementary calculus to be

$$\left.\frac{\alpha}{\gamma}\right|_{\text{optimal}} = \frac{\sigma_{FC}}{\sigma_C^2} \tag{63}$$

in which case $\Delta = -\gamma^2\sigma_{FC}^2/\sigma_C^2$. See Figure 11. The reduction in the flux factor achieved at the optimal relative control strength Eq. (63) is, using Eq. (49) for a $\mathcal{V}_{\mathsf{W}}$ potential,

$$kA\eta(\frac{\gamma\sigma_F}{kA}) - kA\eta(Q\frac{\gamma\sigma_F}{kA}) \tag{64}$$

where

$$Q = \sqrt{1 - \frac{\sigma_{FC}^2}{\sigma_F^2\sigma_C^2}}. \tag{65}$$

If the marginal distribution of the forcing is symmetric, then Eq. (51) yields a similar expression for $\mathcal{V}_{\mathsf{M}}$ potentials. By the Cauchy-Schwartz inequality $\sigma_{FC}^2 < \sigma_F^2\sigma_C^2$ so $0 \le Q \le 1$. From approximation Eq. (53), the maximum reduction Eq. (64) in the flux factor is approximately

$$kA\eta(\frac{\gamma\sigma_F}{kA}) - kA\eta(Q\frac{\gamma\sigma_F}{kA}) \doteq (1-Q)\gamma\sigma_F\eta_\infty \tag{66}$$

for $Q\gamma\sigma_F/(kA) \gg x_o$. This approximation is exact in the limiting case of no damping; that is, the maximum reduction in the flux factor is $(1-Q)\gamma\sigma_F\eta_\infty$ for $k = 0$. Approximation Eq. (66) suggests that $1-Q$ be treated as a measure of the proportional reduction achievable for the flux factor by a given control. The absence of the damping constant k from $1-Q$ implies that while k does play a role in the flux factor, it is secondary to that played by the second order statistics of the forcing and control processes. This is unsurprising since the damping resists the action of both the forcing *and* the control.

4.4. Role of Control Lag

To determine the role of the control lag ℓ vis-à-vis the control strength α we express the components σ_F^2, σ_C^2 and σ_{FC} of Eq. (59) as integrals. Let $c_{\Theta[F]}(t) = \mathrm{Cov}[\Theta[F](\tau +$

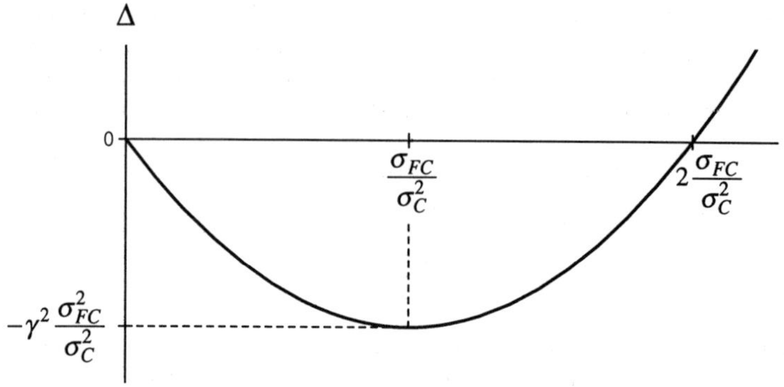

Figure 11. Change Δ in σ^2 as a function of the relative control strength α/γ.

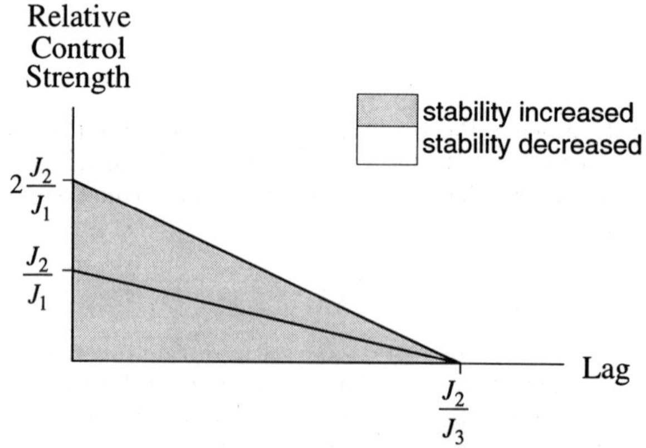

Figure 12. Combinations of relative control strength and lag within the shaded region reduce the flux factor and increase stability of the system within the potential well. Outside the shaded region stability is decreased. The solid line bisecting the shaded region identifies the optimal relative control strength for a given lag.

$t), \Theta[F](\tau)]$ be the autocovariance of the filtered process $\Theta[F](t)$. The Fourier transform $\hat{c}_{\Theta[F]}(\nu) = |\hat{h}(\nu)|^2 |\hat{f}(\nu)|^2$ is the spectral density of $\Theta[F](t)$. Therefore[28]

$$\sigma_F^2 = \frac{1}{2\pi} \int_{-\infty}^{\infty} |\hat{h}(\nu)|^2 |\hat{f}(\nu)|^2 \, d\nu. \tag{67}$$

The spectral density is an even function so

$$\sigma_F^2 = \frac{1}{\pi} \int_0^{\infty} |\hat{h}(\nu)|^2 |\hat{f}(\nu)|^2 \, d\nu \equiv \frac{J_0}{\pi}. \tag{68}$$

For the case in which $F(t) = \dot{W}(t)$ is a white noise process, Eq. (68) admits a useful simplification.

Proposition 2: Suppose the external forcing in Eq. (29) is additive with white WSS forcing process $F(t)$ as in Eq. (31). Then the referred variance $\sigma_F^2 = \mathrm{Var}[\Theta[F](t)]$ of the forcing is, for a $\mathcal{V}_{\mathsf{W}}$ potential, $\sigma_F^2 = A$ where A is the area of the safe region. For a $\mathcal{V}_{\mathsf{M}}$ potential, $\sigma_F^2 = A/2$.

Proof: The forcing is additive so the impulse response of the orbit filter is $h(t) = v_h(-t)$. $\Theta[F](t) = \Theta[\dot{W}](t)$ is a filtered white noise of the type in Eq. (31) so using in Eq. (33)

$$\begin{aligned} \sigma_F^2 &= \mathrm{Var}[\Theta[\dot{W}](t)] \\ &= \int_{-\infty}^{\infty} h^2(t) \, dt \\ &= \int_{-\infty}^{\infty} v_h^2(t) \, dt. \end{aligned} \tag{69}$$

The desired result then follows from in Eqs. (15), (17) and (19).

Let $c_{\Theta[\beta[F]]}(t)$ and $\hat{c}_{\Theta[\beta[F]]}(\nu)$ be the autocovariance and spectral density of the process $\Theta[\beta[F]](t)$ which results from filtering $F(t)$ by β and Θ successively. Then $\hat{c}_{\Theta[\beta[F]]}(\nu) = |\hat{h}(\nu)|^2 |\hat{b}(\nu)|^2 |\hat{f}(\nu)|^2$ and

$$\begin{aligned} \sigma_C^2 &= \mathrm{Var}[\Theta[C](t)] \\ &= \mathrm{Var}[\Theta[\beta[F]](\iota - \ell)] \\ &= \mathrm{Var}[\Theta[\beta[F]](t)] \\ &= \frac{1}{2\pi} \int_{-\infty}^{\infty} |\hat{h}(\nu)|^2 |\hat{b}(\nu)|^2 |\hat{f}(\nu)|^2 \, d\nu \\ &= \frac{1}{\pi} \int_0^{\infty} |\hat{h}(\nu)|^2 |\hat{b}(\nu)|^2 |\hat{f}(\nu)|^2 \, d\nu \\ &\equiv \frac{J_1}{\pi}. \end{aligned} \tag{70}$$

Using Eq. (33) followed by Parseval's theorem[28]

$$\sigma_{FC} = \mathrm{Cov}[\Theta[F](t), \Theta[C](t)]$$

$$
\begin{aligned}
&= \mathrm{Cov}[\Theta[F](t), \Theta[\beta[F]](t-\ell)] \\
&= \int_{-\infty}^{\infty} (h * f)(s)(h * g * f * \delta_\ell)(s)\, ds \\
&= \frac{1}{2\pi}\int_{-\infty}^{\infty} \hat{h}(-\nu)\hat{f}(-\nu)\hat{h}(\nu)\hat{b}(\nu)\hat{f}(\nu)e^{-j\nu\ell}\, d\nu \\
&= \frac{1}{2\pi}\int_{-\infty}^{\infty} |\hat{h}(\nu)|^2|\hat{f}(\nu)|^2\hat{b}(\nu)e^{-j\ell\nu}\, d\nu. \qquad (71)
\end{aligned}
$$

Write $\hat{b}(\nu) = \mathcal{R}(\nu) + j\mathcal{I}(\nu)$ where $\mathcal{R}(\nu)$ and $\mathcal{I}(\nu)$ are the real and imaginary parts of $\hat{b}(\nu)$. The impulse response $b(t)$ of the control filter β is real so the functions $\mathcal{R}(\nu)$ and $\mathcal{I}(\nu)$ are even and odd, respectively. Then

$$
\begin{aligned}
\sigma_{FC} &= \frac{1}{2\pi}\int_{-\infty}^{\infty} |\hat{h}(\nu)|^2|\hat{f}(\nu)|^2[\mathcal{R}(\nu) + j\mathcal{I}(\nu)][\cos\ell\nu - j\sin\ell\nu]\, d\nu \\
&= \frac{1}{2\pi}\int_{-\infty}^{\infty} |\hat{h}(\nu)|^2|\hat{f}(\nu)|^2[\mathcal{R}(\nu)\cos\ell\nu + \mathcal{I}(\nu)\sin\ell\nu]\, d\nu \\
&= \frac{1}{\pi}\int_{0}^{\infty} |\hat{h}(\nu)|^2|\hat{f}(\nu)|^2\mathcal{R}(\nu)\cos\ell\nu\, d\nu \\
&\quad + \frac{1}{\pi}\int_{0}^{\infty} |\hat{h}(\nu)|^2|\hat{f}(\nu)|^2\mathcal{I}(\nu)\sin\ell\nu\, d\nu. \qquad (72)
\end{aligned}
$$

We have $\cos x = 1 + o(x)$ and $\sin x = x + o(x^2)$ in the limit as $x \to 0$ so

$$
\begin{aligned}
\sigma_{FC} &= \frac{1}{\pi}\int_{0}^{\infty} |\hat{h}(\nu)|^2|\hat{f}(\nu)|^2\mathcal{R}(\nu)\, d\nu + \frac{\ell}{\pi}\int_{0}^{\infty} |\hat{h}(\nu)|^2|\hat{f}(\nu)|^2\mathcal{I}(\nu)\nu\, d\nu + o(\ell) \\
&\equiv \frac{J_2}{\pi} - \frac{J_3}{\pi}\ell + o(\ell) \\
&\doteq \frac{J_2}{\pi} - \frac{J_3}{\pi}\ell \qquad (73)
\end{aligned}
$$

for small lags ℓ. In cases of interest $J_2 > 0$ and $J_3 > 0$. Using approximation Eq. (73) for σ_{FC} we have

$$
\begin{aligned}
\sigma^2 &= \gamma^2\sigma_F^2 + \alpha^2\sigma_C^2 - 2\gamma\alpha\sigma_{FC} \\
&\doteq \gamma^2\frac{J_0}{\pi} + \alpha^2\frac{J_1}{\pi} - 2\gamma\alpha\frac{J_2}{\pi} + 2\gamma\alpha\ell\frac{J_3}{\pi}. \qquad (74)
\end{aligned}
$$

If the lag ℓ is small, condition Eq. (62) for increased stability is approximately

$$
\frac{J_1}{2}\frac{\alpha}{\gamma} + J_3\ell < J_2. \qquad (75)
$$

Approximation Eq. (73) from which condition Eq. (75) follows was made for convenience. While higher order terms are readily included in Eq. (73), they do not substantively change the overall result.

Condition Eq. (75) is shown in Figure 12. This condition imposes upper bounds on both of the fundamental control parameters: the relative strength α/γ of the control signal and the lag ℓ. Naturally, to minimize the flux factor and increase stability, the lag ℓ should be small since the smaller the lag in $C(t)$ the better it can cancel the forcing $F(t)$. The upper bound on ℓ indicates that large lags are, as expected, useless and justifies our small-lag approximation. The upper bound on the relative strength α/γ of the control signal is less intuitive. It might be expected that the stronger the control signal the more forcing it could cancel. However, the control $C(t)$ in system Eq. (29) actually has two competing effects. The first, represented by the term $-2\gamma\alpha J_2/\pi$ in Eq. (74), has the effect of decreasing σ^2 and contributes to increased stability through cancellation of $F(t)$. The second effect of $C(t)$ is represented by the terms $\alpha^2 J_1/\pi + 2\gamma\alpha\ell J_3/\pi$ in Eq. (74). For cases of interest, these terms are positive and increase σ^2. Their presence in Eq. (74) is explained by recognizing that $C(t)$ is an additional external forcing on the system and, to the extent that it is mismatched to $F(t)$, increases the flux factor and promotes instability.

Using approximation Eq. (73), the optimal relative control strength is

$$\left.\frac{\alpha}{\gamma}\right|_{\text{optimal}} = \frac{J_2}{J_1} - \ell\frac{J_3}{J_1}. \tag{76}$$

This is the straight line which bisects the shaded region in Figure 12. The degree to which stability is increased is, for the approximate optimal relative control strength Eq. (76), the change Eq. (64) in the flux factor with

$$Q = \sqrt{1 - \frac{(J_2 - \ell J_3)^2}{J_0 J_1}}. \tag{77}$$

For a given relative control strength α/γ and lag ℓ, the effectiveness of the control $\alpha C(t)$ depends on the control filter β and, in particular, on the control filter's gain $|\hat{b}(\nu)| = (\mathcal{R}^2(\nu) + \mathcal{I}^2(\nu))^{1/2}$ and its phase $\text{Tan}^{-1}(\mathcal{I}(\nu)/\mathcal{R}(\nu))$ which governs the filter's intrinsic lag. A useful control is one for which the average control power is small relative to the average forcing power. This suggests a control filter which places spectral power only at the frequencies needed to counteract the filtered forcing $\Theta[F](t)$. This, in addition to minimizing Q in Eq. (77), guides the search for a suitable control filter. Using this procedure, Frey and Simiu demonstrate how a simple "off-the-shelf" parametric filter might be optimized for a given control problem[10].

4.5. Optimal Control Filter

The additive control $\alpha\beta[F](t-\ell)$ in the perturbed Newtonian system Eq. (29) is determined by its strength α, its lag ℓ and the linear time-invariant filter β. The relationship between α and ℓ for fixed β is determined by Eqs. (75) and (76). We now turn to the question of the optimal choice of β. To begin we set $\gamma = \alpha = 1$ in Eq. (29). This entails no loss of generality since there is no restriction on the amplitudes of the control and forcing filters, β and $\mathcal{F}$. We assume that observation of the forcing process is limited to the output of a filter $\mathcal{M} \in \Upsilon$ with impulse response $m(t)$ where Υ is the set of finite-energy (square-integrable impulse response), causal, time-invariant, linear filters. Observation of $F(t)$ may involve a time lag $\ell \geq 0$ so, as in Eq. (57), we posit a lag filter $\mathcal{L}$ with impulse response $\delta(t-\ell)$ where $\delta(t)$ is the Dirac-delta function. Thus we have the observation process $B(t) = \mathcal{M} \circ \mathcal{L}[F](t)$ from which to construct the control signal $C(t)$. We assume $C(t) = \mathcal{G}[B](t)$ for some $\mathcal{G} \in \Upsilon$. Then $C(t) = \mathcal{Q}[F](t)$ where $\mathcal{Q} = \mathcal{M} \circ \mathcal{L} \circ \mathcal{G}$ is the composition of three filters: the measurement filter $\mathcal{M}$, the lag filter $\mathcal{L}$ and a third filter $\mathcal{G} \in \Upsilon$ chosen to minimize the flux factor Ξ for given $\mathcal{M}$, $\mathcal{L}$ and orbit filter Θ. The filters $\mathcal{M}$, $\mathcal{L}$ and $\mathcal{G}$ are all members of Υ so, also, $\mathcal{Q} \in \Upsilon$. The present control model

$$C(t) = \mathcal{Q}[F](t) = (\mathcal{M} \circ \mathcal{L} \circ \mathcal{G})[F](t) \tag{78}$$

is the same as that in Eq. (57) except that, now, the filter β in Eq. (57) has been decomposed into $\beta = \mathcal{G} \circ \mathcal{M}$.

Proposition 1 asserts that for WSS forcing an optimal choice of $\mathcal{G}$ for minimizing the flux factor of a safe region is that which minimizes the referred mean-square cancellation error $E[(\Theta[F-C](t))^2]$. Let $X(t) = \Theta[F](t)$ and $Y = (\Theta \circ \mathcal{M} \circ \mathcal{L})[F](t)$. Then $Y = (\mathcal{M} \circ \mathcal{L})[X](t)$, $\mathcal{G}[Y](t) = \Theta[C](t)$ and the referred mean-square cancellation error $E[(\Theta[F-C](t))^2]$ is

$$\sigma^2 = E[(X(t) - \mathcal{G}[Y](t))^2] \tag{79}$$

The choice of $\mathcal{G} \in \Upsilon$ which minimizes σ^2 in Eq. (79) is the well-known Wiener filter $\mathcal{G} = \mathcal{G}_w$ with transfer function

$$\hat{g}_w(\nu) = \frac{1}{\hat{c}_Y^{(+)}(\nu)} \left[\frac{\hat{c}_{XY}(\nu)}{\hat{c}_Y^{(-)}(\nu)} \right]_{\oplus} \tag{80}$$

where $\hat{c}_Y^{(+)}(\nu)$ and $\hat{c}_Y^{(-)}(\nu)$ are the causal and anticausal parts, respectively, of the spectral factorization[43] of $\hat{c}_Y(\nu)$. $[\hat{\jmath}(\nu)]_{\oplus}$ denotes the Fourier transform of $\jmath(t)u(t)$ where $u(t)$ is the unit step function; $u(t) = 1$ for $t \geq 0$ and $u(t) = 0$ otherwise. Let Υ_* be the set of all causal, time-invariant filters, both linear and nonlinear. It is well-established that if $X(t)$ is Gaussian then $\mathcal{G} = \mathcal{G}_w$ minimizes σ^2 over all $\mathcal{G} \in \Upsilon_*$.

Proposition 3: Let the perturbed Newtonian system Eq. (29) with orbit filter Θ be forced by the WSS process $F(t)$ given in Eq. (31). Assume that the control process $C(t)$ in Eq. (29) has the form $C(t) = \mathcal{G}[B](t)$ where $B(t) = \mathcal{M} \circ \mathcal{L}[F](t)$ is the observed forcing after filtering by measurement and lag filters $\mathcal{M}, \mathcal{L} \in \Upsilon$ with respective transfer functions $\hat{m}(\nu)$ and $e^{-j\nu\ell}$. Then the referred mean-square cancellation error $\sigma^2 = E[(\Theta[F-C](t))^2]$ is minimized over all $\mathcal{G} \in \Upsilon$ by the Wiener filter $\mathcal{G} = \mathcal{G}_w$ with transfer function

$$\hat{g}_w(\nu) = \frac{\left[\hat{c}_X^{(+)}(\nu) e^{j\nu\ell}\right]_{\oplus}}{\hat{c}_X^{(+)}(\nu)\hat{m}(\nu)} \tag{81}$$

where $\hat{c}_X(\nu) = |\hat{h}(\nu)|^2 |\hat{f}(\nu)|^2$ is the spectral power density of $X(t) = \Theta[F](t)$. If the forcing is Gaussian, then $\mathcal{G} = \mathcal{G}_w$ is optimal over all Υ_*.

Proof: It only remains to show that the Wiener filter transfer function $\hat{g}_w(\nu)$ given in Eq. (80) can be expressed as in Eq. (81). From $Y(t) = \mathcal{M} \circ \mathcal{L}[X](t)$ it follows that $\hat{c}_{XY}(\nu) = \hat{c}_X(\nu)\hat{m}(-\nu)\hat{l}(-\nu)$ and

$$\hat{c}_Y(\nu) = \hat{c}_{XY}(\nu)\hat{m}(\nu)\hat{l}(\nu) = \hat{c}_X(\nu)|\hat{m}(\nu)|^2. \tag{82}$$

Spectral factorization of $\hat{c}_Y(\nu)$ yields $\hat{c}_Y(\nu) = \hat{c}_Y^{(+)}(\nu)\hat{c}_Y^{(-)}(\nu)$ where $\hat{c}_Y^{(-)}(\nu) = \hat{c}_Y^{(+)}(-\nu)$ and

$$\hat{c}_Y^{(+)}(\nu) = \hat{c}_X^{(+)}(\nu)\hat{m}(\nu). \tag{83}$$

Thus

$$\frac{\hat{c}_{XY}(\nu)}{\hat{c}_Y^{(-)}(\nu)} = \frac{\hat{c}_X(\nu)\hat{m}(-\nu)\hat{l}(-\nu)}{\hat{c}_X^{(-)}(\nu)\hat{m}(-\nu)} = \hat{c}_X^{(+)}(\nu)e^{j\nu\ell}. \tag{84}$$

Putting this together with Eq. (83) in Eq. (80) yields the desired result.

As an illustration of Proposition 3, we derive the state space flux optimal control signal for system Eq. (29) with the diamond potential Eq. (11) and white WSS forcing, $F(t) = \dot{W}(t)$ (not necessarily Gaussian). For this case $|\hat{f}(\nu)| = 1$ and the orbit filter Θ has impulse response

$$h(t) = \sqrt{2D}e^{-\sqrt{2D}|t|} \tag{85}$$

and transfer function

$$\hat{h}(\nu) = \int_{-\infty}^{\infty} h(t)e^{-j\nu t}\,dt = \frac{4D}{2D+\nu^2}. \tag{86}$$

Let the measurement filter $\mathcal{M}$ have causal impulse response and transfer function

$$m(t) = \frac{1}{\mu}e^{-t/\mu}u(t), \quad \hat{m}(\nu) = \frac{1}{1+j\mu\nu} \tag{87}$$

with time constant $\mu > 0$. We begin the derivation of the Wiener filter $\mathcal{G}$ for this example by observing that

$$\hat{c}_X(\nu) = |\hat{f}(\nu)|^2|\hat{h}(\nu)|^2 = \frac{16D^2}{(2D+\nu^2)^2} = \frac{4A^4}{(A^2+\nu^2)^2} \tag{88}$$

where, recall, $A = \sqrt{2D}$ is the area of the safe region. Spectral factorization of $\hat{c}_X(\nu)$ yields

$$\hat{c}_X^{(+)}(\nu) = \frac{2A^2}{(A+j\nu)^2} \tag{89}$$

so, after some elementary calculation,

$$\left[\hat{c}_X^{(+)}(\nu)e^{j\nu\ell}\right]_{\oplus} = 2A^2e^{-A\ell}\frac{1}{(A+j\nu)^2} + 2A^2\ell e^{-A\ell}\frac{1}{A+j\nu}. \tag{90}$$

The transfer function of $\mathcal{G}_w$ is therefore

$$\begin{aligned}\hat{g}_w(\nu) &= \frac{\left[\hat{c}_X^{(+)}(\nu)e^{j\nu\ell}\right]_\oplus}{\hat{c}_X^{(+)}(\nu)\hat{m}(\nu)} \\ &= e^{-A\ell}(1+j\mu\nu) + \ell e^{-A\ell}(1+j\mu\nu)(A+j\nu) \end{aligned} \tag{91}$$

with corresponding impulse response

$$g_w(t) = (1+A\ell)e^{-A\ell}\delta(t) + (\mu+\ell+A\mu\ell)e^{-A\ell}\dot{\delta}(t) + \mu\ell e^{-A\ell}\ddot{\delta}(t). \tag{92}$$

In terms of the observation process $B(t) = \mathcal{M}\circ\mathcal{L}[F](t)$, the optimal control signal $C(t)$ for this example is, therefore,

$$C(t) = (1+A\ell)e^{-A\ell}B(t) + (\mu+\ell+A\mu\ell)e^{-A\ell}\dot{B}(t) + \mu\ell e^{-A\ell}\ddot{B}(t) \tag{93}$$

and, in the absence of the measurement filter $\mathcal{M}$ (that is, with $\mu = 0$),

$$C(t) = (1+A\ell)e^{-A\ell}B(t) + \ell e^{-A\ell}\dot{B}(t). \tag{94}$$

Potential Eq. (11) is used in this and the next example primarily because of its very simple orbit filter transfer function Eq. (86). However, the general features of the diamond potential are similar to those of the ocean vessel capsizing potential $V_{\text{OV}}(x)$ and to those of the potential $V_{\text{rf}}(x) = -\cos\pi x$ used to model the radio frequency-driven Josephson junction[19]. These potentials, like the diamond potential, are $\mathcal{V}_{\mathcal{M}}$ potentials with a well on the interval $[-1,1]$ bounded by heteroclinic orbits. Their orbit filters have respective transfer functions

$$\hat{h}_{\text{SC}}(\nu) = \frac{\sqrt{2}\pi\nu}{\sinh\frac{\pi\nu}{\sqrt{2}}}, \qquad \hat{h}_{\text{rf}}(\nu) = \frac{2}{\cosh\frac{\nu}{2}}. \tag{95}$$

Employing the power series approximations $\sinh x \doteq x + x^3/6$ and $\cosh x \doteq 1 + x^2/2$ we have

$$\hat{h}_{\text{SC}}(\nu) \doteq 2\frac{1}{1+\frac{1}{6}\left(\frac{\pi\nu}{\sqrt{2}}\right)^2} = \frac{24/\pi^2}{12/\pi^2+\nu^2} \tag{96}$$

and

$$\hat{h}_{\text{rf}}(\nu) \doteq 2\frac{1}{1+\frac{1}{2}\left(\frac{\nu}{2}\right)^2} = \frac{16}{8+\nu^2}. \tag{97}$$

Choose $D = 6/\pi^2$ for approximation Eq. (96) and $D = 4$ for approximation Eq. (97). Each approximation is then identically the transfer function $\hat{h}(\nu)$ in Eq. (86). Thus, for suitable choices of D, the transfer function of $V(x)$ is a rational approximation of both $\hat{h}_{\text{SC}}(\nu)$ and $\hat{h}_{\text{rf}}(\nu)$. See Figure 13. In this sense (the most important one for controlling the flux factor), $V(x)$ approximates both $V_{\text{SC}}(x)$ and $V_{\text{rf}}(x)$.

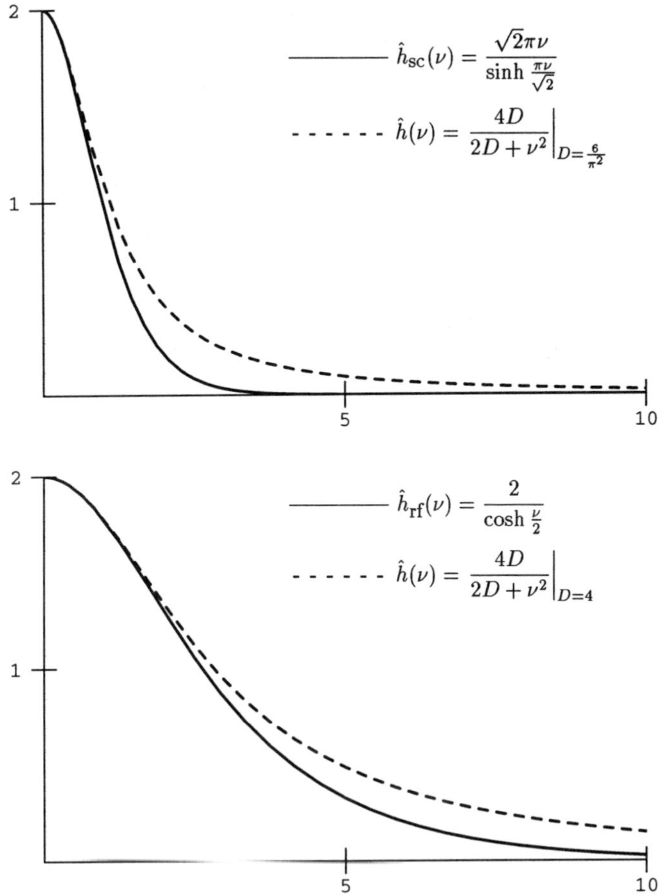

Figure 13. Orbit filter transfer functions $\hat{h}_{sc}(\nu)$ and $\hat{h}_{rf}(\nu)$ for the ship capsizing model and the radio frequency-driven Josephson junction. Also shown are their $\hat{h}(\nu)$ approximations based on the diamond potential.

4.6 Limited Available Power

The average power of a WSS process is its marginal variance so, for example, the average power of $C(t)$ is $\mathrm{Var}[C(t)]$. The Wiener filter solution in Proposition 3 for the optimal control was given in the absence of any constraint on the average power of the control and therefore implicitly assumes the availability of sufficient average power. Suppose $P > 0$ is the available average control power. Then Proposition 3 applies only if $\mathrm{Var}[C(t)] \le P$ for $C(t) = \mathcal{G}_w[B](t)$. This condition can be expressed directly in terms of the orbit filter, the lag, and the forcing.

Proposition 4: Let $P > 0$ be the available average control power. Under the conditions of Proposition 3 with orbit filter transfer function $\hat{h}(\nu)$, lag ℓ and forcing spectral power density $\hat{c}_F(\nu) = |\hat{f}(\nu)|^2$, we have $\mathrm{Var}[C(t)] \le P$ if and only if

$$P \ge \frac{1}{2\pi} \int_{-\infty}^{\infty} \left| \frac{\left[\hat{c}_X^{(+)}(\nu) e^{j\nu\ell}\right]_\oplus}{\hat{h}(\nu)} \right|^2 d\nu \tag{98}$$

where $\hat{c}_X(\nu) = |\hat{h}(\nu)|^2 |\hat{f}(\nu)|^2$. In particular, if $\ell = 0$ then $\mathrm{Var}[C(t)] \le P$ if and only if $P \ge \mathrm{Var}[F(t)]$ where $\mathrm{Var}[F(t)]$ is the average forcing power.

Proof: It follows from $C(t) = \mathcal{G}_w \circ \mathcal{M} \circ \mathcal{L}[F](t)$ that

$$\mathrm{Var}[C(t)] = \frac{1}{2\pi} \int_{-\infty}^{\infty} |\hat{g}_w(\nu)|^2 |\hat{m}(\nu)|^2 |\hat{f}(\nu)|^2 \, d\nu. \tag{99}$$

Equation (98) then follows using Eq. (81) in Proposition 3. Finally, if $\ell = 0$ then

$$\left[\hat{c}_X^{(+)}(\nu) e^{j\nu\ell}\right]_\oplus = \left[\hat{c}_X^{(+)}(\nu)\right]_\oplus = \hat{c}_X^{(+)}(\nu) \tag{100}$$

in which case the right-hand side of Eq. (98) is $\mathrm{Var}[F(t)]$.

For $\ell = 0$ or otherwise the right-hand side of Eq. (98) is bounded above by $\mathrm{Var}[F(t)]$ so the Wiener filter solution in Proposition 3 never requires more average power than that of the forcing. In some applications the average forcing power is small as in, for example, electronic circuits with thermal noise forcing. In these applications the available average control power P would be expected to satisfy Eq. (98), thereby allowing the use of Proposition 3. In ocean vessel capsizing models with extreme beam sea forcing or in earthquake-loaded structures, P would not be expected to satisfy Eq. (98) and Proposition 3 is not applicable. For problems of this latter class, we seek the filter $\mathcal{G} \in \Upsilon$ which minimizes the referred mean-square cancellation error $\sigma^2 = E[(X - \mathcal{G}[Y])^2]$ subject to

$$\sigma^2_{\Theta^{-1} \circ \mathcal{G}[Y]} = \mathrm{Var}[\Theta^{-1} \circ \mathcal{G}[Y]] \le P \tag{101}$$

where Θ^{-1} is the inverse of the orbit filter Θ. It can be shown based on the orthogonality principle for Hilbert space[43] that this minimization problem is solved by $\alpha \mathcal{G}_w$ where $\alpha \le 1$ is a scalar chosen to meet the power constraint Eq. (101), if and only if the orbit filter Θ in Eq. (101) is the identity (with impulse response $\delta(t)$). Since the latter is never the case, something beyond Wiener filter theory is needed to minimize σ^2. While this minimization does not appear to have a closed-form solution, we can express σ^2 in a form which suggests an attractive, albeit less than optimal, solution.

Proposition 5: Let $\mathcal{Q} \in \Upsilon$ with $C(t) = \mathcal{Q}[F](t)$ have impulse response $q(t)$ and transfer function $\hat{q}(\nu)$. Given the orbit filter transfer function $\hat{h}(\nu)$ and the forcing spectral power density $\hat{c}_F(\nu) = |\hat{f}(\nu)|^2$, the referred mean-square cancellation error σ^2 is

$$\sigma^2 = \frac{1}{2\pi}\int_{-\infty}^{\infty} K(\nu)\left[1 - 2|\hat{q}(\nu)|\cos\angle\hat{q}(\nu) + |\hat{q}(\nu)|^2\right] d\nu \tag{102}$$

where $K(\nu) = |\hat{h}(\nu)\hat{f}(\nu)|^2$ and $\angle\hat{q}(\nu)$ is the phase response of $\mathcal{Q}$ satisfying

$$\hat{q}(\nu) = |\hat{q}(\nu)|e^{j\angle\hat{q}(\nu)}. \tag{103}$$

Proof: It follows from $C(t) = \mathcal{Q}[F](t)$ that

$$\begin{aligned}\hat{c}_{F-C}(\nu) &= \hat{c}_F(\nu) - \hat{c}_{FC}(\nu) - \hat{c}_{CF}(\nu) + \hat{c}_C(\nu)\\ &= \hat{c}_F(\nu)\left[1 - 2|\hat{q}(\nu)|\cos\angle\hat{q}(\nu) + |\hat{q}(\nu)|^2\right].\end{aligned} \tag{104}$$

Using this and $K(\nu) = |\hat{h}(\nu)|^2\hat{c}_F(\nu)$ in

$$\sigma^2 = \frac{1}{2\pi}\int_{-\infty}^{\infty} |\hat{h}(\nu)|^2\hat{c}_{F-C}(\nu)\, d\nu \tag{105}$$

gives the desired result.

Suppose the phase response $\angle\hat{q}(\nu)$ of the control filter $\mathcal{Q} = \mathcal{M}\circ\mathcal{L}\circ\mathcal{G}$ is identically zero. This will never be the case but following this line of reasoning we obtain a control filter $\mathcal{Q} \in \Upsilon$ with certain attractive features. Also assume for simplicity that there is no measurement filter or control lag so that $\mathcal{Q} = \mathcal{G}$. With no control lag Proposition 3 applies if $P \geq \mathrm{Var}[F(t)]$. Thus we consider here the case $P < \mathrm{Var}[F(t)]$. From Proposition 4, for $\angle\hat{q}(\nu) = 0$,

$$\sigma^2 = \frac{1}{2\pi}\int_{-\infty}^{\infty} K(\nu)(1 - |\hat{g}(\nu)|)^2\, d\nu. \tag{106}$$

We seek a filter $\mathcal{G} \in \Upsilon$ which minimizes Eq. (106) subject to the constraint $\mathrm{Var}[C(t)] \leq P$ for a given available average power $P > 0$. In the present situation $C(t) = \mathcal{G}[F](t)$ has spectral power density $\hat{c}_C(\nu) = |\hat{g}(\nu)|^2|\hat{f}(\nu)|^2$ and the power constraint $\mathrm{Var}[C(t)] \leq P$ is equivalent to

$$\frac{1}{2\pi}\int_{-\infty}^{\infty} |\hat{g}(\nu)|^2|\hat{f}(\nu)|^2\, d\nu \leq P. \tag{107}$$

To minimize σ^2 in Eq. (106) subject to the power constraint Eq. (107), we apply the method of Lagrange multipliers to the Lagrangian

$$L(|\hat{g}(\cdot)|;\lambda) = \frac{1}{2\pi}\int_{-\infty}^{\infty} K(\nu)(1 - |\hat{g}(\nu)|)^2\, d\nu + \lambda\left(\frac{1}{2\pi}\int_{-\infty}^{\infty} |\hat{g}(\nu)|^2|\hat{f}(\nu)|^2\, d\nu - P\right) \tag{108}$$

with the result that the optimal filter $\mathcal{G} = \mathcal{G}_o$ has amplitude response

$$|\hat{g}_o(\nu)| = \frac{|\hat{h}(\nu)|^2}{\lambda + |\hat{h}(\nu)|^2} \tag{109}$$

where λ is given implicitly by

$$\frac{1}{2\pi}\int_{-\infty}^{\infty} \frac{|\hat{h}(\nu)|^4 |\hat{f}(\nu)|^2}{(\lambda + |\hat{h}(\nu)|^2)^2}\, d\nu = P. \tag{110}$$

The Lagrange multiplier λ is evidently a decreasing function of P with $\lambda \to \infty$ as $P \to 0$ and $\lambda = 0$ for $P = \mathrm{Var}[F(t)]$.

It follows from the amplitude response given in Eq. (109) that the optimal (in the present restricted sense of minimizing Eq. (106)) control has spectral power density

$$\hat{c}_C(\nu) = |\hat{g}_o(\nu)|^2 \hat{c}_F(\nu) = \frac{|\hat{h}(\nu)|^4 |\hat{f}(\nu)|^2}{(\lambda + |\hat{h}(\nu)|^2)^2}. \tag{111}$$

This spectral power density has the intuitively satisfying property that it invests a relatively large fraction of the average available power P at frequencies where both the forcing spectral power density $\hat{c}_F(\nu) = |\hat{f}(\nu)|^2$ is high and $|\hat{h}(\nu)|$ is large. At frequencies where there is little forcing power present or where $|\hat{h}(\nu)|$ is close to zero indicating that the system is relatively uneffected by forcing, $\hat{c}_C(\nu)$ invests relatively little power. Also, consistent with $\mathcal{G}_w$ given in Proposition 3, $\mathcal{G}_o$ has unit amplitude $|\hat{g}_o(\nu)| = 1$ with $g_o(t) = \delta(t)$ for $P = \mathrm{Var}[F(t)]$. Finally, the transfer function $\hat{g}_o(\nu)$ of $\mathcal{G}_o$ depends on the forcing only through λ. This is convenient for purposes of implementation and means $\mathcal{G}_o$ is easily adaptable to changes in average forcing power.

Only the amplitude response $\hat{g}_o(\nu)$ of the filter $\mathcal{G}_o$ is given by Eq. (109). Even so, the corresponding phase response $\angle\hat{g}_o(\nu)$ is determined by the requirement that $\mathcal{G}_o$ be causal. This phase response will not be identically zero as assumed in the derivation of $|\hat{g}_o(\nu)|$. The ultimate effect of this inconsistency on the effectiveness of the filter $\mathcal{G}_o$ is difficult to assess.

To illustrate the procedure for calculating the filter $\mathcal{G}_o$ described above we again consider system Eq. (29) with potential Eq. (11) forced by white noise $F(t) = \dot{W}(t)$ with $\hat{f}(\nu) = 1$. According to Eq. (109), the amplitude response of $\mathcal{G}_o$ for potential Eq. (11) with $\hat{h}(\nu) = 2A^2/(A^2 + \nu^2)$ is

$$|\hat{g}_o(\nu)| = \frac{4A^4}{\lambda(A^2 + \nu^2)^2 + 4A^4}. \tag{112}$$

The causal transfer function $\hat{g}_o(\nu)$ with this amplitude is[13]

$$\hat{g}_o(\nu) = \frac{4/\lambda}{((\lambda_1 + j\nu/A)^2 + \lambda_2^2)^2} \tag{113}$$

where $A = \sqrt{2D}$ and

$$\lambda_1 = \sqrt{\frac{\sqrt{1+4/\lambda}+1}{2}}, \quad \lambda_2 = \sqrt{\frac{\sqrt{1+4/\lambda}-1}{2}}. \tag{114}$$

The impulse response corresponding to $\hat{g}_o(\nu)$ is

$$g_o(t) = \frac{2A}{\lambda_2^3\lambda} e^{-\lambda_1 At} \left[\sin(\lambda_2 At) - \lambda_2 At \cos(\lambda_2 At)\right] u(t). \tag{115}$$

As noted earlier the forcing $F(t)$ is expressed in $g_o(t)$ only through the parameter λ. The functional form of $g_o(t)$ is independent of the forcing's distributional properties.

5. Relative Stability

Stability is a central issue in the design, analysis, and control of many dynamical systems. Stability has many meanings: it may refer to the relative stability of various system states, the relative stability of various system trajectories, the relative stability of a range of states such as those within a particular potential well, or the structural stability[2] of the system itself. A wide range of definitions of stability are recognized, the consensus being that the choice of definition should "depend on pragmatic as well as rigorous goals"[3,32,36]. Prominent conceptions of relative stability include stability in the sense of Lyaponov's "second" or "direct" method[3], Lyaponov exponents[1], relative equilibria[24], probability ratios[38], and other stochastically-oriented indices[20]. For systems modeled as autonomous flows at least three definitions of relative stability can be identified: linear stability, Lyapunov stability and spectral stability[17].

In Newtonian systems for which a potential energy $V(x)$ can be assigned to each system state, stability is usually interpreted in terms of $V(x)$; the system is considered to be in a stable state if its potential energy in that state is a relative minimum. If this minimum is at the bottom of a deep potential well, then, intuitively, the state is highly stable since strong external forcing is needed to drive the system from this state and out of the well. To assess this type of system stability we propose a measure of relative stability, FRRS, based on state space transport. This definition has the same fundamental properties as a more traditional measure of relative stability, PRRS, based on probability ratios. Because FRRS is based on transport out of a potential well it is well-suited for applications in which system failure is identified with escape from a well.

We use FRRS to prove a "blowtorch" theorem for white WSS forcing and, in particular, for white Gaussian forcing. This theorem is consistent with the blowtorch theorem already known for PRRS for the white Gaussian case. This latter theorem states that if the temperature of a section of a potential well is heated (as if by a blowtorch) above that of the surrounding well, then the PRRS of the well is decreased. Proceeding from our definition of FRRS, we obtain the new result that a blowtorch theorem, though valid for white WSS forcing, is not generally valid for colored Gaussian or deterministic harmonic forcing. This has the practical consequence that the addition of the appropriate state-dependent excitation within a well can increase the stability, in our defined sense, of the well against escapes, effectively extending the mean first passage time from the well.

The remainder of this section consists of four subsections. The first of these recalls the definition of PRRS. FRRS is formally introduced in the second subsection. This new measure of relative stability is compared with that of PRRS using the asymmetric Duffing potential and the double square well potential as examples. Proposition 2 facilitates this comparison. In the third subsection, FRRS is proved to satisfy a blowtorch theorem for the case of white WSS forcing. The fourth subsection shows by counterexamples that such a theorem is not, in general, possible for either colored Gaussian forcing or deterministic forcing.

5.1. Probability Ratio Relative Stability

Let $V(x)$ be a bistable potential belonging to $\mathcal{V}_W$ for the reparameterized, linearly damped, additively forced Newtonian system

$$\ddot{x} = -V'(x) + \gamma F(t) - k\dot{x} \tag{116}$$

where $k > 0$, $\gamma > 0$, and the forcing $F(t) = \dot{W}(t)$ is a white Gaussian process. System Eq. (116) can be written as the set of first order stochastic differential equations

$$\begin{aligned} dv(t) &= -[V'(x) + kv(t)]\,dt + \gamma\,dW(t) \\ dx(t) &= v(t)\,dt \end{aligned} \tag{117}$$

where $W(t)$ is a Wiener process.

Energy potentials belonging to $\mathcal{V}_W$ have two wells with the well to the left of $x = 0$ labelled well 1 and that to the right labelled well 2. The PRRS of well 2 with respect to well 1 is[38] the probability ratio $\rho_{21} = p_2/p_1$ where p_1 and p_2 are, respectively, the probabilities of being in well 1 and well 2. According to this definition, well 2 of a bistable system is stable relative to well 1 if the state of the system is more likely to be in well 2 than well 1. The probabilities p_1 and p_2 used in this approach are those of the stationary distribution of the system state. In general, Monte Carlo simulation is necessary to obtain these probabilities. However, for system Eq. (117), the density of the stationary distribution of the state (x, v) can be obtained as a solution of Kramers' equation[14]. This distribution is Boltzmann with density

$$p_s(x, v) = \mathcal{N}_1 \exp\left(-\frac{2kV(x)}{\gamma^2} - \frac{kv^2}{\gamma^2}\right) \tag{118}$$

where $\mathcal{N}_1$ is a normalization constant. The density of the stationary distribution of the variable x alone is then

$$p_s(x) = \mathcal{N}_2 \exp\left(-\frac{2kV(x)}{\gamma^2}\right) \tag{119}$$

where $\mathcal{N}_2$ is another normalization constant. The PRRS of well 2 with respect to well 1 is defined to be the stationary odds ratio

$$\rho_{21} = \lim_{t\to\infty} \frac{P\{x(t) > 0\}}{P\{x(t) < 0\}} \tag{120}$$

provided this limit exists. If the potential $V(x)$ increases sufficiently rapidly for $x \to \pm\infty$ as, for example, is the case with both the double square well potential and the asymmetric Duffing potential, the limit does exist and, from Eq. (119),

$$\rho_{21} = \frac{\int_0^\infty \exp(-2kV(x)/\gamma^2)\,dx}{\int_{-\infty}^0 \exp(-2kV(x)/\gamma^2)\,dx}. \tag{121}$$

The PRRS ρ_{21} of the wells of the asymmetric Duffing potential and of the double square well potential is plotted in Figures 14 and 15 for various values of k, γ and well parameters.

5.2. Flux Ratio Relative Stability

We now introduce a measure ϱ_{21} of relative stability founded on the relative transport of state space out of the safe regions associated with the two wells of $V(x) \in \mathcal{V}_W$. First, express system Eq. (116) in its original form Eq. (29) parameterized by ε:

$$\ddot{x} = -V'(x) + \varepsilon\gamma F(t) - \varepsilon k\dot{x}. \tag{122}$$

The flux ratio relative stability (FRRS) of well 2 with respect to well 1 is defined for this bistable (two well) system to be

$$\varrho_{21} = \frac{\Xi_{1,s}}{\Xi_{2,s}} \tag{123}$$

where Ξ_i, $i = 1, 2$ is the flux factor for transport out of well i's safe region and $\Xi_{i,s} = \Xi_i/A_i$, is its standardized counterpart. A_i is the area of well i's safe region and, as such, is a geometrical measure of the depth and width of well i. According to Eq. (123), well 2 is more stable than well 1 when $\varrho_{21} > 1$. Considering the example of the forced Duffing oscillator in subsection 3.3, $\varrho_{21} > 1$ is intuitively interpreted to mean that escapes from well 2 occur less frequently or at a lower rate than escapes from well 1. The flux factor Ξ_i for well i of system Eq. (122) can be expressed

$$\Xi_i = kA_i\,\eta\left(\frac{\gamma\sigma_i}{kA_i}\right) \tag{124}$$

where σ_i^2 is the variance of the SM distribution of the referred forcing $\Theta_i[F](t)$ and Θ_i is the orbit filter of the homoclinic orbit about the safe region of well i. If $F(t)$ is a zero mean, WSS process then $\sigma_i^2 = \mathrm{Var}[\Theta[F](t)]$. The standardized flux factor corresponding to Eq. (124) is

$$\Xi_{i,s} = k\,\eta(\kappa_i) \tag{125}$$

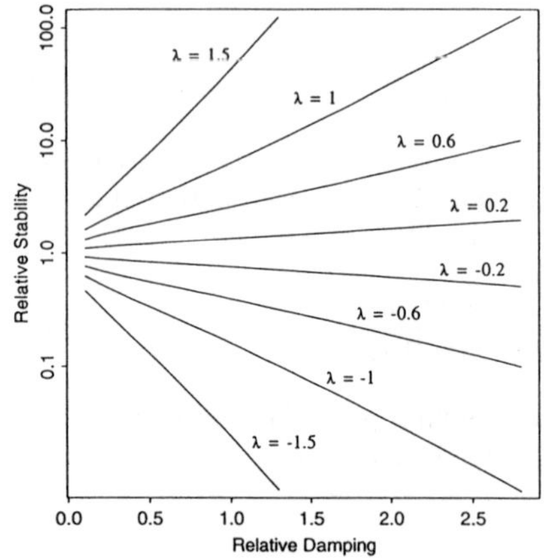

Figure 14. PRRS of the wells of the asymmetric Duffing potential for various values of asymmetry parameter λ and relative damping k/γ^2.

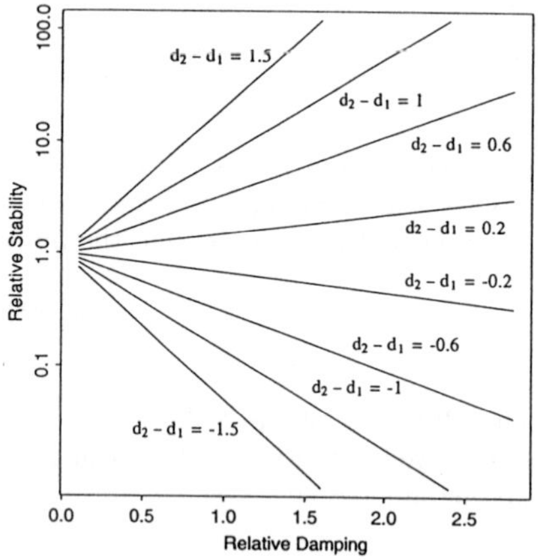

Figure 15. PRRS for the double square well potential for various values of relative depth d_2-d_1 and relative damping k/γ^2.

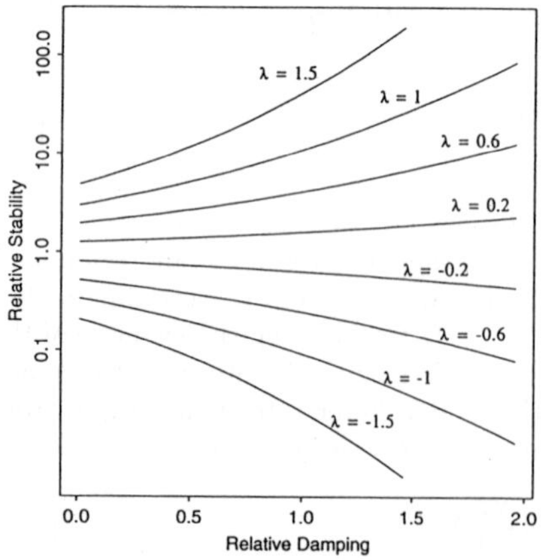

Figure 16. FRRS for the asymmetric Duffing potential for various values of asymmetry parameter λ and relative damping k/γ.

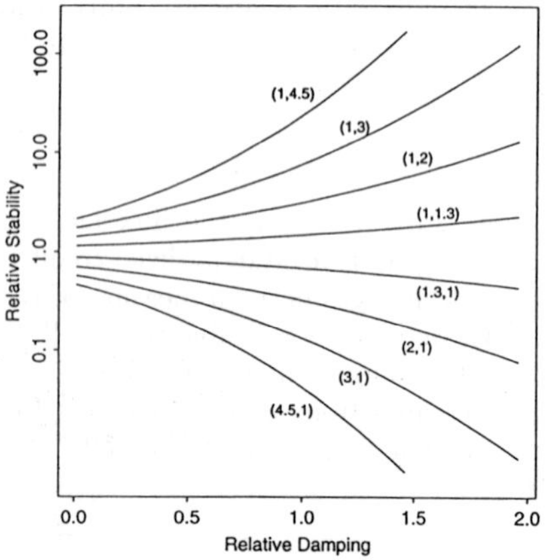

Figure 17. FRRS for the double square well potential for various well sizes ($wd_1^{1/2}$,$wd_2^{1/2}$) and relative damping k/γ.

where

$$\kappa_i = \frac{\gamma\sigma_i}{kA_i}. \tag{126}$$

In the case of WSS forcing, the quantity κ_i is easily shown to be the coefficient of variation (standard deviation divided by mean) of the Melnikov process[8] associated with well i.

Proposition 6: The standardized flux factor $\Xi_{i,s}$ for transport out of well i of system Eq. (122) is a nondecreasing function of κ_i. For white WSS forcing, $\Xi_{i,s}$ is a nonincreasing function of the area A_i. If the support of the SM distribution of the referred forcing $\Theta_i[F](t)$ is the whole real line, then $\Xi_{i,s}$ is an increasing function of κ_i and, for white WSS forcing, $\Xi_{i,s}$ is a decreasing function of A_i.

Proof: The derivative of the function $\eta(x)$ is nonnegative for all $x > 0$ and positive for all $x > 0$ if the SM distribution of $\Theta_i[F](t)$ is the whole real line. (See subsection 3.1.) Therefore $\Xi_{i,s}$ is, as stated, a nondecreasing or increasing function of κ_i. If the forcing $F(t)$ is white WSS then, using Proposition 2, $\sigma_i = \sqrt{A_i}$ and κ_i is inversely proportional to the square root $\sqrt{A_i}$. The remainder of the proposition follows.

The extent of the homoclinic orbit of a $\mathcal{V}_W$ potential well is determined by the size of the well. The wider the well, the greater the extent of the homoclinic orbit in the direction of the state space variable x. The deeper the well, the greater the maximum velocity on the homoclinic orbit and the greater the extent of the orbit in the direction of the state space variable v. Thus the area A_i of well i's safe region reflects both the depth and width of the well; if either the depth or the width of the well is increased then A_i increases. Therefore, Proposition 6 establishes that the flux ratio definition of ϱ_{21} is order consistent; wider, deeper wells are more stable than narrower, shallower wells.

Proposition 7: The relative stability ϱ_{21} of well 2 with respect to well 1 for the perturbed Newtonian system Eq. (122) with colored Gaussian forcing $F(t)$ is

$$\varrho_{21} = \frac{\kappa_1\phi(1/\kappa_1) - 1 + \Phi(1/\kappa_1)}{\kappa_2\phi(1/\kappa_2) - 1 + \Phi(1/\kappa_2)}. \tag{127}$$

Proof: Insert Eq. (54) in Eq. (125) for the standardized flux factor.

Limiting expressions are available for ϱ_{21} in the Gaussian case based on Eq. (127). If the damping k is small relative to $\gamma\sigma_i$ in each well, then $\kappa_i \gg 1$, $i = 1, 2$ and

$$\varrho_{21} \approx \frac{\kappa_1}{\kappa_2} = \frac{A_2}{A_1}\frac{\sigma_1}{\sigma_2}. \tag{128}$$

The Laplace-Feller inequality states[37]

$$0 < \frac{\phi(z)}{z} - 1 + \Phi(z) < \frac{\phi(z)}{z^3} \tag{129}$$

for all $z > 0$. Thus, if the damping k is large relative to $\gamma\sigma_i$ in each well, then $\kappa_i \ll 1$, $i = 1, 2$ and

$$\varrho_{21} \approx \frac{\kappa_1^3\phi(1/\kappa_1)}{\kappa_2^3\phi(1/\kappa_2)}. \tag{130}$$

The relative stability ϱ_{21} is given in terms of κ_i, $i = 1, 2$ in all three Eqs. (127), (128) and (130). For the case of white WSS forcing, we have, using Proposition 2,

$$\kappa_i = \frac{kA_i}{\gamma\sigma_i} = \frac{k}{\gamma}\sqrt{A_i}. \tag{131}$$

Equations (127), (128) and (130) for ϱ_{21} apply for colored Gaussian forcing. Using expressions similar to Eq. (124), the flux ratio relative stability ϱ_{21} finds similar expression for many other processes including, for example, shot noises[9]. Expressions for Ξ_i, and hence ϱ_{21}, can also be obtained when $F(t)$ is a deterministic function[8].

Asymmetric Duffing potential: The velocity components Eq. (25) of the homoclinic orbits are needed for the calculation of the referred variances $\sigma_i^2 = \mathrm{Var}[\Theta_i[F](t)]$ in the case of colored Gaussian forcing. For white Gaussian forcing, these calculations are obviated by Proposition 2. For purpose of comparison with ρ_{21} we assume $F(t)$ is white Gaussian and calculate ϱ_{21} using Eq. (131) with Eq. (26) for A_i in Eq. (127). The results are shown in Figure 16.

Double square well potential: We assume $F(t)$ is a white Gaussian process and calculate ϱ_{21} using Eq. (131) with $A_i = 2w\sqrt{2d_i}$ in Eq. (127). The results are shown in Figure 17.

Comparisons of Figures 14 and 16 and Figures 15 and 17 show that ρ_{21} and ϱ_{21} perform similarly. However, they measure essentially different features of the potential wells of a dynamical system and neither measure can be used to numerically approximate the other. For the PRRS ρ_{21}, the natural relative damping is k/γ^2 since this is the factor that appears in Eq. (121) for ρ_{21}. By contrast, the FRRS is expressed in terms of k/γ as, for example, in Proposition 7. Thus for ϱ_{21} the natural expression of relative damping is k/γ rather than k/γ^2. This difference has significant consequences. First note that as shown in Figure 14 the PRRS ρ_{21} is a monotonic function of k/γ^2; strictly increasing for $\lambda > 0$ and strictly decreasing for $\lambda < 0$. Figure 16 shows that FRRS is likewise a monotonic function of k/γ. Suppose λ is positive and $k = \gamma = 1$. Then $k/\gamma^2 = k/\gamma = 1$. Now consider increasing k to 8 and γ to 4. Then k/γ^2 *decreases* from 1 to 1/2 while k/γ *increases* from 1 to 2. Here is an example where a change in system parameters results in a decrease in PRRS and an increase in FRRS. This is not paradoxical; rather, it underscores the fact, already noted, that ϱ_{21} and ρ_{21} measure different quantities. The PRRS reflects the mean time to escape from well 2 based on the stationary distribution Eq. (118) of system states within the competing wells. FRRS is both tighter in its focus and narrower in its scope and is based on the fraction of a certain portion of state space which eventually escapes the potential well. This portion of state space is that within the so-called stochastic layer[23]; i.e, the infinitesimally narrow strip of state space along the pseudo-separatrix. This portion of state space consists of exactly those states with nearly sufficient energy (kinetic plus potential in the Newtonian system model) to escape the potential well. Thus the difference between the two measures of stability is akin to that between conditional and unconditional probabilities. PRRS is based on the stationary well occupancy probability averaged over all energy levels while FRRS is more closely related to the conditional probability of escape given that the system has been excited to an energy level at which very little additional excitation may be required for escape. This suggests that PRRS be applied to time-invariant systems with stationary forcing and that FRRS be used for problems involving arbitrarily varying forcing with less than fully-known higher-order statistics such as, for example, problems of earthquake loading on structures or ship capsizing in extreme seas.

5.3. Blowtorch Theorem

The blowtorch theorem, as first formulated, concerns the PRRS ρ_{21} of the potential wells of a bistable system subjected to state-dependent white Gaussian forcing[21,38]. The theorem states that if in some interval I belonging to well 2 the effective temperature is increased above the background temperature outside of I, then the PRRS of well 2 is decreased with respect to well 1. The temperature to which the theorem refers is that of the "heat bath" which is the source of the diffusion of the system state. In the case of system Eq. (117) the role of temperature is played by the forcing strength γ. FRRS also satisfies a blowtorch theorem. The theorem stated here for FRRS treats forcing with the same state-dependency as that of the probability ratio blowtorch theorem but includes all white WSS forcing processes. The probability ratio blowtorch theorem treats only white Gaussian forcing.

Theorem 1: Let system Eq. (122) be bistable with potential energy $V(x) \in \mathcal{V}_{\mathsf{W}}$ and suppose $F(t)$ is a zero mean, white WSS process. Well 2 of $V(x)$ is assumed to be contained within the interval $(0, \infty)$ while well 1 is contained within $(-\infty, 0)$. Let $I \subseteq (0, \infty)$ be a subinterval of well 2 and let the forcing strength γ be state-dependent,

$$\gamma(x) = \gamma_0 + \gamma_1 1_I(x), \tag{132}$$

where $\gamma_0 > 0$, $\gamma_1 \geq 0$ and $1_I(\cdot)$ is the indicator function of the interval I; $1_I(x) = 1$ if $x \in I$ and $1_I(x) = 0$ otherwise. Then the FRRS ϱ_{21} of well 2 with respect to well 1 is a nonincreasing function of γ_1 for $\gamma_1 > 0$. If the forcing is a zero-mean, white Gaussian process then ϱ_{21} is a decreasing function of γ_1.

Proof: The FRRS

$$\varrho_{21} = \frac{\Xi_{1,s}}{\Xi_{2,s}} = \frac{A_2}{A_1}\frac{\Xi_1}{\Xi_2} \tag{133}$$

is a function of γ_1 only through the functional dependence of the flux factor $\Xi_2 = \Xi_2(\gamma_1)$ on γ_1. From Eq. (124) this flux factor is

$$\Xi_2 = kA_2\,\eta(\frac{\sigma_2}{kA_2}) \tag{134}$$

where σ_2^2 is the variance of the SM distribution of the referred forcing $\Theta_2[F](t)$. The forcing is multiplicative so the orbit filter Θ_2 is time-invariant and linear. Equation (134) for Ξ_2, though derived in Section 3 for additive excitations, is equally valid for a multiplicative excitation. The filtered counterpart of the white WSS process $F(t)$ is the stochastic integral

$$\mathcal{I}(t) = \Theta_2[F](t) = \int_{-\infty}^{\infty} h_2(t-s)\,dW(s) \tag{135}$$

where $\vec{z}_{h2}(t) = (x_{h2}(t), v_{h2}(t))$ is the homoclinic orbit of well 2 and

$$h_2(t) = v_{h2}(-t)\gamma(x_{h2}(-t)) \tag{136}$$

is the impulse response of Θ_2 associated with the multiplicative excitation $\gamma(x(t))F(t)$. Then

$$\begin{aligned}
\sigma_2^2 &= \int_{-\infty}^{\infty} \dot{x}_2^2(t)\gamma^2(x_2(t))\,dt \\
&= \int_{-\infty}^{\infty} \dot{x}_2^2(t)(\gamma_0 + \gamma_1 1_J(t))^2\,dt \\
&= \gamma_0^2 \int_{-\infty}^{\infty} \dot{x}_2^2(t)\,dt + (2\gamma_0\gamma_1 + \gamma_1^2)\int_{-\infty}^{\infty} \dot{x}_2^2 1_J(t)\,dt \\
&= \gamma_0^2 \int_{-\infty}^{\infty} \dot{x}_2^2(t)\,dt + (2\gamma_0\gamma_1 + \gamma_1^2)\int_J \dot{x}_2^2(t)\,dt
\end{aligned} \tag{137}$$

where $J = \{t \in \Re : x_2(t) \in I\}$. The referred variance σ_2^2 is an increasing function of γ_1 by inspection of Eq. (137). The flux factor Ξ_2 in Eq. (134) is, in turn, by Proposition 6 a nondecreasing function of σ_2^2. The result for WSS forcing then follows from Eq. (133). For Gaussian forcing Ξ_2 in Eq. (134) is an increasing function of σ_2^2, hence the result for Gaussian forcing.

5.4. Counterexamples

The previous section established that FRRS, like PRRS, satisfies a blowtorch theorem for white Gaussian forcing and also, more generally, for white WSS forcing. It is of interest to determine whether a blowtorch theorem is possible for other forms of forcing—for example, colored Gaussian forcing or deterministic forcing. Because of its intractability, this question is at present unanswered for PRRS. By contrast, the convenient expressions established for FRRS do admit answers. We show by way of counterexamples that a flux ratio blowtorch theorem does not generally apply to either colored Gaussian forcing or sinusoidal deterministic forcing. Perhaps most important, these examples show that it is possible to further stabilize a potential well by additional state-dependent forcing.

Colored Gaussian forcing: Consider as in Theorem 1 system Eq. (122) with double-well potential $V(x) \in \mathcal{V}_W$ and state-dependent forcing strength $\gamma(x) = \gamma_0 + \gamma_1 1_I(x)$. Let $I = \{x > 0 : |x - \delta| < \ell\}$ with $\delta \geq \ell > 0$. Let $F(t)$ be a colored Gaussian process taking the form of a filtered white Gaussian process $F(t) = (f * \dot{W})(t)$ such that $f(t)$ is square-integrable. Assume solely for convenience that $V(x) = V(-x)$ for all x. Then for $\gamma_1 = 0$ the two wells of $V(x)$ are equally stable and $\varrho_{21} = 1$. We show that if $\ell \neq \delta$ then $\varrho_{21} > 1$ for a range of $\gamma_1 > 0$. This rules out a general flux ratio blowtorch theorem for colored Gaussian forcing.

For the present example we have just as in the proof of Theorem 1 that the dependence of the FRRS ϱ_{21} on γ_1 hinges on the dependence of Ξ_2 on γ_1: if Ξ_2 is a decreasing function of γ_1 then ϱ_{21} is an increasing function of γ_1. Also, for Gaussian forcing the flux Ξ_2 is an increasing function of the variance $\sigma_2^2 = \mathrm{Var}[(h_2 * F)(t)]$. Here, as before, $h_2(t) = v_{h2}(-t)\gamma(x_{h2}(-t))$ is the impulse response of the orbit filter Θ_2. To demonstrate that $\varrho_{21} > 1$ for $\gamma_1 > 0$, it suffices to show that σ_2^2 can be decreased by choosing $\gamma_1 > 0$.

Let $\mathcal{I}(t) = \Theta_2[F](t) = (h_2 * F)(t)$. Then $\mathcal{I}(t) = (h_2 * f * \dot{W})(t)$ and $\mathcal{I}(t)$ can be expressed as the Wiener integral

$$\mathcal{I}(t) = \int_{-\infty}^{\infty} (h_2 * f)(t-s)\, dW(s). \tag{138}$$

Then, using Eq. (33),

$$\sigma_2^2 = \mathrm{Var}[\mathcal{I}(t)] = \int_{-\infty}^{\infty} (h_2 * f)^2(t)\, dt. \tag{139}$$

The convolution of $f(t)$ and the impulse response $h_2(t) = v_{h2}(-t)\gamma(x_{h2}(-t))$ of Θ_2 is

$$\begin{aligned}
(h_2 * f)(t) &= \int_{-\infty}^{\infty} f(t-s) v_{h2}(-s)\gamma(x_{h2}(-s))\, ds \\
&= \int_{-\infty}^{\infty} f(t+s) v_{h2}(s)[\gamma_0 + \gamma_1 1_I(x_{h2}(s))]\, ds \\
&= \gamma_0 \int_{-\infty}^{\infty} f(t+s) v_{h2}(s)\, ds + \gamma_1 \int_J f(t+s) v_{h2}(s)\, ds \\
&= \gamma_0 \Psi(\Re, t) + \gamma_1 \Psi(J, t)
\end{aligned} \tag{140}$$

where $\Re = (-\infty, \infty)$, $J = \{t \in \Re : x_{h2}(t) \in I\}$ and

$$\Psi(A, t) = \int_A f(s+t) v_{h2}(s)\, ds \tag{141}$$

for any interval A. From Eqs. (139) and (140) we have that

$$\begin{aligned}
\sigma_2^2 &= \int_{-\infty}^{\infty} (\gamma_0 \Psi(\Re, t) + \gamma_1 \Psi(J, t))^2\, dt \\
&= \gamma_0^2 \int_{-\infty}^{\infty} \Psi^2(\Re, t)\, dt + \gamma_1^2 \int_{-\infty}^{\infty} \Psi^2(J, t)\, dt \\
&\quad + 2\gamma_0\gamma_1 \int_{-\infty}^{\infty} \Psi(\Re, t)\Psi(J, t)\, dt.
\end{aligned} \tag{142}$$

Note that if $f(t)$ is the Dirac-δ function so that $F(t)$ is white Gaussian noise then Eq. (142) recovers Eq. (137). The first term in Eq. (142) is the variance of $(h_2 * F)(t)$ for $\gamma_1 = 0$. Thus the second and third terms of Eq. (142) represent the change in variance due to $\gamma_1 > 0$. The second term of Eq. (142) is always nonnegative hence the sign of the third term is pivotal; if the integral

$$\int_{-\infty}^{\infty} \Psi(\Re, t)\Psi(J, t)\, dt \tag{143}$$

is negative then no matter what its magnitude, choosing γ_1 in the interval

$$0 < \gamma_1 < -2\gamma_0 \frac{\int_{-\infty}^{\infty} \Psi(\Re, t)\Psi(J, t)\, dt}{\int_{-\infty}^{\infty} \Psi^2(J, t)\, dt} \tag{144}$$

produces some reduction in the variance $\mathrm{Var}[(h_2 * F)(t)]$ below its value for $\gamma_1 = 0$. It is a matter of elementary calculus to show that the choice of γ_1 in the interval Eq. (144) which maximally reduces $\mathrm{Var}[(h_2 * F)(t)]$ is the interval's midpoint

$$\gamma_1 = -\gamma_0 \frac{\int_{-\infty}^{\infty} \Psi(\Re, t)\Psi(J, t)\, dt}{\int_{-\infty}^{\infty} \Psi^2(J, t)\, dt}. \tag{145}$$

The decrease in $\mathrm{Var}[(h_2 * F)(t)]$ corresponding to this choice of γ_1 is

$$-\gamma_0 \frac{\left[\int_{-\infty}^{\infty} \Psi(\Re, t)\Psi(J, t)\, dt\right]^2}{\int_{-\infty}^{\infty} \Psi^2(J, t)\, dt}. \tag{146}$$

Thus it is established that the sign of the integral Eq. (143) is critical and it remains to be determined if there are instances in which this sign is negative. Let

$$\phi_f(\tau) = \int_{-\infty}^{\infty} f(t)f(t+\tau)\, dt \tag{147}$$

and

$$\phi_J(\tau) = \int_{-\infty}^{\infty} v_{h2}(t)v_{h2}(t+\tau)1_J(t+\tau)\, dt \tag{148}$$

be, respectively, the autocorrelation of $h(t)$ and the crosscorrelation[22] of $v_{h2}(t)$ and $v_{h2}(t)1_J(t)$. Then

$$\begin{aligned}
\int_{-\infty}^{\infty} \Psi(\Re, t)\Psi(J, t)\, dt &= \int_{-\infty}^{\infty} \left[\int_{-\infty}^{\infty} f(s+t)v_{h2}(s)\, ds\right] \left[\int_{-\infty}^{\infty} f(r+t)v_{h2}(r)1_J(r)\, dr\right] dt \\
&= \int_{-\infty}^{\infty}\int_{-\infty}^{\infty}\int_{-\infty}^{\infty} f(s+t)f(r+t)v_{h2}(s)v_{h2}(r)1_J(r)\, dr\, ds\, dt \\
&= \int_{-\infty}^{\infty}\int_{-\infty}^{\infty}\int_{-\infty}^{\infty} f(s+t)f(r+t)\, dt\, v_{h2}(s)v_{h2}(r)1_J(r)\, dr\, ds \\
&= \int_{-\infty}^{\infty}\int_{-\infty}^{\infty} \phi_f(r-s)v_{h2}(s)v_{h2}(r)1_J(r)\, dr\, ds \\
&= \int_{-\infty}^{\infty} \phi_f(\tau)\phi_J(\tau)\, d\tau.
\end{aligned} \tag{149}$$

Letting $\hat{\phi}_f(\nu)$ and $\hat{\phi}_J(\nu)$ be the Fourier transforms of $\phi_f(t)$ and $\phi_J(t)$, respectively, we then have[28]

$$\int_{-\infty}^{\infty} \Psi(\Re, t)\Psi(J, t)\, dt = \frac{1}{2\pi}\int_{-\infty}^{\infty} \hat{\phi}_f(\nu)\hat{\phi}_J(\nu)\, d\nu. \tag{150}$$

The Fourier transform $\hat{\phi}_f(\nu)$ is the energy density spectrum of the function $f(t)$. Equivalently, it is the mean-square spectral density of the Gaussian process $F(t)$. In any case, $\hat{\phi}_f(\nu)$ is real and nonnegative and, moreover, by suitable choice of $f(t)$, the spectral density $\hat{\phi}_f(\nu)$ can be so constructed that its support is a subset of any desired interval. Thus if an interval can be found throughout which $\hat{\phi}_J(\nu)$ is negative, then it will be established that the pivotal integral Eq. (143) can be negative.

Let $V(x)$ be the double square well potential Eq. (27) with width w and $d_1 = d_2 = 1/2$ and suppose $I = \{x > 0 : |x - \delta| < \ell\} \subset [0, w]$. For $d_2 = 1/2$ the speed $|v_{h2}(t)|$ on the homoclinic orbit of well 2 is unity for $x_{h2}(t) > 0$. Then

$$J = \{t \in \Re : |t - \delta| < \ell \text{ or } |t + \delta| < \ell\}. \tag{151}$$

The crosscorrelation $\psi_J(t)$ can be written as a convolution of the functions $-v_{h2}(t)$ and $v_{h2}(t)1_J(t)$ using the fact that both of these functions are odd. Therefore

$$\hat{\psi}_J(\nu) = -\hat{v}_{h2}(\nu)\hat{g}(\nu) \tag{152}$$

where $\hat{g}(\nu)$ and $\hat{v}_{h2}(\nu)$ are, respectively, the Fourier transforms of $g(t) \equiv v_{h2}(t)1_J(t)$ and $v_{h2}(t)$. Because well 2 is square with depth $d_2 = 1/2$ the velocity component $v_{h2}(t)$ of the well's homoclinic orbit is

$$v_{h2}(t) = 1_{(-w,0)}(t) - 1_{(0,w)}(t). \tag{153}$$

Then

$$g(t) = 1_{(-\ell-\delta,-\ell+\delta)}(t) - 1_{(\ell-\delta,\ell+\delta)}(t). \tag{154}$$

The Fourier transforms $\hat{v}_{h2}(\nu)$ and $\hat{g}(\nu)$ are found by elementary means with the result that

$$\hat{\psi}_J(\nu) = \frac{16}{\nu^2}\sin^2\frac{w\nu}{2}\sin\ell\nu\sin\delta\nu. \tag{155}$$

For $\ell \neq \delta$ intervals of angular frequencies ν exist throughout which $\hat{\psi}_J(\nu) < 0$. If the spectral energy of $F(t)$ is concentrated in these intervals then the integrals in Eq. (150) will be negative and, as already reasoned above, the variance σ_2^2 will for suitable choice of γ_1 be reduced from its $\gamma_1 = 0$ value. Transport from well 2 is thereby diminished and the relative stability of the well is increased.

Deterministic sinusoidal forcing: Consider as before system Eq. (122) with double-well potential $V(x)$ and state-dependent forcing strength $\gamma(x) = \gamma_0 + \gamma_1 1_I(x)$ but now take $F(t)$ to be $F(t) = \cos\nu t$. Again let $V(x)$ be the double square well potential with well width w and depths $d_1 = d_2 = 1/2$. Take $I = \{x > 0 : |x - \delta| < \ell\}$ such that $I \in [0, w]$. In this example we show that for the double square well potential and given δ and ℓ, the flux Ξ_2 can be reduced to zero by suitable choice of $\gamma_1 > 0$. Hence here the relative stability of well 2 can be made to be $\varrho_{21} = \infty$. This rules out a general blowtorch theorem for deterministic forcing and indicates the possibility of reducing or completely suppressing instability due to deterministic sinusoidal forcing by additional state-dependent forcing.

We begin by calculating the SM distribution $\mu(\cdot)$ of the filtered forcing $\Theta_2[F](t)$. The filtered counterpart of $F(t) = \cos\nu t$ is

$$\begin{aligned}\Theta_2[F](t) &= \int_{-\infty}^{\infty} \dot{x}_2(-s)\gamma(x_2(-s))\cos\nu(t-s)\,ds \\ &= \sin\nu t\int_{-\infty}^{\infty} \dot{x}_2(-s)\gamma(x_2(-s))\sin\nu s\,ds \\ &= \mathcal{A}\sin\nu t \end{aligned}\tag{156}$$

where the amplitude $\mathcal{A}$ is, using Eq. (153),

$$\mathcal{A} = \frac{4}{\nu}\left[\gamma_0 \sin^2 \frac{\nu w}{2} + \gamma_1 \sin \nu\delta \sin \nu\ell\right]. \tag{157}$$

In the present case $\Theta_2[F](t)$ is nonrandom and periodic with period $T_o = 2\pi/\nu$ so, for any subinterval $A \subseteq \Re$,

$$\begin{aligned}
\mu(A) &= \lim_{T\to\infty} \frac{1}{2T}\int_{-T}^{T} E[1_A(\Theta_2[F](t))]\,dt \\
&= \frac{1}{T_o}\int_0^{T_o} 1_A(\Theta_2[F](t))\,dt \\
&= \frac{1}{T_o}\int_0^{T_o} 1_A(\mathcal{A}\cos\nu t)\,dt \\
&= E[1_A(\mathcal{A}\cos\nu U)] \\
&= P\{\mathcal{A}\cos\nu U \in A\}
\end{aligned} \tag{158}$$

where U is a random variable uniformly distributed over the interval $[0, T_o]$. Thus the SM distribution of $\Theta_2[F](t)$ is the same as the distribution of the random variable $\mathcal{A}\cos\nu U$. For $\mathcal{A} \neq 0$, we find after an elementary calculation that the density of $\mu(\cdot)$ is

$$f_\mu(y) = \frac{1}{\pi|\mathcal{A}|}\left(1 - \frac{y^2}{\mathcal{A}^2}\right)^{-1/2} \tag{159}$$

for $|y| < |\mathcal{A}|$ and $f_\mu(y) = 0$ otherwise. For $\mathcal{A} = 0$, $\mu(\cdot)$ is a point mass at $y = 0$. Thus in this case $\sigma_2^2 = 0$ and $\Xi_2 = kA_2\eta(\sigma_2/(kA_2)) = 0$. Choose, for example, the interval I with

$$\ell = \frac{\pi}{2\nu}, \qquad \delta = \frac{3\pi}{2\nu}. \tag{160}$$

Then $\sin\nu\delta\sin\nu\ell = -1$. For

$$\gamma_1 = \frac{\gamma_0}{2}\sin^2 \nu w \tag{161}$$

the amplitude $\mathcal{A}$ in Eq. (157) is then zero and we have demonstrated for the double square well potential with sinusoidal forcing that additional state-dependent forcing can completely stabilize a well. This shows that there can be no general blowtorch theorem for deterministic forcing. Finally, note that given Eq. (160) for ℓ and δ it is not necessary to have γ_1 as large as indicated in Eq. (161). Because the support of $\mu(\cdot)$ is, as shown in Eq. (159), the interval $(-|\mathcal{A}|, |\mathcal{A}|)$ it suffices to have $|\mathcal{A}| < kA_2$ for $\Xi_2 = 0$. The condition $|\mathcal{A}| < kA_2$ requires only

$$\left|\gamma_1 - \frac{\gamma_0}{2}\sin^2 \nu w\right| < \frac{\nu k A_2}{8} \tag{162}$$

and is a weaker condition than Eq. (161).

6. Final Remarks

Only two-dimensional dynamical systems with Newtonian velocity field $\vec{\pi}(\vec{z}) = \vec{\pi}(x,v) = (v, -V'(x))$ are treated here, in part for specificity but primarily because of the direct and intuitive connection between potential wells and regions bounded by separatrices which exists for these systems. Many of the results obtained for these systems can without doubt be extended to both more general two-dimensional vector fields and to Newtonian vector fields with more than one degree-of-freedom. Extensions to higher dimensional non-Newtonian dynamical systems are more problematic. For such systems with more than two dimensions, the intersection of stable and unstable invariant manifolds which form the separatrices in two-dimensional Newtonian systems may have codimension greater than one. In such cases these intersections do not partition the state space and the definition of safe regions is unclear. Conditions for codimension one intersection and related issues are addressed by Wiggins[42].

Throughout we have considered only linear damping as represented by the term $-k\dot{x}$ in Eq. (29). This term can be replaced with other forms of damping with no substantive change in the analysis beyond modification of the constant A in Eq. (39), Eq. (41) and subsequent expressions. There is also a possibility that damping outside the weak regime can be treated using available Melnikov-theoretic results for this situation[29].

The Wiener filter theory exploited here to identify the optimal filter for controlling the flux factor relies on the wide-sense stationarity of the forcing and knowledge of the forcing spectral density. This seems a natural formulation for the present problem since without these assumptions and the related assumption of forcing ergodicity, the total flux factor Ξ cannot be expressed as in Eqs. (44) and (45). Nevertheless, for application purposes a Kalman-Bucy formulation[43] of the filtering problem may be more attractive. This formulation would require neither stationary forcing nor knowledge of the forcing spectrum but would require that the orbits, which present themselves so naturally in terms of orbit filters, be expressed as a state equation. Unlike the Wiener theory which leads to a filter as a solution, the Kalman-Bucy theory yields a differential equation for the optimal control opening the way for a recursive implementation. However, for a Kalman-Bucy approach to be meaningful, work remains to develop usable expressions for Ξ comparable to Eqs. (44) and (45) for the nonstationary case.

We have introduced FRRS as a new measure of relative stability for potential wells based on state space transport. FRRS is attractive for several reasons. First, it exhibits reciprocal symmetry ($\varrho_{21} = 1/\varrho_{12}$) and meets the essential test of consistent order established in Proposition 6; wider, deeper wells are more stable. Both of these properties are evident in Figures 16 and 17. Second, it has the expected blowtorch property for white Gaussian forcing as stated in Theorem 1. Moreover, the results obtained for colored or nonGaussian forcing suggest that whiteness rather than Gaussianity is critical to the blowtorch theorem. Third, FRRS has a clear geometrical foundation in terms of state space transport, tying a well's stability to the flux factor for transport from the well. Fourth, it admits closed-form expressions for very general forms of damping and external forcing—both stochastic and deterministic. This holds true for additive forcing and more generally for multiplicative forcing. No other definition of relative stability has this scope. Its tractability makes it particularly attractive for problems of system design and analysis and dynamical control. By contrast, closed-form expressions are available for PRRS in only the simplest cases. PRRS is not even meaningful for deterministic forcing because no probability is involved. (Though for deterministic forcing one might, in lieu of stationary probabilities, make ergodic assignments in definition Eq. (120) of ρ_{21}.)

It was observed in Subsection 5.2 that the flux factor is perhaps analogous to a conditional probability of exit given that the system is at or near the separatrix bounding the safe region. It may be possible using either subharmonic Melnikov theory[2,29] or approximating wells to create for a potential well a sequence of nested safe regions indexed by energy level, well depth or other equivalent indices. The flux factors of these safe regions might be summed as conditional probabilities to produce something like an unconditional probability of escape. Alternatively, the indexed sequence of flux factors might be used to craft more capable controls. These and other questions remain to be addressed.

7. Acknowledgements

This work is a result of valuable and stimulating discussions with Emil Simiu of the U.S. National Institute of Standards and Technology. Support by the Ocean Engineering Division of the U.S. Office of Naval Research under Grant N00014-94-0284 is also gratefully acknowledged.

8. References

1. S.T. Ariaratnam and W.-C. Xie, *J. of Applied Mechanics* **60** (1993) 677–682.

2. D.K. Arrowsmith and C.M. Place, *An Introduction to Dynamical Systems* (Cambridge University Press, New York, 1990).

3. S. Barnett and R.G. Cameron, *Introduction to Mathematical Control Theory*, Second Edition (Clarendon Press, Oxford, U.K., 1985).

4. D. Beigie, A. Leonard and S. Wiggins, *Nonlinearity* **4**(3) (1991) 775–819.

5. V. Brunsden, J. Cortell and P.J. Holmes, *J. of Sound and Vibration* **130**(1) (1989) 1–25.

6. H. Cramer and M.R. Leadbetter, *Stationary and Related Stochastic Processes* (Wiley, New York, 1967).

7. L. Edelstein-Keshet, *Mathematical Models in Biology* (Random House, New York, 1988).

8. M.R. Frey and E. Simiu, *Physica D* **63** (1993) 321–340.

9. M.R. Frey and E. Simiu, in *Computational Stochastic Mechanics*, eds. A.H-D. Cheng and C.Y. Yang (Elsevier Science Publishers, Essex, U.K., 1993).

10. M.R. Frey and E. Simiu, *Physica D* in press (1996).

11. M.R. Frey and E. Simiu, in *Proc. of the Second Inter. Conf. on Stochastic Computational Mechanics*, Athens, Greece, 1994.

12. M.R. Frey and E. Simiu, *Physics Letters A* in review (1995).

13. M.R. Frey, *IEEE Trans. on Automatic Control* **41**(2) (1996) 216–223.

14. C.W. Gardiner, *Handbook of Stochastic Methods for Physics, Chemistry and the Natural Sciences*, Second Edition (Springer Verlag, New York, 1990).

15. R.M. Gray, *Probability, Random Processes, and Ergodic Properties* (Houghton Mifflin, Boston, 1972).

16. J. Guckenheimer and P. Holmes, *Nonlinear Oscillations, Dynamical Systems, and Bifurcations of Vector Fields* (Springer-Verlag, New York, 1983).

17. J.E. Howard, *Celestial Mechanics and Dynamical Astronomy* **48** (1990) 267–288.

18. S.-R. Hsieh, A.W. Troesch and S.W. Shaw, *Proc. Royal Soc. London, Series A* **446**(7) (1994) 195–211.

19. R.L. Kautz, *J. of Applied Physics* **58**(1) (1985).

20. R.Z. Khasminskii, *Stochastic Stability of Differential Equations* (Sitjhoff & Noordhoff, The Netherlands, 1980).

21. R. Landauer, *Physical Review A* **12**(2) (1975) 636–638.

22. B.P. Lathi, *Signals, Systems and Communication* (Wiley, New York, 1965).

23. A.J. Lichtenberg and M.A. Lieberman, *Regular and Stochastic Motion* (Springer, New York, 1983).

24. J.H. Maddocks, *IMA J. of Applied Mathematics* **46** (1991) 71–99.

25. V.K. Melnikov, *Trans. Moscow Math. Society* **12** (1963) 1–57.

26. K.R. Meyer and C.R. Sell, *Trans. Amer. Math. Society* **314**(1) (1989) 63–105.

27. F. Moon, *Chaotic Vibrations* (Wiley, New York, 1987).

28. A. Papoulis, *The Fourier Integral and Its Applications* (McGraw-Hill, New York, 1988).

29. L. Perko, *Differential Equations and Dynamical Systems* (Springer-Verlag, New York, 1991).

30. C. Pezeshki and E.H. Dowell, *Physica D* **32** (1988) 194.

31. V. Rom-Kedar and S. Wiggins, *Arch. Rat. Mech. Analysis* **109** (1990) 239–298.

32. D. Schultz and J. Melsa, *State Functions and Linear Control Systems* (McGraw-Hill, New York, 1967).

33. E. Simiu and C. Hagwood, in *Proc. of the Second Inter. Conf. on Stochastic Computational Mechanics*, (Athens, Greece, 1994).

34. E. Simiu and M.R. Frey, *J. of Engineering Mechanics* in press (1995).

35. E. Simiu and M. Franaszek, in *Proc. of the Symp. on Vibration and Control of Stochastic Dynamical Systems*, ed. L.A. Bergman, (Design Engineering Technical Conference, Boston, MA, 1995).

36. R.F. Stengel, *Stochastic Optimal Control, Theory and Application* (Wiley, New York, 1986).

37. Y.L. Tong, *The Multivariate Normal Distribution* (Springer-Verlag, New York, 1990).

38. N.G. Van Kampen, *IBM J. of Research and Development* **32**(1) (1988) 107–111.

39. S. Wiggins, *Physics Letters A* **24**(3) (1987) 138–142.

40. S. Wiggins, *Introduction to Applied Nonlinear Dynamical Systems and Chaos* (Springer-Verlag, New York, 1990).

41. S. Wiggins, *Chaotic Transport in Dynamical Systems* (Springer-Verlag, New York, 1992).

42. S. Wiggins, *Physica D* **44** (1990) 471–501.

43. E. Wong and B. Hajek, *Stochastic Processes in Engineering Systems* (Springer-Verlag, New York, 1985).

Uncertainty Modeling in Vibration and Fuzzy Analysis of Structural Systems, pp. 101–146
edited by B. M. Ayyub, A. Guran and A. Haldar
Series on Stability, Vibration and Control of Systems Series B: Vol. 10

DYNAMICS OF NONLINEAR STOCHASTIC SYSTEMS: A FREQUENCY DOMAIN APPROACH

AHSAN KAREEM
Dept. of Civil Engrg. & Geo. Sci., Univ. of Notre Dame,
Notre Dame, IN 46556-0767, USA

JUN ZHAO
Barnett & Casbarian, Inc.,
Houston, TX 77024-1596, USA

MICHAEL A. TOGNARELLI
Dept. of Civil Engrg. & Geo. Sci., Univ. of Notre Dame,
Notre Dame, IN 46556-0767, USA

KURTIS R. GURLEY
Dept. of Civil Engrg. & Geo. Sci., Univ. of Notre Dame,
Notre Dame, IN 46556-0767, USA

ABSTRACT

The method of equivalent statistical linearization, commonly used in frequency domain analyses of nonlinear stochastic systems, fails to capture the non-Gaussianity of the response of such systems in terms of its higher-order statistics. In addition, response energy in frequency ranges outside that of the input spectrum is not observed using this technique. Herein, methods of equivalent statistical quadratization and cubicization are proposed, whereby nonlinearities in the excitation of practical systems are cast in an equivalent quadratic or cubic form rather than in an equivalent linear form. The present techniques take advantage of the Gaussianity of the first order response (when the system input is Gaussian) to simplify the recasting of the nonlinearity in its approximate polynomial form. A Volterra series approach in the frequency domain leads to the development of transfer functions from which the response spectrum as well as statistics of the response may be obtained. Response cumulants, computed up to the fourth order via direct integration or the Kac-Siegert technique, reveal the non-Gaussian character of the response and, when used in the framework of some available non-Gaussian probability density function models, indicate acceptable agreement with time-domain simulations of the original nonlinear differential equations. In addition, approximations for the response power spectral densities reveal spectral peaks outside the range of input frequencies which are brought about by nonlinear effects that are undetectable using linearization techniques. This, too, corroborates information gained from time-domain simulations.

1. Introduction

Many practical examples of nonlinear systems exist in all disciplines of modern engineering. For example, the tension leg platform (TLP) is the most promising concept among different structural systems being considered for tapping into deep-water oil resources. The compliant nature of TLP motions in the horizontal plane makes these platforms sensitive to low frequency oscillations due to wind and wave drift forces. The matter of analyzing the response characteristics of such a system is complicated by the fact that both the wind loads and the wave loads acting on TLPs are nonlinear. For example, the expression for wind drag force contains the square of the fluctuating velocity term and that for wave drag force in Morison equation[1] contains a nonlinear term involving the water particle velocity. Furthermore, the hydrodynamic loads due to potential effects (diffraction and radiation) contain inherent quadratic load effects (e.g. Refs. 2 & 3). Historically, analyses of nonlinear systems in the frequency domain have been based on the statistical linearization approach (e.g., Refs. 4-6). The linearization approach, however, fails to adequately represent important features of the response in the presence of such nonlinearities. Particularly, the response power spectral density function spans only the range of excitation frequencies while the energy near the system resonant frequency components is nonexistent. Further, since higher-order cumulant values vanish the approximate response probability density function remains Gaussian, giving rise to underprediction of the response extremes which are very important for design considerations.

Obviously, the next level of improvement involves statistical quadratization. In this chapter, one type of system nonlinearity is expressed in terms of an equivalent polynomial in which up to quadratic terms are included. The concept of quadratization or polynomial approximation of nonlinear effects has been used in the study of hydrodynamic loads on offshore structures (e.g., Refs. 7-29). Addressing a selection of these works specifically, Borgman[7] was first in formulating the spectral and statistical representation of nonlinear drag forces on cylindrical structures utilizing a memoryless transformation. His work has been followed by several investigators. Most of these studies are limited to obtaining the second-order statistics of the loading or response process while some are extended to extremes of wave force statistics and higher-order response cumulants. Li and Kareem[14] employed a trivariate Hermite polynomial expansion up to the second order to represent the nonlinear drag-induced hydrodynamic load on a TLP. This analysis included the influence of wave free-surface fluctuations on the response statistics of a TLP. Kareem and Hsieh[18] used a similar approach in the analysis of jacket-type platforms for non-Gaussian waves. Spanos and Donley[15,16,17] formulated a more general quadratization technique for the treatment of arbitrary statistically asymmetric nonlinear systems. This approach was employed to address the response of a TLP subjected to wave and current loads, with the inclusion of the fluid-structure interaction term in the Morison equation. The investigation established frequency domain moment expressions up to the third order, but considered fourth-order moment computations prohibitive, and developed a probability density function estimation based on the Gram-Charlier expansion. Kareem and Zhao[19,20,21] formulated an alternative quadratization method for the analysis of nonlinear wind and wave loads on TLPs which capitalized, in terms of computational efficiency, on the Gaussianity of the first-order response solution. A procedure which was developed in their studies for factoring the multi-fold integrals found in the cumulant expressions helped to make calculations of fourth-order response statistics feasible, and more accurate probability density approximations possible.

For the case of a statistically symmetric nonlinearity, the method of equivalent statistical quadratization degenerates, in effect, to the method of equivalent statistical linearization. Example situations of such nonlinearities which arise in practical scenarios are Morison drag force with no current or a structural system which may be modelled as having cubic Duffing stiffness. The use of a cubic polynomial approximation has been considered by some to express drag force nonlinearity[7]. These types of formulations, however, ignored relative fluid-structure velocity, thereby permitting the use of a static or memoryless transformation, and only sought to make statistical characterizations up to the second order. Hu and Lutes[22], using a nonlinear static transformation to represent drag force, developed an expression for the fourth-order cumulant of the force. Borgman[23] and Olagnon, et al.[24] implemented a Volterra series approach to describe the polynomial drag force encountered in offshore applications. Both of these investigations, however, were limited to considerations of the spectral densities of force, response or both and did not explicitly address the evaluation of higher-order statistical properties. Additional examples focussing on the utility of third-order polynomial representations of nonlinearities in evaluating higher-order response statistics via moment closure techniques may be found in the literature[11,25,26,27]. Bendat[28] considered analyses of nonlinear systems via higher-order spectra.

To address more completely the needs of a system for which a memoryless transformation is not applicable, e.g., drag force formulations containing relative velocity terms, a method of equivalent statistical cubicization has been developed[19,20,29] which is useful for the general case of a statistically symmetric nonlinearity in the loading or the system itself. This technique casts such a nonlinearity as a cubic polynomial so that the system response may be treated as the sum of first-order and third-order Volterra systems. The response statistics up to the fourth order are obtainable as well via this approach.

This chapter addresses the response of a TLP to wind and wave loads. The nonlinear loading is expressed via quadratization or cubicization in terms of an equivalent polynomial that contains terms up to quadratic or cubic order. Subsequently, frequency domain integral expressions for the response power spectral density and cumulants up to the fourth order are obtained based on application of the Volterra theory to the approximated system according to the work of Bedrosian and Rice[30]. It is important to note that this approach to obtaining the response and its statistics requires only that the input process be given. It is not a matter of system identification for which a priori knowledge of the response power spectral density is also necessary. Since closed-form solutions for the cumulant integrals are not attainable, a direct numerical integration method is utilized to evaluate the first four cumulants, as well as a scheme to obtain the cumulants by the Kac-Siegert[31] technique. Both analyses include the effects of the relative fluid structure velocity. The probability density function and crossing rates may be approximated based on the higher-order cumulants utilizing a moment-based Hermite method[32] and the Maximum Entropy method[33,34] which alleviate within their framework some of the inherent difficulties encountered in the use of the Gram Charlier series expansion. Comparisons of all findings to results based on numerical Monte Carlo simulation of the original governing nonlinear equations verify the validity of the aforementioned procedures.

2. Theoretical Background

Historically, there exist two fundamental approaches to solving nonlinear stochastic problems. One is based on the theory of Markov processes and the associated Fokker-Planck equation, initially applied in the analysis of Brownian motion in statistical physics. The second is based on spectral analysis which has been extensively applied in the field of communications and electrical engineering. Herein, the latter approach will be pursued by assimilating a nonlinear system to a quadratic or cubic form utilizing the method of equivalent statistical quadratization or cubicization and applying the Volterra theory for system analysis in the frequency domain. Thereby, a statistical description may be made for the system response up to the fourth order to more accurately characterize its non-Gaussian nature. The statistical quantities thus obtained may be employed within several current frameworks for estimating probability distributions and crossing rates of non-Gaussian processes, with the ultimate goal of effectively describing the extremes of the response process for use in future design considerations.

2.1 Volterra Series

A Volterra series expansion may be viewed as a regular expansion in power series, "with memory"[30]. Systems with polynomial nonlinear transformations may be treated as Volterra equivalent systems. A general second-order Volterra equivalent system may be described as the following,

$$x(t) = \int_{-\infty}^{\infty} h_1(\tau)u(t-\tau)d\tau + \frac{1}{2}\int_{-\infty}^{\infty}\int h_2(\tau_1, \tau_2)u(t-\tau_1)u(t-\tau_2)d\tau_1 d\tau_2 = x_1(t) + \frac{1}{2}x_2(t) \tag{1}$$

where $u(t)$ is an input process, $h_1(\tau)$ and $h_2(\tau_1, \tau_2)$ are linear and second-order impulse response functions, respectively, and $x_1(t)$ and $x_2(t)$ are linear and second-order response components. Notice that the first-order kernel is simply the impulse response function of a linear system, while the higher-order kernels can be viewed as higher-order impulse response functions which serve to characterize the various orders of nonlinearities. The general formulation of the kernels is not available, but when the nonlinear transformation is in the polynomial form, the kernels can be evaluated. The application of this series to nonlinear systems was first investigated by Wiener [35] in 1942. Later, Barrett[36] reported on a systematic study of the utility of the Volterra series for analyzing physical systems.

The first term of Eq. (1), representing the output of a first-order Volterra system, is the same as the response of a linear, time-invariant system where the impulse-response function, $h_1(\tau)$, is subject to conditions of causality and stability. Physically, the causality condition stipulates that the response cannot depend on any future input, i.e. $h_1(\tau) = 0$, for $\tau < 0$. The stability condition requires that

$$\int_{-\infty}^{\infty} |h_1(\tau)|^2 d\tau < \infty .$$

The frequency domain transfer function for the first-order Volterra system may be derived by finding the system response to an input, $e^{i\omega t}$,

$$x_1(t) = \int_{-\infty}^{\infty} h_1(\tau) e^{i\omega(t-\tau)} d\tau = H_1(\omega) e^{i\omega t}, \tag{2}$$

where $H_1(\omega)$, the transfer function, is the Fourier transform of $h_1(\tau)$. The inverse relationship is given by

$$h_1(\tau) = \frac{1}{2\pi} \int_{-\infty}^{\infty} H_1(\omega) e^{i\omega\tau} d\omega . \tag{3}$$

From this development, it is important to note that the response of a linear time-invariant system will not contain frequency components which are not present in the input. This is not true for nonlinear systems.

The second term in Eq. (1) is a two-dimensional convolution which represents the output of a second-order Volterra system where $h_2(\tau_1, \tau_2)$ is the second-order Volterra kernel and is generally assumed symmetric. This assumption is reasonable since there exists a procedure whereby any asymmetric kernel may be symmetrized without loss of generality[37]. Thus, the order of appearance of τ and σ is unimportant. This kernel must satisfy the causality and stability conditions mentioned earlier and may be related to its corresponding frequency domain transfer function via the two-dimensional Fourier transform,

$$H_2(\omega_1, \omega_2) = \int_{-\infty}^{\infty}\int_{-\infty}^{\infty} h_2(\tau_1, \tau_2) e^{-i\omega_1\tau_1} e^{-i\omega_2\tau_2} d\tau_1 d\tau_2 . \tag{4}$$

The inverse relationship is thus,

$$h_2(\tau_1, \tau_2) = \frac{1}{(2\pi)^2} \int_{-\infty}^{\infty}\int_{-\infty}^{\infty} H_2(\omega_1, \omega_2) e^{(i\omega_1\tau_1 + i\omega_2\tau_2)} d\omega_1 d\omega_2 . \tag{5}$$

Clearly, then, the response of the second-order Volterra system will contain energy at frequencies which are, in fact, the sums and differences of frequencies contained in the input.

In certain particular circumstances, which will be described in the upcoming subsection, the casting of a nonlinearity in a form amenable to the application of the second-order Volterra series adds no additional information to our analysis. In such situations, the nonlinearity may be cast in a cubic form such that it is represented by the sum of the outputs of a first-order and a third-order Volterra system,

$$x = x_1 + \frac{1}{6}x_3 = \int_{-\infty}^{\infty} h_{x_1}(\tau_1) f(t-\tau_1) d\tau_1 +$$

$$\frac{1}{6}\int_{-\infty}^{\infty}\int_{-\infty}^{\infty}\int_{-\infty}^{\infty} h_{x_3}(\tau_1, \tau_2, \tau_3) f(t-\tau_1) f(t-\tau_2) f(t-\tau_3) d\tau_1 d\tau_2 d\tau_3 .$$

The corresponding third-order transfer function in the frequency domain may be obtained according to,

$$H_3(\omega_1, \omega_2, \omega_3) = \int_{-\infty}^{\infty}\int_{-\infty}^{\infty}\int_{-\infty}^{\infty} h_3(\tau_1, \tau_2, \tau_3) e^{-i\omega_1\tau_1} e^{-i\omega_2\tau_2} e^{-i\omega_3\tau_3} d\tau_1 d\tau_2 d\tau_3 . \tag{6}$$

The inverse relation is,

$$h_3(\tau_1, \tau_2, \tau_3) = \frac{1}{(2\pi)^3}\int_{-\infty}^{\infty}\int_{-\infty}^{\infty} H_3(\omega_1, \omega_2, \omega_3) e^{(i\omega_1\tau_1 + i\omega_2\tau_2 + i\omega_3\tau_3)} d\omega_1 d\omega_2 d\omega_3 . \tag{7}$$

2.2 Types of Nonlinearities

In this study, a nonlinear system will be cast in the form of the Volterra functional series up to either the second or third order for the purpose of statistical analysis. It is first important to note that the statistical characteristics of a given nonlinearity make it more or less conducive to analysis by a particular order Volterra system. Indeed, a *statistically symmetric* nonlinear function, *g(u)*, defined as a function for which $E[g(u)^{2n-1}] = 0$ for all *n*, is not treatable by the present quadratization technique, but may be addressed when the approximating system is the sum of a first- and a third-order Volterra system. Conversely, a *statistically asymmetric* nonlinear function, *g(u)*, for which $E[g(u)^{2n-1}]$ is nonzero for all *n*, may be effectively approximated by the present technique. Given a zero-mean, Gaussian process, *u(t)*, some examples of statistically symmetric nonlinearities are $g(u) = u|u|$ and $g(u) = u^3$. The functions $g(u) = (u+a)|u+a|$ and $g(u) = u^2$ are examples of statistically asymmetric nonlinearities[19].

2.3 Modelling of Wind and Wave Loads

The TLP surge response, x, is modelled as an equivalent single-degree-of-freedom system given by

$$M\ddot{x} + C\dot{x} + Kx = F , \tag{8}$$

where M is the structural mass and added mass; C is structural damping; K is structural stiffness; and F represents one or a combination of wind, wave and current forces.

Because of the compliant nature of TLP motions in the horizontal plane, their surge response is sensitive to wind-induced drag force fluctuations. Some works covering the second-order statistical characterization of the response of a TLP under the dynamic effects of wind are found in Refs. 13 & 38-43. Kareem and Zhao[21] in 1994 extended the analysis to include up to fourth-order response statistics using an equivalent statistical quadratization technique as well as Kac-Siegert approach. Their study was thus able to producc non-Gaussian probability densities for the response processes as well as crossing rates and peak distributions. The wind force acting in the surge direction is expressed in terms of the relative wind-structure velocity by (e.g., Ref. 44),

$$F_{wind} = K_w(w + W - \dot{x})^2, \tag{9}$$

where $K_w = \frac{1}{2}\rho_a C_a A$; where ρ_a is the air density, C_a is the TLP shape coefficient, A is the total above-water area of the TLP exposed to wind, W is the mean wind velocity at a reference height of ten meters and w is the fluctuating wind velocity. More extensive development of forces on a TLP due to wind involving the complex structure of the platform above the water level are given in Ref. 44.

Often, for land-based structures, the square of the fluctuating wind velocity is ignored. While this practice does not significantly affect root-mean-square response statistics prediction, it can introduce sizable errors in the peak distribution estimation since the parent distribution is no longer Gaussian (e.g., Refs. 13 & 45-47). Thus, to adequately characterize the extreme response the velocity-squared term must be retained in the computation of higher-order cumulants. Notably, the inclusion of the fluid-structure interaction in Eq. (9), rather than solely the fluctuating wind component, introduces additional non-Gaussianity in the excitation, since now not only the Gaussian w, but also the inherently non-Gaussian velocity response, $\dot{x}$, to the nonlinear loading are squared[21]. It has been seen that the inclusion of this interaction plays a significant role in much more accurately approximating the higher-order response cumulants than analyses for which the interaction is not considered.

Now, consider the wave loads on a TLP which, in general, are more dominant than the wind loads. Typical TLP structures, depending on their submerged geometry and size, experience a combination of wave-induced viscous and potential loads. The viscous effects are generally described through the drag term of Morison equation and the potential effects, i.e., diffraction and radiation forces, are based on diffraction of waves by the structure. For example,

$$F_u(t, \dot{x}) = \int_{S_v} 0.5\rho C_d |u(y, z, t) - \dot{x}(t)|(u(y, z, t) - \dot{x}(t))ds, \tag{10}$$

where $F_u(t, \dot{x})$ is the viscous force, S_v is the submerged surface area, ρ and C_d are water density and a force coefficient, respectively, and u is the water particle velocity at a position given by y and z and time, t. The water particle velocity in a simplistic formulation may be based on the undisturbed far field or may be derived from diffraction analysis. The first-order diffraction loads are given by the convolution of the wave surface elevation, $\eta(t)$, with an appropriate convolution kernel,

$$f_d(t) = \int_{-\infty}^{\infty} h_d(\tau)\eta(t-\tau)d\tau, \tag{11}$$

where,

$$h_d(\tau) = \frac{1}{2\pi}\int_{-\infty}^{\infty} H_d(\omega)\exp(-i\omega\tau)d\omega,$$

and $H_d(\omega)$ is the diffraction force transfer function which can be derived from a linear diffraction analysis. The wave radiation force is given in terms of frequency-dependent added mass and radiation coefficients which can be obtained from diffraction analysis.

The higher-order effects resulting from hydrodynamic loads of viscous and potential origins introduce nonlinearity with the consequence of non-Gaussian statistical features. These higher-order effects are attributable to the following sources: (i) nonlinearity in Bernoulli's equation; (ii) nonlinearity in Morison drag term; (iii) nonlinearity in the free surface wave profile; (iv) displacement and velocity dependence of wave-induced forces; and (v) nonlinear diffraction (e.g., Refs. 2, 13-15, 18 & 48-56). The second-order forces can be expressed as

$$f_d^{(2)}(t) = \int_{-\infty}^{\infty}\int_{-\infty}^{\infty} h_f^{(2)}(\tau_1, \tau_2)\eta(t-\tau_1)\eta(t-\tau_2)d\tau_1 d\tau_2, \tag{12}$$

where the second-order kernel, $h_f^{(2)}(\tau_1, \tau_2)$, is the Fourier transform of the second-order quadratic function $H_d^{(2)}(\omega_1, \omega_2)$. The hydrodynamic loads of potential origin can be conveniently expressed in the above format. The quadratization technique is necessary to express the second- order viscous load effects on TLPs. Since the potential effects do not require the quadratization procedure, the objective of this study could be met by just treating the drag-induced viscous loads. For this reason, computationally efficient Morison equation was used for hydrodynamic loads instead of a combination of drag and potential effects from Morison equation and diffraction theory.

The wave force is expressed in terms of the relative velocity by a modified form of Morison equation for the drag force acting on the TLP in the surge direction[2],

$$F_{wave} = K_m\dot{u} + K_d|u + U - \dot{x}|(u + U - \dot{x}), \tag{13}$$

where $K_m = \rho C_m V_e$; $K_d = \frac{1}{2}\rho C_d A_e$. In this formulation, the first term represents inertial force and the second describes the viscous effects. The relative importance of these two terms can be quantified in terms of the non-dimensional drag to inertial force ratio, R_{di}, according to[57],

$$R_{di} = \frac{K_d\sigma_u^2}{K_m\sigma_{\dot{u}}},$$

where u, and $\dot{u}$ are water particle velocity and acceleration, respectively, and U is the current speed. Here, ρ is the water density, C_m and C_d are the inertia and drag coefficients which are usually determined from experimental data, and A_e and V_e are related to the area and volume of the submerged portion of the platform. The ocean environment is described by a model wave elevation spectrum, e.g., JONSWAP, Pierson-Moskowitz, etc. (e.g., Refs. 2 & 3) which can be related to the power spectrum of the fluctuating water particle velocity, u, by a linear transfer function.

Unlike the wind force, the viscous wave force is not cast in a purely polynomial form, but must be so approximated for the implementation of the Volterra theory. Again, it is important to note that although the wave process may be assumed as Gaussian, the structural velocity in the preceding equations, due to its nature as a response to a nonlinear forcing function, is no longer Gaussian.

2.4 Equivalent Statistical Quadratization

In a quadratization approach, a statistically asymmetric nonlinear system having an arbitrary form is approximated by a second-order polynomial expression for analysis within the Volterra framework. In the approximate quadratic representation of the system the linear component is analogous to conventional statistical linearization while the retention of a second-order component gives the method its name.

The governing equation of a single-degree-of-freedom nonlinear system containing statistically asymmetric nonlinearities in both the system characteristics and the excitation may be written,

$$M\ddot{x} + C\dot{x} + Kx + g(x, \dot{x}) = f_L(\upsilon) + f_N(\upsilon, \dot{x}), \tag{14}$$

where M, C, and K are the structural mass, damping and stiffness, respectively, $g(x, \dot{x})$ represents a system nonlinearity, $f_L(\upsilon)$ and $f_N(\upsilon, \dot{x})$ represent the linear and nonlinear forcing terms, respectively, and υ is the input wind velocity or water particle velocity process. For arbitrary nonlinear functions, however, the response of the system may not be written as in Eq. (1). The forthcoming discussion will outline some approximations, culminating in the quadratization procedure itself, which will assist in making any statistically asymmetric nonlinear functions amenable to analysis via Volterra series.

2.4.1 Slow Drift Approximation

To eliminate the computational difficulty imposed by the non-Gaussian structural velocity, a slowly varying drift approximation is invoked. For a system with low natural frequency, the slowly varying drift motion plays an important role. This leads to a reasonable assumption, i.e., the higher-order nonlinear velocity terms can be neglected. As an example, the nonlinear terms in the wind and wave induced drag descriptions expanded in Taylor series in terms of the second-order component, $\dot{x}_2$, about the linear, Gaussian component, $\dot{x}_1$, are given by

$$(w + W - \dot{x})^2 \cong (w + W - \dot{x}_1)^2 - 2E[w + W - \dot{x}_1]\frac{\dot{x}_2}{2}, \tag{15}$$

$$|u+U-\dot{x}|(u+U-\dot{x}) \cong |u+U-\dot{x}_1|(u+U-\dot{x}_1) - 2E[|u+U-\dot{x}_1|]\frac{\dot{x}_2}{2}. \tag{16}$$

where $E[\ ]$ denotes mathematical expectation. It is assumed as well that the second-order response itself is small compared to the first-order response and terms involving its higher-order powers may thus be neglected.

2.4.2 Splitting Technique

Returning to Eq. (14), the nonlinear functions are expanded in Taylor series in terms of the quadratic response and its velocity, and apply the assumptions of the previous section,

$$g(x,\dot{x}) = g(x_0+x_1,\dot{x}_1) + \frac{\partial g(x_0+x_1,\dot{x}_1)}{\partial x}\frac{x_2}{2} + \frac{\partial g(x_0+x_1,\dot{x}_1)}{\partial \dot{x}}\frac{\dot{x}_2}{2} + O(x_2^2,\dot{x}_2^2)$$

$$g(x,\dot{x}) \approx g(x_0+x_1,\dot{x}_1) + \mu_{gx}\frac{x_2}{2} + \mu_{gv}\frac{\dot{x}_2}{2}, \tag{17}$$

where x_0 is a static response and μ_{gx} and μ_{gv} are the expected values of the terms they replace. Similarly,

$$f_N(\upsilon,\dot{x}) = f_N(\upsilon,\dot{x}_1) + \frac{\partial f_N(\upsilon,\dot{x}_1)}{\partial \dot{x}}\frac{\dot{x}_2}{2} + O(\dot{x}_2^2),$$

$$f_N(\upsilon,\dot{x}) \approx f_N(\upsilon,\dot{x}_1) - \mu_{fv}\frac{\dot{x}_2}{2}. \tag{18}$$

By expanding in this way, two nonlinear functions of Gaussian processes (the initial terms on the right-hand sides of Eq. (17) and Eq. (18)) remain along with two additional damping terms and an additional stiffness term.

2.4.3 Quadratization Procedure

Since the two nonlinearities discussed in the previous section still have arbitrary forms, they must finally be approximated in terms of quadratic polynomials for the Volterra series technique to be effective. To this end, second-order approximations for both terms are written as,

$$g(x_0+x_1,\dot{x}_1) \approx \beta_0 + \beta_1 x_1 + \beta_2\dot{x}_1 + \frac{1}{2}(\beta_3 x_1^2 + \beta_4 x_1\dot{x}_1 + \beta_5\dot{x}_1^2)$$

$$f_N(\upsilon,\dot{x}_1) \approx \sum_i \int_0^{L_i} dz\alpha_0 + \sum_i \int_0^{L_i} dz\alpha_1(\upsilon-\dot{x}_1) + \frac{1}{2}\sum_i \int_0^{L_i} dz\alpha_2(\upsilon-\dot{x}_1)^2, \tag{19}$$

where the summation indicates the total excitation acting on each element of the structure. The unknown coefficients in Eq. (19) are solved for by mean-square minimization of the following error terms,

$$\varepsilon_g = E\left[\left(g(x_0 + x_1, \dot{x}_1) - \left(\beta_0 + \beta_1 x_1 + \beta_2 \dot{x}_1 + \frac{1}{2}(\beta_3 x_1^2 + \beta_4 x_1 \dot{x}_1 + \beta_5 \dot{x}_1^2)\right)\right)^2\right],$$

$$\varepsilon_f = E\left[\left(f_N(\upsilon, \dot{x}_1) - \left(\sum_i \int_0^{L_i} dz\alpha_0 + \sum_i \int_0^{L_i} dz\alpha_1(\upsilon - \dot{x}_1) + \frac{1}{2}\sum_i \int_0^{L_i} dz\alpha_2(\upsilon - \dot{x}_1)^2\right)\right)^2\right], \quad (20)$$

which yields linear systems of algebraic equations. An advantage of the present technique is that none of the expected values computed in Eq. (20) involve non-Gaussian processes.

Now, letting,

$$a_0 = \sum_i \int_0^{L_i} dz\alpha_0, \; a_1 = \sum_i \int_0^{L_i} dz\alpha_1 \text{ and } a_2 = \sum_i \int_0^{L_i} dz\alpha_2,$$

an equivalent set of Volterra system equations can be written as[37],

$$x_0 = \frac{a_0 - \beta_0}{K}$$

$$M\ddot{x}_1 + (C + a_1 + \beta_2)\dot{x}_1 + (K + \beta_1)x_1 = f_L(\upsilon) + a_1\upsilon,$$

$$M\ddot{x}_2 + (C + \mu_{gv} + \mu_{fv})\dot{x}_2 + (K + \mu_{gx})x_2 = a_2(\upsilon - \dot{x}_1)^2 - (\beta_3 x_1^2 + \beta_4 x_1 \dot{x}_1 + \beta_5 \dot{x}_1^2), \quad (21)$$

for which the following transfer functions are derivable,

$$H_x^{(1)}(\omega) = H_1(\omega) H_f^{(1)}(\omega),$$

$$H_{\dot{x}}^{(1)}(\omega) = i\omega H_x^{(1)}(\omega),$$

$$H_v(\omega) = H_\upsilon(\omega) - i\omega H_x^{(1)}(\omega), \text{ (fluid-structure interaction term)}$$

$$H_x^{(2)}(\omega_1, \omega_2) = H_2(\omega_1 + \omega_2) H_f^{(2)}(\omega_1, \omega_2),$$

$$H_{\dot{x}}^{(2)}(\omega_1, \omega_2) = i(\omega_1 + \omega_2) H_x^{(2)}(\omega_1, \omega_2), \quad (22)$$

where,

$$H_f^{(1)}(\omega) = H_L(\omega) + a_1 H_\upsilon(\omega),$$

$$H_x^{(2)}(\omega_1, \omega_2) = a_2 H_v(\omega_1) H_v(\omega_1) -$$

$$\left(\beta_3 + \beta_4 \frac{i(\omega_1 + \omega_2)}{2} + \beta_5 \omega_1 \omega_2\right) H_x^{(1)}(\omega_1) H_x^{(1)}(\omega_2)$$

and

$$H_1(\omega) = [(K+\beta_1) - \omega^2 M + i\omega(C + a_1 + \beta_2)]^{-1},$$

$$H_2(\omega) = [(K+\mu_{gx}) - \omega^2 M + i\omega(C + \mu_{gv} + \mu_{fv})]^{-1}.$$

In future developments, the following relationships will be helpful,

$$H_x^{(1)*}(\omega) = H_x^{(1)}(-\omega),$$

$$H_x^{(2)*}(\omega_1, \omega_2) = H_x^{(2)}(-\omega_1, -\omega_2) \ . \tag{23}$$

2.4.4 Response Statistics

For the purpose of addressing the response statistics, all of the random processes considered herein are assumed stationary and ergodic. In conventional linear analysis, the response statistics are described by the second-order statistics of the response. In the present case the response distribution is no longer Gaussian, therefore, higher-order moments or cumulants are needed to describe the response statistics. The response statistics are considered in terms of the response cumulants rather than the moments. The cumulants are the coefficients of the expansion of the log-characteristic function of a random process and are chosen over moments as an analysis tool because their expressions are simpler and they are directly related to the departure from Gaussianity of a given process (e.g. Ref. 58). Indeed, for a Gaussian process, cumulants of order higher than second order vanish, while nonzero higher-order even moments may still exist. The first-order cumulant is the mean of the response and the second-order cumulant is equal to its variance. The third- and fourth-order cumulants are descriptors of the skewness and kurtosis, respectively, of the process, quantifying its departure from Gaussianity. Mathematically the cumulants, k_i, and moments, m_i, are related as follows:

$$k_1 = m_1,$$

$$k_2 = m_2 - m_1^2,$$

$$k_3 = m_3 - 3m_1 m_2 + 2m_1^3,$$

$$k_4 = m_4 - 3m_2^2 - 4m_1 m_3 + 12 m_1^2 m_2 - 6m_1^4. \tag{24}$$

The skewness and kurtosis representing a departure from Gaussianity are given by

$$\gamma_3 = \frac{k_3}{k_2^{3/2}}; \text{ and } \gamma_4 = \frac{k_4}{k_2^2}. \tag{25}$$

The power spectrum and the first four cumulants can be obtained from the following[30],

$$D_{xx}(\omega) = k_1\delta(\omega) + \left|H_x^{(1)}(\omega)\right|^2 D(\omega) + \frac{1}{2}\int_{-\infty}^{\infty} \left|H_x^{(2)}(\theta, \omega-\theta)\right|^2 D(\theta)D(\omega-\theta)d\theta, \tag{26}$$

where the transfer functions are as defined in the previous section, $D_{xx}(\omega)$ is the two-sided spectrum of $x(t)$; and $D(\omega)$ represents the two-sided spectrum of $\upsilon(t)$.

For brevity's sake, let $H_x^{(1)}(1)$, $H_x^{(2)}(1,2)$, $D(1)$ represent $H_x^{(1)}(\omega_1)$, $H_x^{(2)}(\omega_1,\omega_2)$ and $D(\omega_1)\ d\omega_1$, respectively. The associated cumulants are given below

$$k_1 = x_0 + \frac{1}{2}\int_{-\infty}^{\infty} H_x^{(2)}(1,-1)D(1),$$

$$k_2 = \int_{-\infty}^{\infty} H_x^{(1)}(1)H_x^{(1)}(-1)D(1) + \frac{1}{2}\int_{-\infty}^{\infty}\int_{-\infty}^{\infty} H_x^{(2)}(1,2)H_x^{(2)}(-1,-2)D(1)D(2),$$

$$k_3 = 3\int_{-\infty}^{\infty}\int_{-\infty}^{\infty} H_x^{(1)}(1)H_x^{(1)}(2)H_x^{(2)}(-1,-2)D(1)D(2)$$

$$+\int_{-\infty}^{\infty}\int_{-\infty}^{\infty}\int_{-\infty}^{\infty} H_x^{(2)}(1,2)H_x^{(2)}(-1,3)H_x^{(2)}(-2,-3)D(1)D(2)D(3),$$

$$k_4 = 12\int_{-\infty}^{\infty}\int_{-\infty}^{\infty}\int_{-\infty}^{\infty} H_x^{(1)}(1)H_x^{(1)}(2)H_x^{(2)}(-1,3)H_x^{(2)}(-2,-3)D(1)D(2)D(3)$$

$$+3\int_{-\infty}^{\infty}\int_{-\infty}^{\infty}\int_{-\infty}^{\infty}\int_{-\infty}^{\infty} H_x^{(2)}(1,2)H_x^{(2)}(-1,3)H_x^{(2)}(-2,4)H_x^{(2)}(-3,-4)D(1)D(2)D(3)D(4),$$

(27)

where x_0 in the expression for k_1 is the static response.

While most of the information about a random process is contained in the first two cumulants, the latter two yield important information necessary for evaluating the extremes of the response. In general, expressions for the response cumulants of Volterra systems are not available. Inspecting Eq. (27), even for the present case in which only terms up to the second order are treated, the higher order cumulant expressions become increasingly complex. In the sections to follow, two methods for addressing the issue of the excessive computational effort required in obtaining higher-order cumulants will be discussed.

2.4.5 *Direct Integration Method*

The calculation of the fourth-order cumulant was considered prohibitive, not only because of the behavior of the integrand, but also due to very extensive computational effort needed in evaluating the multi-fold integrals. Bedrosian and Rice[30] stated that the four-fold integral in the above equations cannot be carried out because of its complexity.

Recently, Spanos and Donley (e.g., Ref. 15) reported a similar difficulty. The present paper presents the evaluation of the four-fold integral in the above equations by reducing it into a three-fold integral after some mathematical manipulations.

To further assist in factorizing the more complicated integrals, the following one- and two-dimensional transfer functions may be developed,

$$C_{10}(\omega) = H_x^{(1)}(\omega),$$

$$C_{11}(\omega) = \int_{-\infty}^{\infty} H_x^{(1)}(\alpha)H_x^{(2)}(-\alpha, \omega)D(\alpha)d\alpha,$$

$$C_{20}(\omega_1, \omega_2) = H_x^{(2)}(\omega_1, \omega_2),$$

$$C_{22}(\omega_1, \omega_2) = \int_{-\infty}^{\infty} H_x^{(2)}(\alpha, \omega_1)H_x^{(2)}(-\alpha, \omega_2)D(\alpha)d\alpha, \tag{28}$$

noting that $C_{11}(-\omega) \neq C^*_{11}(\omega)$ and $C_{22}(\omega_1, \omega_2)$ is Hermitian, i.e., $C_{22}(\omega_1, \omega_2) = C^*_{22}(\omega_2, \omega_1)$. Then, the cumulant expressions of Eq. (27) can be recast in the following manner,

$$k_1 = x_0 + \frac{1}{2}\int_{-\infty}^{\infty} C_{20}(1, -1)D(1),$$

$$k_2 = \int_{-\infty}^{\infty} |C_{10}(1)|^2 D(1) + \frac{1}{2}\int_{-\infty}^{\infty}\int_{-\infty}^{\infty} |C_{20}(1, 2)|^2 D(1)D(2),$$

$$k_3 = 3\int_{-\infty}^{\infty} C_{10}(1)C_{11}(-1)D(1) + \int_{-\infty}^{\infty}\int_{-\infty}^{\infty} C_{20}(-1, -2)C_{22}(1, 2)D(1)D(2),$$

$$k_4 = 12\int_{-\infty}^{\infty} |C_{11}(1)|^2 D(1) + 3\int_{-\infty}^{\infty}\int_{-\infty}^{\infty} |C_{22}(1, 2)|^2 D(1)D(2). \tag{29}$$

By this approach, the solution of the fourth-order cumulant involves an effort equal to that needed for solving the third-order cumulant without any compromise on the accuracy.

2.4.6 Kac-Siegert Technique

A second approach to evaluating response cumulants is named after Kac and Siegert who first applied it to the theory of noise in radio receivers with square law detectors in 1947[31]. In fact, it may be taken as the generalized Fourier series representation method[59], based on the theory of linear integral equations. Its application to ocean engineering was first given by Neal[60]. Later, Vinje[61] discussed its application to the statistical distribution of second-order forces and motions. The Kac-Siegert technique has also been used extensively by Naess[62] and Naess and Johnsen[63] for offshore problems. Kareem and Zhao[20] and Langley and McWilliams[64] have also utilized the Kac-Siegert approach in offshore applications. The following is a complete formulation of the response cumulants up to fourth-order employing the Kac-Siegert technique, including the fluid-structure interaction which introduces nonlinear damping to the system.

The generalized Fourier series expansion can be written in one- and two-dimensional forms.

$$Q_1(\omega) = \sum_{i=1}^{\infty} \rho_i \phi_i(\omega),$$

$$Q_2(\omega_1, \omega_2) = \sum_{i=1}^{\infty} \sum_{j=1}^{\infty} \beta_{ij} \phi_i(\omega_1) \phi^*{}_j(\omega_2). \tag{30}$$

The above expression is sometimes called the degenerate kernel or separable kernel within the theory of integral equations. In the case when the second-order transfer function, $Q_2(\omega_1, \omega_2)$, is Hermitian, the basis functions, $\phi_i(.)$, are chosen as the characteristic functions of the following Fredholm homogeneous integral equation of the second kind,

$$\int_a^b Q_2(\omega_1, \omega_2) \phi(\omega_2) dy = \lambda \phi(\omega_1). \tag{31}$$

The nontrivial eigenvalues of Eq. (31) are real and corresponding eigenfunctions are orthogonal to one another, i.e., $(\phi_i, \phi_j) = \delta_{ij}$. Due to this orthogonality, $Q_2(\omega_1, \omega_2)$ may be recast as,

$$Q_2(\omega_1, \omega_2) = \sum_{i=1}^{\infty} \lambda_i \phi_i(\omega_1) \phi_i^*(\omega_2), \tag{32}$$

the convergence of which can be shown by Hilbert's and Mercer's theorems[59].

For the present application, the second order system function is,

$$Q_2(\omega_1, \omega_2) = G(\omega_1) H_x^{(2)}(\omega_1, -\omega_2) G(\omega_2), \tag{33}$$

where $G(\omega_1) = D(\omega_1)^{1/2}$, and $Q_2(\omega_1, \omega_2)$ is Hermitian. Both ω_1 and ω_2 represent frequencies, which are discretized as $\{\omega_i, i{=}1{,}2{,}\ldots, N\}$, with equal intervals Δ.

The discrete form of the homogeneous integral equation constitutes a linear algebraic eigenvalue problem as follows,

$$\sum_{j=1}^{N} Q_2(\omega_i, \omega_j) W_j \phi(\omega_j) \Delta + \sum_{j=1}^{N} Q_2(\omega_i, -\omega_j) W_j \phi(-\omega_j) \Delta = \lambda \phi(\omega_i), \tag{34}$$

$$\sum_{j=1}^{N} Q_2(-\omega_i, \omega_j) W_j \phi(\omega_j) \Delta + \sum_{j=1}^{N} Q_2(-\omega_i, -\omega_j) W_j \phi(-\omega_j) \Delta = \lambda \phi(-\omega_i), \tag{35}$$

where W_j are the weighting factors determined by the numerical method used to evaluate these equations. For slowly varying drift response applications, Naess[62] employed Newman's approximation[48] and ignored the interaction terms at sum frequencies, thus the second term in Eq. (34) and the first term in Eq. (35) may be eliminated. However, this assumption, though valid for slow drift response, is not applicable in the case of response due to wind loads. The preceding equations can be solved numerically to obtain the eigenvalues and the corresponding eigenvectors as follows.

Essentially, the result is a linear eigenvalue problem of dimension N which is expressible in matrix form as,

$$[A][W]\{\underset{\sim}{\phi}\} + [B][W]\{\underset{\sim}{\phi}\}^* = \lambda\{\underset{\sim}{\phi}\}, \tag{36}$$

where

$$A_{ij} = G(\omega_i) H_x^{(2)}(\omega_i, -\omega_j) G(\omega_j) \Delta,$$

$$B_{ij} = G(\omega_i) H_x^{(2)}(\omega_i, \omega_j) G(\omega_j) \Delta, \tag{37}$$

and $[W]$ is a diagonal matrix, whose elements are determined by the numerical integration method chosen. As an example, the weighting factor elements for the composite Simpson's rule are,

$$W_{1,1} = \frac{1}{3}, W_{2,2} = \frac{4}{3}, W_{3,3} = \frac{2}{3}, \ldots, W_{N-2,N-2} = \frac{2}{3}, W_{N-1,N-1} = \frac{4}{3}, W_{N,N} = \frac{1}{3}.$$

Noting that Eq. (36) is not Hermitian due to the involvement of the matrix of weighting factors, another diagonal matrix, $[V]$, is introduced, where $V_{i,i} = \sqrt{W_{i,i}}$. Then, a new vector may be specified as,

$$\{\underset{\sim}{\Phi}\} = [V]\{\underset{\sim}{\phi}\}. \tag{38}$$

Multiplying both sides of Eq. (36) by $[V]$ leaves,

$$[V][A][V]\{\underset{\sim}{\Phi}\} + [V][B][V]\{\underset{\sim}{\Phi}\}^* = \lambda\{\underset{\sim}{\Phi}\}. \tag{39}$$

To solve Eq. (39), $[A]$, $[B]$, and $\{\underset{\sim}{\Phi}\}$ are rewritten in terms of their real and imaginary parts, i.e.,

$$[A] = [A_R] + i[A_I],$$

$$[B] = [B_R] + i[B_I], \text{ and}$$

$$\{\underset{\sim}{\Phi}\} = \{\underset{\sim}{\Phi}_R\} + i\{\underset{\sim}{\Phi}_I\}.$$

Then the real matrix form which is equivalent to Eq. (39) is expressible as,

$$\begin{bmatrix} [V][A_R+B_R][V] & [V][B_I-A_I][V] \\ [V][A_I+B_I][V] & [V][A_R-B_R][V] \end{bmatrix} \begin{bmatrix} \underset{\sim}{\Phi}_R \\ \underset{\sim}{\Phi}_I \end{bmatrix} = \lambda \begin{bmatrix} \underset{\sim}{\Phi}_R \\ \underset{\sim}{\Phi}_I \end{bmatrix}, \tag{40}$$

which is symmetric when $[A_R]$, $[B_R]$, and $[B_I]$ are symmetric and $[A_I]^T = -[A_I]$.

Solving Eq. (40) yields $2N$ real eigenvalues and $2N$ eigenvectors. Then, normalizing the eigenvectors and recombining the complex vectors gives,

$$\{\underset{\sim}{\phi}\} = [V]^{-1}\{\underset{\sim}{\Phi}\} = [V]^{-1}\{\underset{\sim}{\Phi}_R + i\underset{\sim}{\Phi}_I\}. \tag{41}$$

According to the work by Kac and Siegert[31], the nth-order cumulant, k_n, is obtained from,

$$k_n = \sum_{i=1}^{N}\left(\frac{n!}{2}\rho_i^2\lambda_i^{n-2}(1-\delta_{n1}) + \frac{(n-1)!}{2}\lambda_i^n\right), \tag{42}$$

where ρ_i is the first-order system coefficient, given by

$$\rho_i = \sum_{j=1}^{N} H_x^{(1)}(\omega_j)G(\omega_j)\phi_i^*(\omega_j)W_j\Delta. \tag{43}$$

and δ_{ij} is the Kronecker delta function.

Specifically, the first four response cumulants are:

$$k_1 = x_0 + \frac{1}{2}\sum_{i=1}^{M}\lambda_i,$$

$$k_2 = \sum_{i=1}^{M}\rho_i^2 + \frac{1}{2}\sum_{i=1}^{M}\lambda_i^2,$$

$$k_3 = 3\sum_{i=1}^{M}\rho_i^2\lambda_i + \sum_{i=1}^{M}\lambda_i^3, \text{ and}$$

$$k_4 = 12\sum_{i=1}^{M}\rho_i^2\lambda_i^2 + 3\sum_{i=1}^{M}\lambda_i^4, \tag{44}$$

where the series expressions are truncated after M terms. It has been observed that the number of terms, M, required for convergence is related to the system damping[21]. Indeed, for larger damping fewer terms need to be retained.

A notable advantage of the Kac-Siegert approach is that it provides information on the cumulants higher than the fourth order. Usually, however, only the third- and fourth-order cumulants are significant and possess physical interpretations in terms of skewness and kurtosis.

2.4.7 Combined Wind, Wave and Current Effects

Typically, wind and wave loadings impinge concurrently on a TLP in a given ocean environment. Indeed, wind plays a part in generating waves but the exact correlation between wind velocity fluctuations and wave elevations has not yet been formalized. Herein, the combined effect of wind and waves is viewed through structural response motions. Assuming: (i) that the total response is simply the sum of the response due to wave loading and the response due to wind loading; (ii) the wind loading is small relative to the wave loading; (iii) the response velocity due to wind loading is small relative to the response velocity due to wave loading, the equations of motion due to combined wind, wave and current loadings are given by

$$M\ddot{x}^{wind} + (C + C_{wave})\dot{x}^{wind} + Kx^{wind} = K_w(w + W - \dot{x}^{wind})^2,$$

$$M\ddot{x}^{wave} + (C + C_{wind})\dot{x}^{wave} + Kx^{wave} = K_m\dot{u} + K_d|u + U - \dot{x}^{wave}|(u + U - \dot{x}^{wave}), \quad (45)$$

where

$$C_{wind} = -E\left[\frac{\partial f_{wind}}{\partial \dot{x}}\right] \text{ and } C_{wave} = -E\left[\frac{\partial f_{wave}}{\partial \dot{x}}\right]$$

denote the damping introduced by the wind loads and wave loads, f_{wind} and f_{wave}, respectively. The solution of the preceding equations is obtained following the two procedures outlined earlier in the text. Based on the assumptions made earlier, the total response cumulants can be obtained by a simple summation of the response cumulants due to all loadings, (e.g. Ref. 65)

$$k_m^{total} = k_m^{wind} + k_m^{wave}, \quad m = 1, 2, 3, 4, \ldots \quad (46)$$

2.5 Equivalent Statistical Cubicization

The SDOF system equation of motion is, again, Eq. (14) where a statistically symmetric nonlinearity in the system or the excitation is now considered, e.g., for the case when the current speed, $U = 0$ m/s. The output of Eq. (14) is in this case cast as the sum of the outputs of a first-order and a third-order Volterra system. It is further assumed that the nonlinear response and its velocity are small such that terms of $O(x_3^2)$ and $O(\dot{x}_3^2)$ may be neglected.

2.5.1 Splitting Technique

Since the Volterra functional series now chosen consists of a linear and a cubic response, it is necessary to represent the symmetric nonlinearities in the system as well as in the excitation in a manner amenable to this technique. To do so, the nonlinearity is expanded as a Taylor series including up to first-order terms of its cubic part as follows,

$$g(x, \dot{x}) = g(x_1, \dot{x}_1) + \frac{\partial g}{\partial x}(x_1, \dot{x}_1)\frac{x_3}{6} + \frac{\partial g}{\partial \dot{x}}(x_1, \dot{x}_1)\frac{\dot{x}_3}{6} + \ldots$$

$$g(x, \dot{x}) \approx g(x_1, \dot{x}_1) + \mu_{gx}\frac{x_3}{6} + \mu_{gv}\frac{\dot{x}_3}{6}. \tag{47}$$

where μ_{gx} and μ_{gv} are as defined previously and, because of symmetry, there is no static offset, x_0. Analogously, a symmetric nonlinearity in the excitation may be treated[19],

$$f_N(\upsilon, \dot{x}) = f_N(\upsilon, \dot{x}_1) + \frac{\partial f_N(\upsilon, \dot{x}_1)}{\partial \dot{x}}\frac{\dot{x}_3}{6} + O(\dot{x}_3^2)$$

$$f_N(\upsilon, \dot{x}) \approx f_N(\upsilon, \dot{x}_1) - \mu_{fv}\frac{\dot{x}_3}{6}. \tag{48}$$

These formulations again leave behind nonlinear leading terms which involve only the Gaussian first-order system response and linear second terms involving the third-order response.

2.5.2 Cubicization Procedure

When the initial terms in Eq. (47) and Eq. (48) are not in polynomial form, they must be so approximated for use in the Volterra framework. Thus, the cubic polynomial approximations,

$$g(x_1, \dot{x}_1) \approx \beta_1 x_1 + \beta_2 \dot{x}_1 + \frac{1}{6}(\beta_6 x_1^3 + \beta_7 x_1^2 \dot{x}_1 + \beta_8 x_1 \dot{x}_1^2 + \beta_9 \dot{x}_1^3), \tag{49}$$

$$f_N(\upsilon, \dot{x}_1) \approx a_1(\upsilon - \dot{x}_1) + \frac{1}{6}a_3(\upsilon - \dot{x}_1)^3. \tag{50}$$

yield desirable forms, where the unknown coefficients are determined by minimizing the mean-square of the error. Again, because the expected values taken in the mean-square error minimization procedure are functions only of the first-order response and its velocity, their Gaussian character offers some computational convenience.

Following the cubicization procedure, the equivalent Volterra representation for Eq. (14), when $g(x, \dot{x})$ is a statistically symmetric nonlinearity is,

$$M\ddot{x}_1 + (C + \beta_2 + a_1)\dot{x}_1 + (K + \beta_1)x_1 = f_L(\upsilon) + a_1\upsilon,$$

$$\begin{aligned} &M\ddot{x}_3 + (C + \mu_{gv} + \mu_{fv})\dot{x}_3 + (K + \mu_{gx})x_3 = \\ &a_3(\upsilon - \dot{x}_1)^3 - (\beta_6 x_1^3 + \beta_7 x_1^2 \dot{x}_1 + \beta_8 x_1 \dot{x}_1^2 + \beta_9 \dot{x}_1^3) \end{aligned} \tag{51}$$

having transfer functions,

$$H_x^{(1)}(\omega) = H_1(\omega)H_f^{(1)}(\omega),$$

$$H_{\dot{x}}^{(1)}(\omega) = i\omega H_x^{(1)}(\omega),$$

$$H_v(\omega) = H_\upsilon(\omega) - i\omega H_x^{(1)}(\omega),$$

$$H_x^{(3)}(\omega_1, \omega_2, \omega_3) = H_3(\omega_1 + \omega_2 + \omega_3) H_f^{(3)}(\omega_1, \omega_2, \omega_3),$$

$$H_{\dot{x}}^{(3)}(\omega_1, \omega_2, \omega_3) = i(\omega_1 + \omega_2 + \omega_3) H_x^{(3)}(\omega_1, \omega_2, \omega_3), \tag{52}$$

where,

$$H_1(\omega) = [K + \beta_1 - \omega^2 M + i\omega(C + \beta_2 + a_1)]^{-1},$$

$$H_3(\omega) = [(K + \mu_{gx}) - \omega^2 M + i\omega(C + \mu_{gv} + \mu_{fv})]^{-1},$$

$$H_f^{(1)}(\omega) = H_L(\omega) + a_1 H_\upsilon(\omega),$$

and,

$$H_f^{(3)}(\omega_1, \omega_2, \omega_3) = a_3 H_v(\omega_1) H_v(\omega_2) H_v(\omega_3)$$

$$-(\beta_6 + i\beta_7 \frac{\omega_1 + \omega_2 + \omega_3}{3} - \beta_8 \frac{\omega_1\omega_2 + \omega_1\omega_3 + \omega_2\omega_3}{3}$$

$$-i\beta_9 \omega_1\omega_2\omega_3) \cdot H_x^{(1)}(\omega_1) H_x^{(1)}(\omega_2) H_x^{(1)}(\omega_3).$$

It is again useful to note that,

$$H_x^{(1)*}(\omega) = H_x^{(1)}(-\omega),$$

$$H_x^{(3)*}(\omega_1, \omega_2, \omega_3) = H_x^{(3)}(-\omega_1, -\omega_2, -\omega_3). \tag{53}$$

2.5.3 *Response Statistics*

It may be shown that for the case of the symmetric nonlinearity with a zero-mean input process, the first- and third-order response cumulants are zero. Using the transfer functions derived earlier within the framework of a direct numerical integration scheme, the second- and fourth-order response cumulants are[19,20,30],

$$k_2 = \int_{-\infty}^{\infty} |H_x^{(1)}(1)|^2 D(1) + \int_{-\infty}^{\infty}\int_{-\infty}^{\infty} H_x^{(1)}(1) H_x^{(3)}(-1, 2, -2) D(1) D(2)$$

$$+ \frac{1}{4} \int_{-\infty}^{\infty}\int_{-\infty}^{\infty}\int_{-\infty}^{\infty} H_x^{(3)}(1, 2, -2) H_x^{(3)}(-1, 3, -3) D(1) D(2) D(3)$$

$$+\frac{1}{6}\int_{-\infty}^{\infty}\int_{-\infty}^{\infty}\int_{-\infty}^{\infty}\left|H_x^{(3)}(1,2,3)\right|^2 D(1)D(2)D(3),$$

$$k_4 = 4\int_{-\infty}^{\infty}\int_{-\infty}^{\infty}\int_{-\infty}^{\infty} H_x^{(1)}(1)H_x^{(1)}(2)H_x^{(1)}(3)H_x^{(3)}(-1,-2,-3)D(1)D(2)D(3)$$

$$+3\int_{-\infty}^{\infty}\int_{-\infty}^{\infty}\int_{-\infty}^{\infty}\int_{-\infty}^{\infty} H_x^{(1)}(1)H_x^{(1)}(2)H_x^{(3)}(-1,3,-3)H_x^{(3)}(-2,4,-4)$$

$$\cdot D(1)D(2)D(3)D(4)$$

$$+6\int_{-\infty}^{\infty}\int_{-\infty}^{\infty}\int_{-\infty}^{\infty}\int_{-\infty}^{\infty} H_x^{(1)}(1)H_x^{(1)}(2)H_x^{(3)}(-1,-2,3)H_x^{(3)}(-3,4,-4)$$

$$\cdot D(1)D(2)D(3)D(4)$$

$$+3\int_{-\infty}^{\infty}\int_{-\infty}^{\infty}\int_{-\infty}^{\infty}\int_{-\infty}^{\infty} H_x^{(1)}(1)H_x^{(1)}(2)H_x^{(3)}(-1,3,4)H_x^{(3)}(-2,-3,-4)$$

$$\cdot D(1)D(2)D(3)D(4) + O(x_3^3), \tag{54}$$

where for the sake of brevity, $H_x^{(1)}(1) = H_x^{(1)}(\omega_1)$, $H_x^{(3)}(1,2,3) = H_x^{(3)}(\omega_1,\omega_2,\omega_3)$, and $D(1) = D(\omega_1)d\omega_1$ in the cumulant expressions. The response power spectral density may be estimated as,

$$D_{xx}(\omega) = \left|H_x^{(1)}(\omega) + \frac{1}{2}\int_{-\infty}^{\infty} H_x^{(3)}(\omega,\theta,-\theta)D(\theta)d\theta\right|^2 D(\omega)$$

$$+\frac{1}{6}\int_{-\infty}^{\infty}\int_{-\infty}^{\infty}\left|H_x^{(3)}(\theta,\tau,\omega-\theta-\tau)\right|^2 D(\theta)D(\tau)D(\omega-\theta-\tau)d\theta d\tau. \tag{55}$$

2.6 Probability Distribution of Response

Following the evaluation of the first four moments or cumulants of response, the non-Gaussian distribution of response processes can be obtained with a subsequent estimation of the extreme value distribution. In this study, Gram-Charlier Series, Hermite Moment Approach and Maximum Entropy method are utilized. A short description of each follows.

2.6.1 Gram-Charlier Series Distribution

This method is based on expanding the distribution of a non-Gaussian random variable, x, in a Hermite series with a Gaussian "parent function." The simplest form of this method is expressible as (e.g., Ref. 66)

$$p_G(x) = \frac{1}{\sqrt{2\pi}\sigma}\exp\left[-\frac{(x-m_1)^2}{2\sigma^2}\right]\left\{1+\frac{\gamma_3}{3!}H_3\left(\frac{x-m_1}{\sigma}\right)+\frac{\gamma_4}{4!}H_4\left(\frac{x-m_1}{\sigma}\right)+\dots\right\}, \quad (56)$$

where the Hermite polynomials are given as

$$H_3(z) = z^3 - 3z,$$

$$H_4(z) = z^4 - 6z^2 + 3, \quad (57)$$

and m_1, σ, γ_3 and γ_4 are the mean, standard deviation, skewness and kurtosis of x, respectively.

The Gram-Charlier series distribution, however, has an inherent shortcoming in its limited ability to characterize the tail regions of the distribution and, thus, the extremes of a given process. In fact, this type of series expansion can exhibit negative probabilities in the tail regions.

2.6.2 Moment-Based Hermite Transformation Method

In this method, the non-Gaussian variable is expanded in Hermite polynomials in terms of a standardized Gaussian process[32, 67]. This transformation is valid for a non-normal process, $x(t)$, which is expressible as a monotonic function of a standard normal process, $u(t)$. For the softening processes dealt with in this study, the relationship between a standardized form of $x(t)$ and $u(t)$ truncated after the third-order term is,

$$\frac{x-m_1}{\sigma} = \alpha\{u + h_3(u^2-1) + h_4(u^3-3u)\}, \quad (58)$$

where

$$\alpha = (1 + 2h_3^2 + 6h_4^2)^{-1/2},$$

$$h_3 = \frac{\gamma_3}{4+2\sqrt{1+1.5\gamma_4}}, \text{ and}$$

$$h_4 = \frac{\sqrt{1+1.5\gamma_4}-1}{18}.$$

Having made this transformation, the probability density function of x may be derived as,

$$p_H(x) = \frac{1}{\sqrt{2\pi}}\exp\left[-\frac{u^2(x)}{2}\right]\frac{du(x)}{dx} \tag{59}$$

where for this truncated expansion, *u(x)* is,

$$u(x) = [\sqrt{\xi^2(x)+c}+\xi(x)]^{1/3} - [\sqrt{\xi^2(x)+c}-\xi(x)]^{-1/3} - a, \tag{60}$$

where

$$\xi(x) = 1.5b\left(a+\frac{x-m_1}{\alpha\sigma}\right)-a^3,\ a = \frac{h_3}{3h_4},\ b = \frac{1}{3h_4},\ \text{and } c = (b-1-a^2)^3.$$

2.6.3 Maximum Entropy Method

In statistical mechanics, the entropy of a given state is directly related to its probability of occurrence. Increasing entropy is associated with increasing disorder. Jaynes[68] approached estimation of the probability density function (PDF) of the states of a system using an entropy functional, choosing as the best distribution the one having maximum entropy among all the distributions which satisfied the observed moment constraints. Sobczyk and Trebicki[34] employed stochastic differential equations in tandem with the entropy functional to estimate the PDF of a given system response when the first N moments of the response were known.

According to the principle of maximum entropy in one dimension, an appropriate probability density function, $p(x)$, must maximize the entropy functional,

$$H = -\int p(x)\ln p(x)dx, \tag{61}$$

while satisfying constraints specified via moment equations or moment values. Applying the constraints, in the present case the moment values themselves from our Volterra system analysis, via the Lagrange multiplier technique of variational calculus, the maximum entropy distribution is expressible as,

$$p_M(x) = \exp\left(-\sum_{k=0}^{N}\lambda_k x^k\right). \tag{62}$$

where N is the number of moments given and the coefficients (Lagrange multipliers) may be determined by matching the moments according to,

$$\int_{\infty}^{-\infty} x^n p_M(x)dx = m_n,\ n = 0, 1, \ldots, N. \tag{63}$$

The existence of a maximum entropy-based distribution is limited to a range of possible values of the moments for which the system of equations Eq. (63) is solvable (e.g., Refs. 34 & 69). For the case when the first four moments are known, Tagliani[69] developed a set of criteria for the existence of such a distribution based on the response statistics themselves,

$$\sigma > 0; \;\; \gamma_3 > \sigma + \frac{1}{\sigma}; \;\; \text{and } \gamma_4 + 3 > 1 + \gamma_3^2, \tag{64}$$

where σ, γ_3, and γ_4 are the standard deviation, skewness and kurtosis of the response process.

Solving Eq. (63) for the Lagrange multipliers requires the implementation of an iterative procedure. From the normalization condition for any PDF,

$$\int_{-\infty}^{\infty} p_0(x)dx = \exp(\lambda_0), \tag{65}$$

where, for $N = 4$,

$$p_0(x) = \exp(-\lambda_1 x - \lambda_2 x^2 - \lambda_3 x^3 - \lambda_4 x^4). \tag{66}$$

Defining

$$I_0 = \exp(\lambda_0), \tag{67}$$

yields

$$I_n = m_n I_0, \text{ for } n = 1, 2, 3, 4, \tag{68}$$

where,

$$I_n = \int_{-\infty}^{\infty} x^n p_0(x)dx. \tag{69}$$

Applying Newton's method, the iteration then takes the following form, where k is the iteration index,

$$[F'(\underset{\sim}{\lambda}^{(k)})]\{\underset{\sim}{\lambda}^{(k+1)} - \underset{\sim}{\lambda}^{(k)}\} = \{-\underset{\sim}{F}(\underset{\sim}{\lambda}^{(k)})\}, \tag{70}$$

where,

$$\{\underset{\sim}{\lambda}^{(k)}\} = \begin{bmatrix} \lambda_1^{(k)} \\ \lambda_2^{(k)} \\ \lambda_3^{(k)} \\ \lambda_4^{(k)} \end{bmatrix},$$

$$F_i(\underset{\sim}{\lambda}^{(k)}) = I_i - m_i I_0, \text{ and}$$

$$F'_{ij}(\underset{\sim}{\lambda}^{(k)}) = -I_{i+j} + m_i I_j. \tag{71}$$

It has been noted, in addition, that since this system is highly nonlinear, the determination of these unknown coefficients or Lagrange multipliers is sensitive to the initial values chosen[21]. Constraints on the initial values are difficult to derive theoretically, and while Sobczyk and Trebicki[34] have provided a set of constraining inequalities for the Lagrange multipliers, these, too, are sometimes difficult to implement in practical analysis. Hence, a scheme has been developed to estimate the initial values using the Gram-Charlier series as a first approximation to the PDF.

Assuming that the corresponding Gram-Charlier distribution gives a reasonable approximation to the maximum entropy distribution which is ultimately sought, an error may be quantified as,

$$\varepsilon = \int_{y_{min}}^{y_{max}} [\ln p_M(y) - \ln p_G(y)]dy, \tag{72}$$

where,

$$p_G(y) = \frac{1}{\sqrt{2\pi}\sigma}\exp\left(-\frac{y^2}{2\sigma^2}\right)\left[1 + \frac{\gamma_3}{3!}H_3\left(\frac{y}{\sigma}\right) + \frac{\gamma_4}{4!}H_4\left(\frac{y}{\sigma}\right)\right], \tag{73}$$

is the Gram-Charlier distribution with H_3, H_4, γ_3, and γ_4 as described in the main text and

$$p_m(y) = \exp(-\lambda_0 - \lambda_1 y - \lambda_2 y^2 - \lambda_3 y^3 - \lambda_4 y^4), \tag{74}$$

is the maximum entropy distribution. The bounds of integration, y_{min} and y_{max}, are chosen as wide as possible such that the tails of the Gram-Charlier distribution remain positive.

Minimization of the error yields the following linear system for the five unknowns:

$$\begin{bmatrix} a_1 & a_2 & a_3 & a_4 & a_5 \\ a_2 & a_3 & a_4 & a_5 & a_6 \\ a_3 & a_4 & a_5 & a_6 & a_7 \\ a_4 & a_5 & a_6 & a_7 & a_8 \\ a_5 & a_6 & a_7 & a_8 & a_9 \end{bmatrix} \begin{bmatrix} \lambda_0 \\ \lambda_1 \\ \lambda_2 \\ \lambda_3 \\ \lambda_4 \end{bmatrix} = \begin{bmatrix} p_0 \\ p_1 \\ p_2 \\ p_3 \\ p_4 \end{bmatrix}, \tag{75}$$

where

$$a_n = \frac{y_{max}^n - y_{min}^n}{n}, \tag{76}$$

and

$$p_n = -\int_{y_{min}}^{y_{max}} y^n \ln p_G(y)dy. \tag{77}$$

Numerical experimentation has shown that the λ_i values obtained using this procedure are suitable starting values for the iteration procedure which determines the maximum entropy distribution parameters[21].

2.7 Mean Upcrossing Rate and Distribution of Maxima

The distribution of the maxima of a process can be approximated in terms of its mean upcrossing rate. Mathematically, the mean upcrossing rate is,

$$\nu(x) = \int_0^{\infty} \dot{x} p_{x,\dot{x}}(x, \dot{x})d\dot{x}. \tag{78}$$

This expression involves the joint probability density function of a random process and its first time derivative, which is difficult to obtain for an arbitrary non-Gaussian process. However, the crossing rate may be easily derived from the crossings of nonlinear transformations of Gaussian processes[67]. In the case of a moment-based Hermite transformation wherein a non-Gaussian process, *x(t)*, is related to a Gaussian process, *u(t)*, the crossing rate may be written as,

$$\nu(x) = \nu_0 \exp\left(\frac{-u^2(x)}{2}\right), \tag{79}$$

where *u(x)* is a standard normal process, related to the non-Gaussian process, *x(t)*, by Eq. (60) and ν_0 is the zero-crossing rate given by $\sigma_{\dot{u}}/2\pi$. The variance of the velocity of the parent Gaussian process is expressible in terms of the variances of the non-Gaussian process and its velocity according to,

$$\sigma_{\dot{u}}^2 = \frac{\sigma_{\dot{x}}^2}{\alpha^2\sigma_x^2(1 + 4h_3^2 + 18h_4^2)}. \tag{80}$$

Substituting this expression into Eq. (79), gives for the crossing rate,

$$\nu(x) = \frac{1}{2\pi}\frac{\sigma_{\dot{x}}}{\alpha\sigma_x\sqrt{1 + 4h_3^2 + 18h_4^2}}\exp\left(\frac{-u^2(x)}{2}\right). \tag{81}$$

An approximate distribution of the maxima of a non-Gaussian response may also be obtained using the Hermite method as,

$$p_E(x) = u(x)\exp\left(-\frac{u^2(x)}{2}\right)\frac{du(x)}{dx}, \tag{82}$$

where *u(x)* is related to *x* according to Eq. (60).

2.8 Time-Domain Simulation

The procedures laid out herein have been verified via a Monte Carlo simulation technique given a prescribed power spectrum, $S(\omega)$, for a random process, $\zeta(t)$. Sample time histories of either the wind or the wave process, given the appropriate spectral representation, may be generated according to,

$$\zeta(t) = \sum_{j=1}^{N} \sqrt{2S(\omega_j)\Delta\omega_j}\cos(\omega_j t + \theta_j), \tag{83}$$

where θ_j are independent random phases distributed uniformly between 0 and 2π, and $\omega_j = j\Delta\omega$. Consideration should be given to appropriately choosing $\Delta\omega$ to suitably discretize the particular spectrum of the process being simulated, paying close attention to the trade-off in terms of time resolution and the overall system dynamics. The efficiency of this simulation procedure may be boosted significantly by employing a fast Fourier transform method (e.g., Refs. 70 & 71).

Note that while fluctuating wind velocity spectra are readily available, wave processes are frequently represented by the power spectral density of the surface elevation rather than the water particle velocity or acceleration. Therefore, in the present study, time histories of the horizontal water particle velocity and acceleration must be obtained from the wave surface elevation process, $\eta(t)$, according to, (e.g. Ref. 2)

$$u(x, z, t) = \omega\frac{\cosh kz}{\sinh kd}\exp(-ikx)\eta(t),$$

$$\dot{u}(x, z, t) = \omega^2\frac{\cosh kz}{\sinh kd}\exp(-ikx)\eta(t), \tag{84}$$

where x is the horizontal coordinate of the location of interest, z is its vertical coordinate measured upward from the sea floor, d is the water depth and k is the wave number which is related to the frequency, ω, by the linear dispersion relation,

$$\omega^2 = gk\tanh kd, \tag{85}$$

where g is gravitational acceleration.

Having the underlying process time histories, developing the force processes is a simple matter. Subsequently, the response, $x(t)$, may be evaluated using a step-by-step numerical integration of the governing differential equation possessing a desirable accuracy. From these simulations, the response moments are obtained as,

$$m_i = \frac{1}{N}\sum_{j=0}^{N-1}(x_j - \bar{x})^i, \tag{86}$$

where $i = 2, 3, \ldots$ and $\bar{x}$ is the response mean.

3. Examples

To illustrate the nonlinear effects introduced by wind and wave loads, an idealized TLP model is utilized. Since the wind force is already in a quadratic form, it is readily cast as an equivalent Volterra system[21] and only the wave drag force is treated rigorously here. First, the splitting technique is performed whereby the nonlinearity in the right side of Eq. (13) is expanded as follows in a Taylor series in terms of second order response velocity,

$$|u+U-\dot{x}|(u+U-\dot{x}) = |u+U-\dot{x}_1|(u+U-\dot{x}_1) - 2|u+U-\dot{x}_1|\frac{\dot{x}_2}{2} + O(\dot{x}_2^2). \quad (87)$$

The slow drift approximation is employed, and in order to eliminate a time-dependence of the additional damping term on the right-hand side of Eq. (87), the coefficient is approximated by its expected value, i.e.,

$$|u+U-\dot{x}|(u+U-\dot{x}) \approx |u+U-\dot{x}_1|(u+U-\dot{x}_1) - 2E[|u+U-\dot{x}_1|]\frac{\dot{x}_2}{2}. \quad (88)$$

Since the initial term on the right-hand side in Eq. (88) is not cast in a polynomial form and as such is not yet tractable by the Volterra approach, the quadratization procedure is now invoked to approximate it in terms of the relative fluid-structure velocity as follows,

$$|u+U-\dot{x}_1|(u+U-\dot{x}_1) \approx a_0 + a_1(u-\dot{x}_1) + \frac{a_2}{2}(u-\dot{x}_1)^2. \quad (89)$$

The polynomial approximation of Eq. (89) may then be tailored by minimizing the mean-square of the following error term,

$$\varepsilon = |u+U-\dot{x}_1|(u+U-\dot{x}_1) - a_0 - a_1(u-\dot{x}_1) - \frac{a_2}{2}(u-\dot{x}_1)^2. \quad (90)$$

This minimization produces a system of three equations for the unknowns, a_i,

$$\begin{bmatrix} 1 & 0 & \sigma^2 \\ 0 & \sigma^2 & 0 \\ \sigma^2 & 0 & 3\sigma^4 \end{bmatrix} \begin{bmatrix} a_0 \\ a_1 \\ a_2 \end{bmatrix} = \begin{bmatrix} E[|u+U-\dot{x}_1|(u+U-\dot{x}_1)] \\ E[(u-\dot{x}_1)|u+U-\dot{x}_1|(u+U-\dot{x}_1)] \\ 2E[(u-\dot{x}_1)^2|u+U-\dot{x}_1|(u+U-\dot{x}_1)] \end{bmatrix}, \quad (91)$$

which when solved yields,

$$a_0 = 2U\sigma(rb_1+b_2);\ a_1 = 4\sigma(rb_1+b_2), \text{and } a_2 = 4b_1, \quad (92)$$

where

$$b_1 = \frac{1}{\sqrt{2\pi}}\int_0^r exp\left(-\frac{y^2}{2}\right)dy,\ b_2 = \frac{1}{\sqrt{2\pi}}exp\left(-\frac{r^2}{2}\right);$$

$$r = \frac{U}{\sigma}\text{; and } \sigma^2 = E[(u - \dot{x}_1)^2].$$

Turning attention back the system Eq. (91), an important advantage of the present technique may be noted in the fact that all of the expected values taken involve only Gaussian quantities and functions thereof.

A more detailed treatment of the expected values appearing in the right-hand side vector of Eq. (91) is now given. Letting $v = u - \dot{x}_1$, v is a zero-mean Gaussian process with standard deviation, σ. The probability density function of v is thus,

$$f_V(v) = \frac{1}{\sqrt{2\pi}\sigma}\exp\left(-\frac{v^2}{2\sigma^2}\right). \tag{93}$$

Proceeding, then, the first expectation is rewritten,

$$E[|v+U|(v+U)] = \int_{-\infty}^{\infty} |v+U|(v+U)\frac{1}{\sqrt{2\pi}\sigma}\exp\left(-\frac{v^2}{2\sigma^2}\right)dv \tag{94}$$

$$= \int_{-U}^{\infty} (v+U)^2\frac{1}{\sqrt{2\pi}\sigma}\exp\left(-\frac{v^2}{2\sigma^2}\right)dv + \int_{-U}^{-\infty} (v+U)^2\frac{1}{\sqrt{2\pi}\sigma}\exp\left(-\frac{v^2}{2\sigma^2}\right)dv\,.$$

Expanding the polynomial in v and employing the properties of the even and odd functions in the expansion yields,

$$E[|v+U|(v+U)] = 2U^2\int_0^U \frac{1}{\sqrt{2\pi}\sigma}\exp\left(-\frac{v^2}{2\sigma^2}\right)dv + \frac{4U\sigma}{\sqrt{2\pi}}\exp\left(-\frac{U^2}{2\sigma^2}\right)$$

$$+\int_0^U v^2\frac{1}{\sqrt{2\pi}\sigma}\exp\left(-\frac{v^2}{2\sigma^2}\right)dv\,.$$

Letting $y = v/\sigma$, and integrating the last term above by parts finally gives,

$$E[|v+U|(v+U)] = 2(U^2+\sigma^2)\int_0^{U/\sigma} \frac{1}{\sqrt{2\pi}}\exp\left(-\frac{y^2}{2}\right)dy + \frac{2U\sigma}{\sqrt{2\pi}}\exp\left(-\frac{U^2}{2\sigma^2}\right). \tag{95}$$

Continuing, as above, with the second expectation:,

$$E[v|v+U|(v+U)] = \int_{-\infty}^{\infty} v|v+U|(v+U)\frac{1}{\sqrt{2\pi}\sigma}\exp\left(-\frac{v^2}{2\sigma^2}\right)dv \tag{96}$$

$$= \int_{-U}^{\infty} v(v+U)^2 \frac{1}{\sqrt{2\pi}\sigma} \exp\left(-\frac{v^2}{2\sigma^2}\right) dv + \int_{-U}^{-\infty} v(v+U)^2 \frac{1}{\sqrt{2\pi}\sigma} \exp\left(-\frac{v^2}{2\sigma^2}\right) dv \,.$$

Again, expanding the polynomial in v and employing even and odd function properties-yields,

$$E[v|v+U|(v+U)] = \frac{2U^2\sigma}{\sqrt{2\pi}} \exp\left(-\frac{U^2}{2\sigma^2}\right) + 2\int_{U}^{\infty} v^3 \frac{1}{\sqrt{2\pi}\sigma} \exp\left(-\frac{v^2}{2\sigma^2}\right) dv$$

$$+ 4\int_{0}^{U} Uv^2 \frac{1}{\sqrt{2\pi}\sigma} \exp\left(-\frac{v^2}{2\sigma^2}\right) dv \,.$$

Integrating the latter two terms by parts leaves,

$$E[v|v+U|(v+U)] = 4U\sigma^2 \int_{0}^{U/\sigma} \frac{1}{\sqrt{2\pi}} \exp\left(-\frac{y^2}{2}\right) dy + \frac{4\sigma^3}{\sqrt{2\pi}} \exp\left(-\frac{U^2}{2\sigma^2}\right). \tag{97}$$

Finally, treating the third expectation:

$$E[v^2|v+U|(v+U)] = \int_{-\infty}^{\infty} v^2|v+U|(v+U) \frac{1}{\sqrt{2\pi}\sigma} \exp\left(-\frac{v^2}{2\sigma^2}\right) dv \,, \tag{98}$$

$$= \int_{-U}^{\infty} v^2(v+U)^2 \frac{1}{\sqrt{2\pi}\sigma} \exp\left(-\frac{v^2}{2\sigma^2}\right) dv + \int_{-U}^{-\infty} v^2(v+U)^2 \frac{1}{\sqrt{2\pi}\sigma} \exp\left(-\frac{v^2}{2\sigma^2}\right) dv \,.$$

Following the same procedure as laid out previously, multiplying out the polynomial in v and taking advantage of the even or odd character of each term in the expansion yields,

$$E[v^2|v+U|(v+U)] = 2\int_{0}^{U} v^4 \frac{1}{\sqrt{2\pi}\sigma} \exp\left(-\frac{v^2}{2\sigma^2}\right) dv + 2U^2 \int_{0}^{U} v^2 \frac{1}{\sqrt{2\pi}\sigma} \exp\left(-\frac{v^2}{2\sigma^2}\right) dv$$

$$+ 4U \int_{U}^{\infty} v^3 \frac{1}{\sqrt{2\pi}\sigma} \exp\left(-\frac{v^2}{2\sigma^2}\right) dv \,.$$

Integrating each of the terms above by parts one or more times gives,

$$E[v^2|v+U|(v+U)] = 2\sigma^2(U^2+3\sigma^2) \int_{0}^{U/\sigma} \frac{1}{\sqrt{2\pi}} \exp\left(-\frac{y^2}{2}\right) dy + \frac{2U\sigma^3}{\sqrt{2\pi}} \exp\left(-\frac{U^2}{2\sigma^2}\right). \tag{99}$$

Now the equations of motion for wave excitation can be expressed as

$$M\ddot{x}_1 + (C + a_1)\dot{x}_1 + Kx_1 = K_m\dot{u} + a_1 u,$$

$$M\ddot{x}_2 + (C + a_1)\dot{x}_2 + Kx_2 = a_2(u - \dot{x}_1)^2, \tag{100}$$

where now, $a_0 = K_d a_0$; $a_1 = K_d a_1$; and $a_2 = K_d a_2$. The static response of this system may be given as,

$$x_0 = \frac{a_0}{K}. \tag{101}$$

It is then desired to characterize the time-varying system response in the frequency domain. Thus, the following transfer functions are developed to relate x_1, $(u - \dot{x}_1)$, and x_2, respectively to the input water particle velocity spectrum

$$H_x^{(1)}(\omega) = (K_m i\omega + a_1)H(\omega),$$

$$H_v(\omega) = 1 - i\omega H_x^{(1)}(\omega),$$

$$H_x^{(2)}(\omega_1, \omega_2) = a_2 H(\omega_1 + \omega_2)H_v(\omega_1)H_v(\omega_2), \tag{102}$$

where $H(\omega) = [K - \omega^2 M + i\omega(C + a_1)]^{-1}$. The cumulants of the response based on these frequency domain formulations are given in the earlier discussion.

Similarly, the equations of motion under the wind force are given by

$$M\ddot{x}_1 + (C + a_1)\dot{x}_1 + Kx_1 = a_1 w,$$

$$M\ddot{x}_2 + (C + a_1)\dot{x}_2 + Kx_2 = a_2(w - \dot{x}_1)^2, \tag{103}$$

where $a_0 = K_w W^2$; $a_1 = 2K_w W$; and $a_2 = 2K_w$. The static response has the same form as Eq. (101) and the transfer function for x_1, now has the form,

$$H_x^{(1)}(\omega) = a_1 H(\omega). \tag{104}$$

The transfer functions for $(w - \dot{x}_1)$ and x_2 maintain the same form as H_v and $H_x^{(2)}$ in Eq. (102). Finally, note in Eq. (100) and Eq. (103) the presence of terms containing $(u - \dot{x}_1)^2$ and $(w - \dot{x}_1)^2$ which are squares of Gaussian fluid-structure interaction processes and include nonlinear damping terms.

As an example of a situation in which application of cubicization is necessary, let us now treat the case where the system excitation is again given by Eq. (13), but the current, U, is zero. The nonlinear drag term in this case is symmetric.

First, the splitting technique is employed by expanding the nonlinearity in a Taylor series and neglecting higher-order terms involving the third-order response velocity,

$$|u - \dot{x}|(u - \dot{x}) = |u - \dot{x}_1|(u - \dot{x}_1) + 2|u - \dot{x}_1|\frac{\dot{x}_3}{6} + O(\dot{x}_3^2)$$

$$\approx |u - \dot{x}_1|(u - \dot{x}_1) - \mu_{fv}\frac{\dot{x}_3}{6}, \tag{105}$$

where,

$$\mu_{fv} = -2E[|u - \dot{x}_1|] = -\sqrt{\frac{8}{\pi}}\sigma_v .$$

Then, applying cubicization to the nonlinear leading term in Eq. (105) yields,

$$|u - \dot{x}_1|(u - \dot{x}_1) \approx a_1(u - \dot{x}_1) + \frac{a_3}{6}(u - \dot{x}_1)^3, \tag{106}$$

for which mean-square minimization of the error term yields,

$$a_1 = \sqrt{\frac{2}{\pi}}\sigma_v \text{ and } a_3 = \frac{1}{\sigma_v}\sqrt{\frac{8}{\pi}},$$

and the relative velocity, $v = u - \dot{x}_1$.

The equivalent Volterra system representing the original nonlinear equation thus becomes,

$$M\ddot{x}_1 + (C + a_1)\dot{x}_1 + Kx_1 = K_m\ddot{u} + a_1 u,$$

$$M\ddot{x}_3 + (C + 2a_1)\dot{x}_3 + Kx_3 = a_3(u - \dot{x}_1)^3, \tag{107}$$

where the cubicization coefficients are redefined as $a_1 = K_d a_1$ and $a_3 = K_d a_3$. This representation yields transfer functions for the first- and third-order responses in the frequency domain as follows,

$$H_x^{(1)} = (K_m i\omega + a_1)H_1(\omega),$$

$$H_x^{(3)}(\omega_1, \omega_2, \omega_3) = a_3 H_3(\omega_1 + \omega_2 + \omega_3)H_v(\omega_1)H_v(\omega_2)H_v(\omega_3), \tag{108}$$

where,

$$H_v(\omega) = 1 - i\omega H_x^{(1)}(\omega),$$

$$H_1(\omega) = [K - \omega^2 M + i\omega(C + a_1)]^{-1},$$

$$H_3(\omega) = [K - \omega^2 M + i\omega(C + 2a_1)]^{-1}.$$

These expressions may be used in tandem with the cumulant relations given earlier in this chapter to develop the statistics of the response.

The TLP is modelled as a single degree of freedom system with structural and added mass, $M = 7.1286\times10^7$ kg, stiffness, $K = 2.8143\times10^5$ N/m and a structural damping ratio, $C/2M\omega_N$, of 0.05. Also, $K_m = 4\times10^7$ and $K_d = 6\times10^5$ are the inertia and drag coefficients, respectively. First consider the system, Eq. (100), only. The input is described by a Pierson-Moskowitz spectrum characterized by a significant wave height of

12 m and a peak frequency of 0.363 rad/s, well above the resonance region for the TLP surge mode. Nonetheless, Figures 1 and 2 indicate a surge response peak due to the second order forces in the resonance region of the TLP which is captured by the quadratization technique in the first figure, but is not seen in the response obtained from linearization. The same figures also illustrate that the presence of currents increases the quadratic contribution. Figure 2, which illustrates the case when no current is present and $K_d = 7.5\times10^5$, reveals the limitation of the quadratization technique. That is, the procedure degenerates to linearization when the nonlinearity becomes statistically symmetric. Nevertheless, this type of higher-order response energy may be captured by a cubicization approach[19,29] as will be illustrated momentarily. Figure 3 outlines the relationship between the current speed and the additional hydrodynamic damping imparted to the system. Figures 4 and 5 offer a comparison of the higher-order cumulants obtained via the present technique to those obtained via numerical simulation for a range of current speeds. Discrepancies in these figures may be accounted for by the realization that the quadratization technique does introduce an approximation as well as the fact that at lower current speeds, the wave drag force behaves in a more statistically symmetrical manner. Hence, at lower current speeds, quadratization becomes an increasingly less effective approximating technique.

It has been further observed that application of Kac-Siegert approach, and Newman's approximation[62], to obtain the higher-order statistics of the response to Eq. (100) yields results similar to those observed in Figures 4 and 5. The similarity of all results indicates that Newman's approximation, which ignores sum frequency contributions of the second-order force, is adequate for analyzing TLP surge response due to wave loads. Figure 6 illustrates the important result that the inertia coefficient controls the linear response, while the drag coefficient has more significant bearing on the quadratic response. As a result, Figure 7 shows that as the drag coefficient increases the higher-order response cumulants increase, as is expected when the response becomes increasingly non-Gaussian.

Figure 8 is a schematic of a TLP under the influence of concurrent wind and wave loadings, i.e. a combination of Eq. (100) and Eq. (103) as described in the earlier discussion. For this example, a mean wind speed of 20 m/s and $K_w = 1250$ are assumed. The wind spectrum is modeled according to Ref. 44 with a stress coefficient of 0.15. The combined response cumulants due to both wind and wave loads compare acceptably to those obtained via numerical simulation in Figures 9 and 10, thus supporting the use of the simplified, wind-wave combination model proposed in this study. Comparing the spectrum of Figure 12 to that of Figure 1, it is observed that the dynamics of the wind produce a significant low frequency peak in the response of the combined system. Finally, Figure 12 presents the interesting result that the low frequency peak diminishes with increasing current speed, indicating the importance of the hydrodynamic damping introduced by larger currents on the wind load response of the system.

The response statistics obtained through application of the quadratization technique are useful in characterizing the overall distribution and crossing rates of the non-Gaussian response. Figure 13 is the probability distribution of the response due to the combination wind and wave loading. In this figure, the departure from Gaussianity is readily evident in

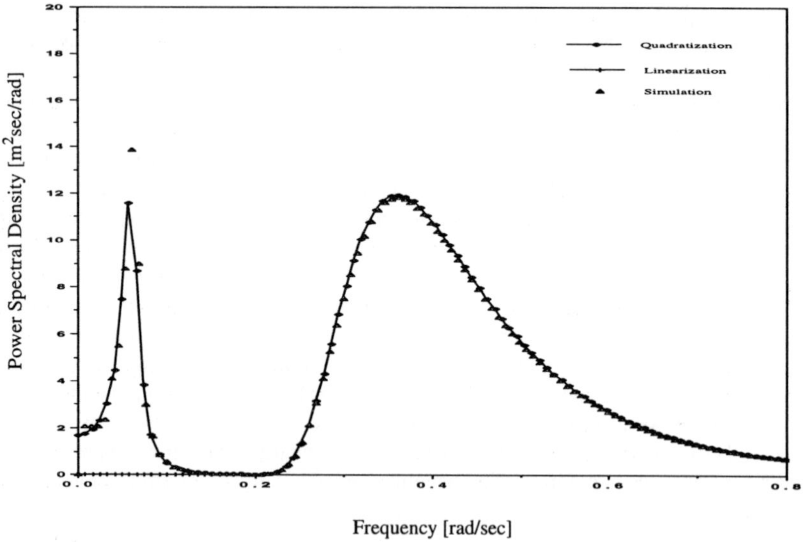

Figure 1: TLP response spectrum due to wave loading (U=1.0 m/s)

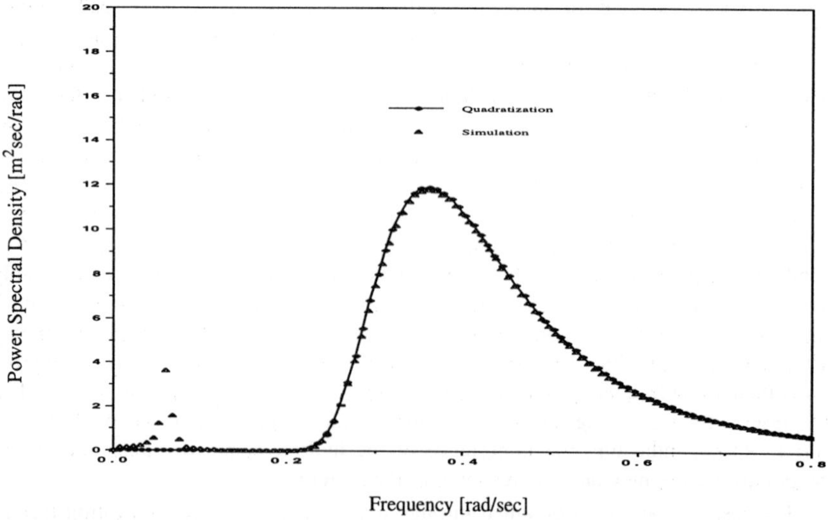

Figure 2: TLP response spectrum due to wave loading (U=0.0 m/s)

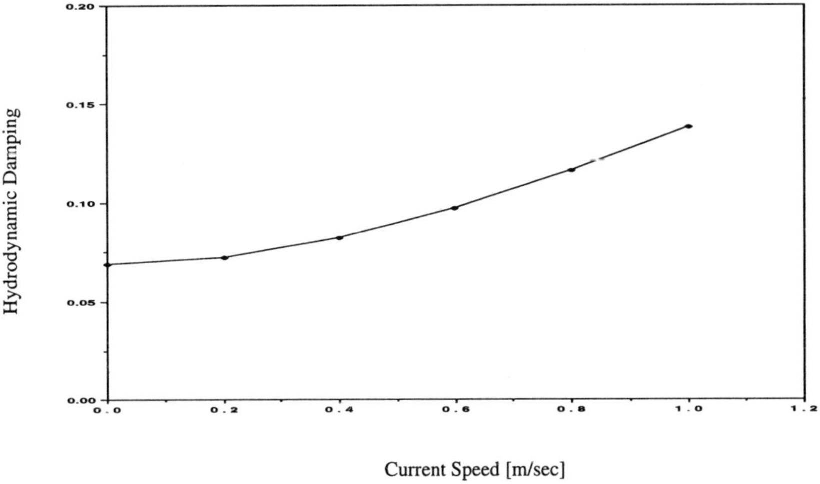

Figure 3: Variation of hydrodynamic damping with current speed

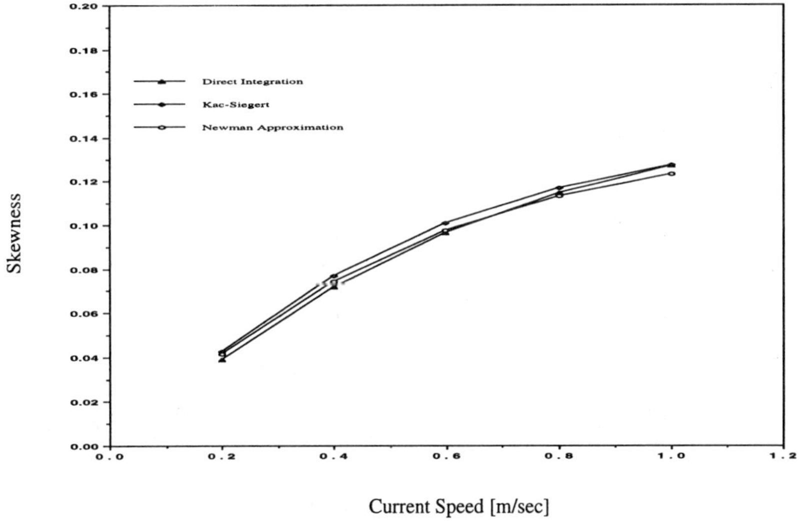

Figure 4: Skewness of TLP response due to wave loading as it varies with current speed

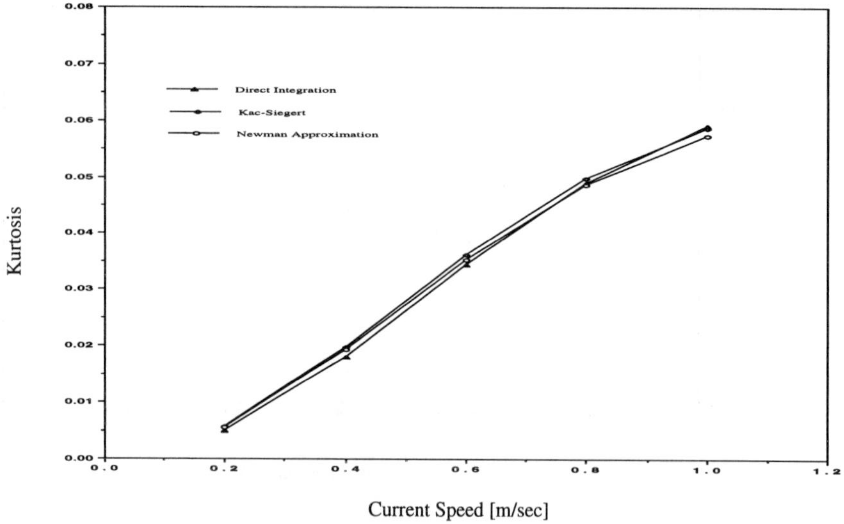

Figure 5: Kurtosis of TLP response due to wave loading as it varies with current speed

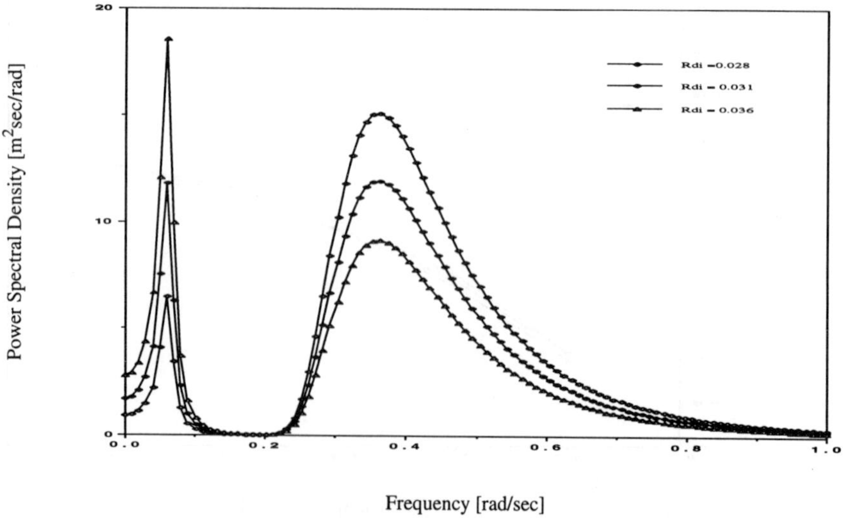

Figure 6: The influence of the drag-to-inertial force ratio on the TLP response spectrum

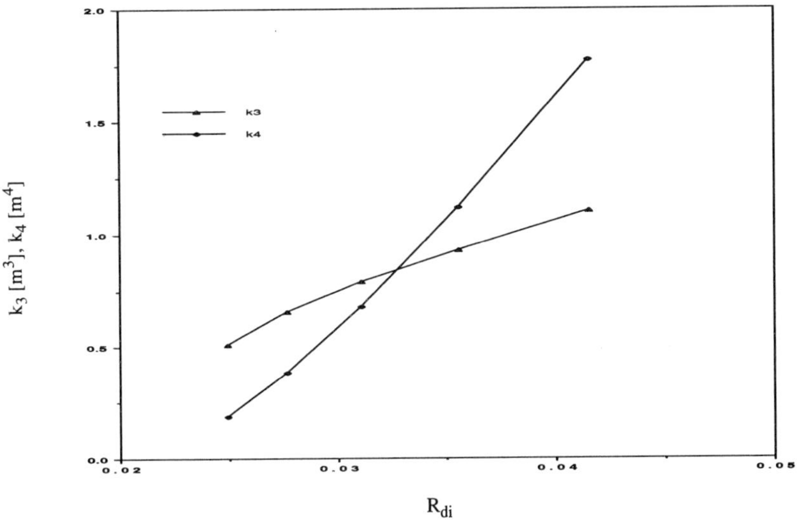

Figure 7: The influence of the drag-to-inertial force ratio on the TLP response cumulants

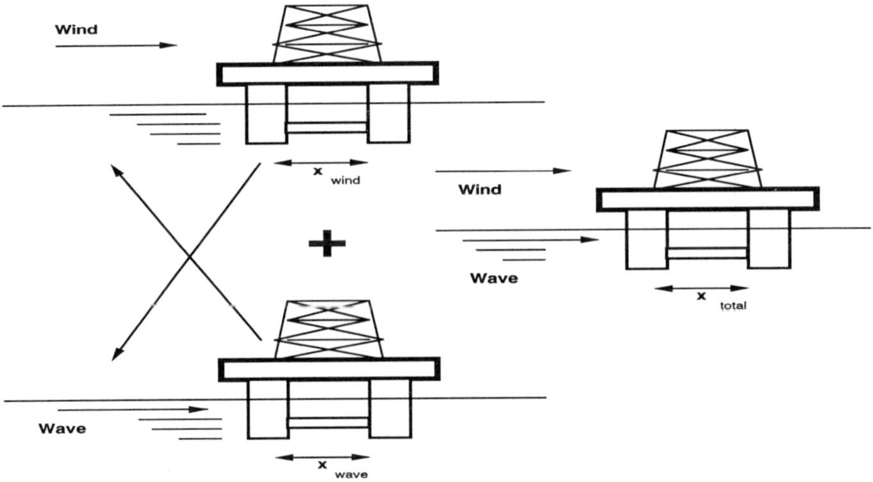

Figure 8: Schematic of wind-wave interaction on a TLP

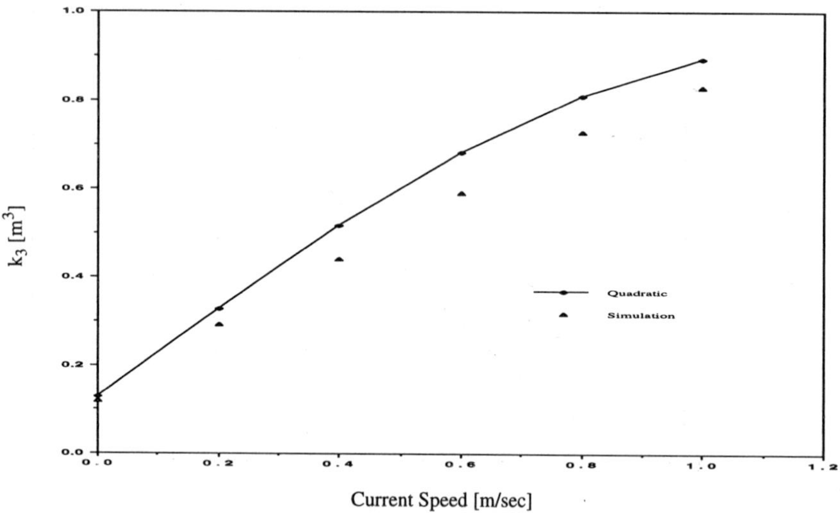

Figure 9: Third-order cumulant of TLP response due to combined wind-wave loading

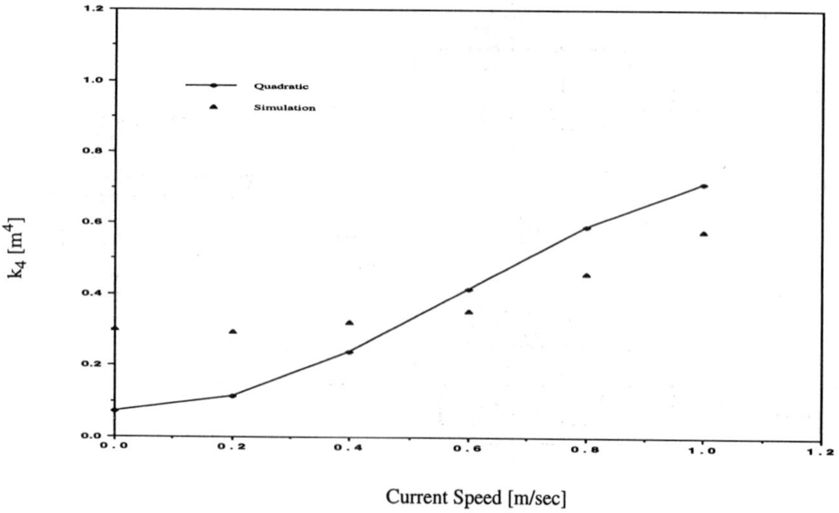

Figure 10: Fourth-order cumulant of TLP response due to combined wind-wave loading

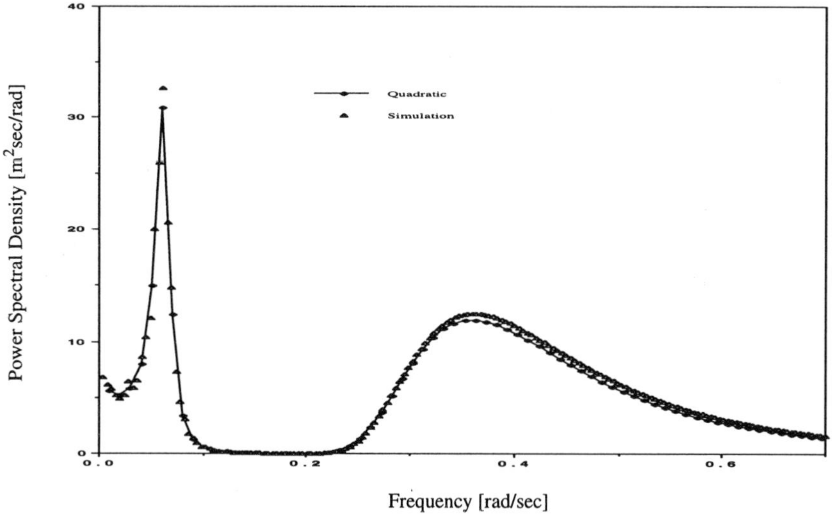

Figure 11: TLP response spectrum due to combined wind-wave loading (U=0.4 m/s)

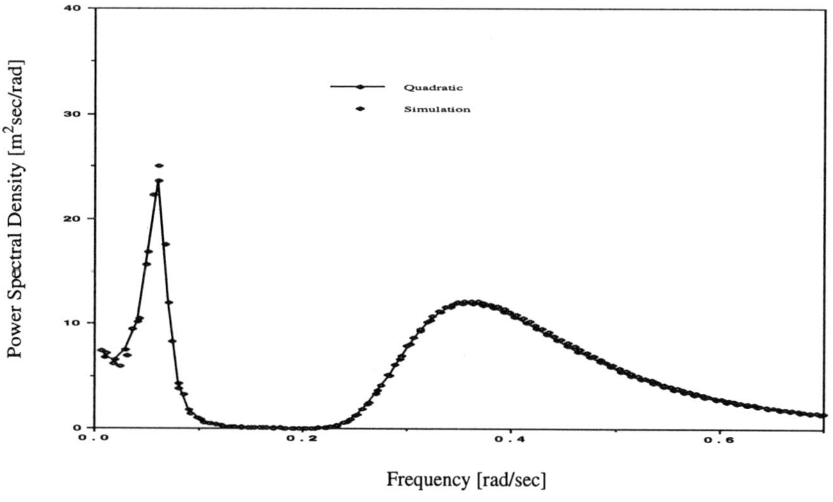

Figure 12: TLP response spectrum due to combined wind-wave loading (U=1.0 m/s)

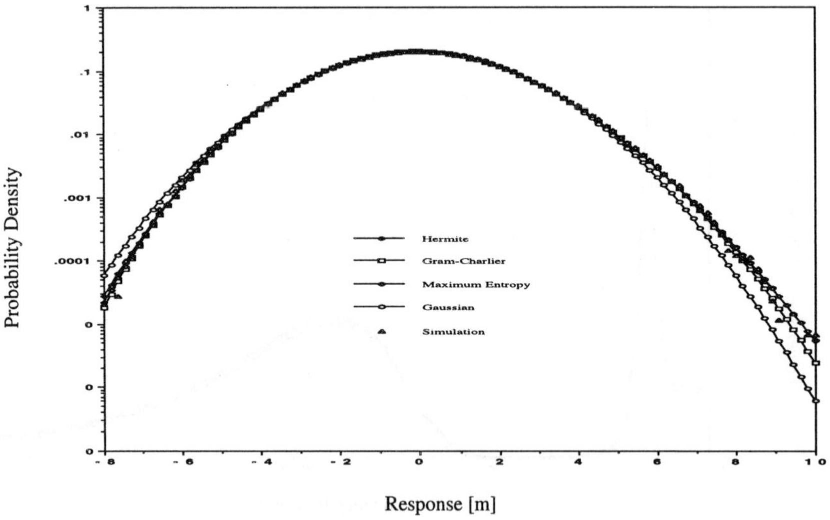

Figure 13: PDF of TLP response due to combined wind-wave loading

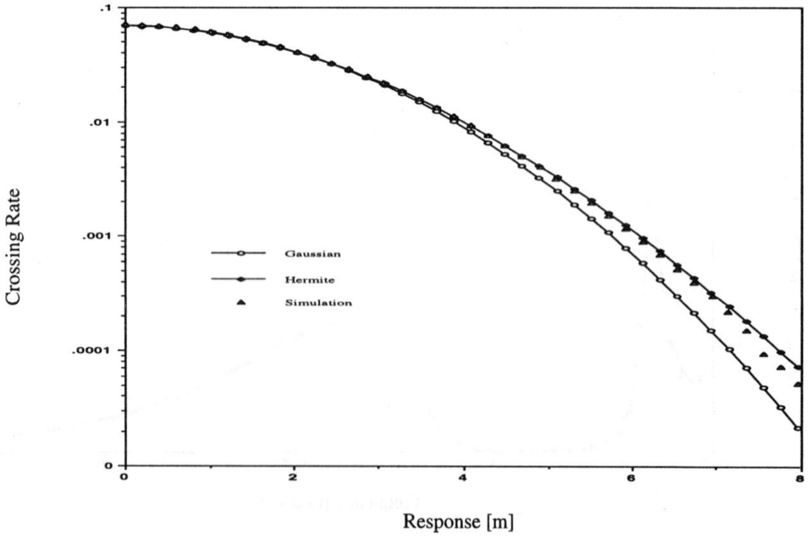

Figure 14: Crossing rates of the TLP response process due to combined wind-wave loading

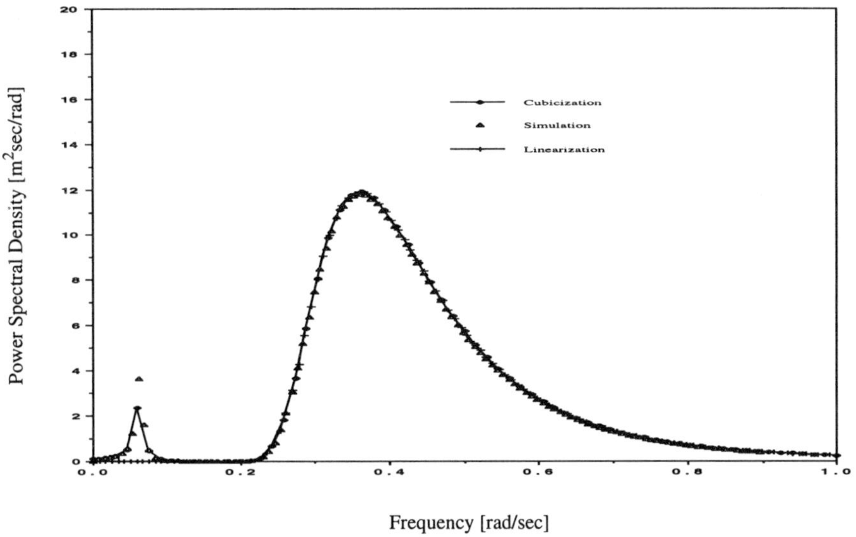

Figure 15: TLP response spectrum due to wave loading (U=0.0 m/s)

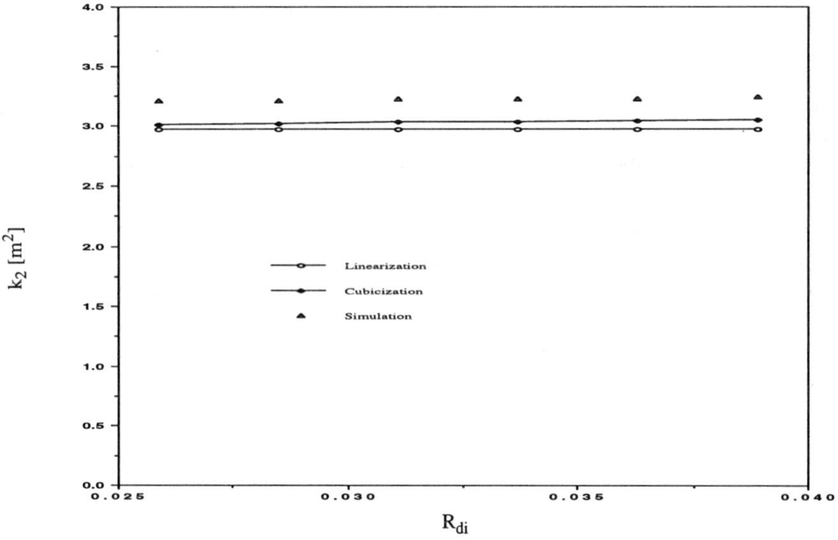

Figure 16: Second-order response cumulant due to wave loading (U=0.0 m/s)

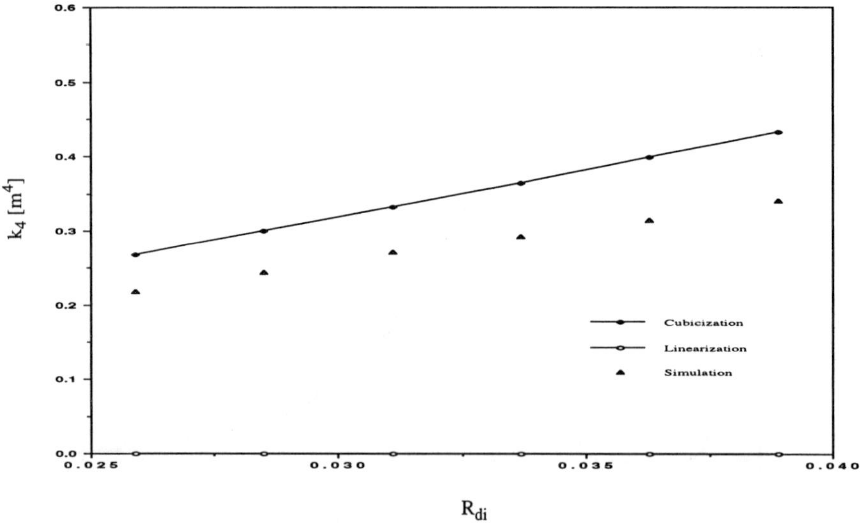

Figure 17: Fourth-order response cumulant due to wave loading (U=0.0 m/s)

the tails of the distribution and is captured by including the higher-order cumulant information gleaned from the present techniques in the probability density approximation. The best approximations seem to be given by the moment-based Hermite method and the maximum entropy method. The ability to characterize the tail contributions of these probability density functions will greatly enhance the accuracy of the prediction of response extremes. Indeed, Figure 14 illustrates favorable approximation of the crossing rate of the system response to wind-wave loads by the moment-based Hermite approximation.

Finally, Figures 15-17 illustrate the usefulness of equivalent statistical cubicization fora addressing statistically symmetric nonlinearities. Figure 15 indicates that for the same TLP characteristics treated in Figure 2, the cubicization technique is able to closely match the low frequency peak in the power spectral density of the response which was not captured by the cubicization technique. The final two figures highlight the improvements over linearization introduced by the equivalent statistical cubicization technique in determining the second- and fourth-order response cumulants over a range of drag-to-inertia force ratios.

4. Concluding Remarks

The approaches presented in this chapter address the treatment in the frequency domain of two classes of nonlinearities that arise from wind and wave loadings on TLPs. For several illustrative examples, the results obtained in terms of TLP response power spectral densities and cumulants are in good agreement with simulation results. The higher-order cumulants are used to determine the response probability distributions and crossing rates using available approximating techniques. The subsequently derived response distributions are also in good agreement with numerically simulated results. It is finally important to note that because of additional approximations introduced within the cubicization framework, this technique may be more sensitive to the degree of nonlinearity present in the particular problem considered. For this reason, it is strongly encouraged that the reader carefully assess each specific scenario before applying any approximate technique.

5. Acknowledgments

The support for this study was provided in part by NSF grant BCS-9096274 and ONR Grant No. 00014-93-1-0761. The third and fourth authors were supported in part by Department of Education GAANN Fellowships during the preparation of this chapter.

6. References

1. J.R. Morison, M.P. O'Brien, J.W. Johnson and S.A. Schaaf, *Petroleum Transactions, AIME* **189** (1950) 149-57.
2. Chakrabarti, S.K. *Hydrodynamics of Offshore Structures* (Springer-Verlag, New York, 1987).
3. T. Sarpkaya and M. Issacson, *Mechanics of Wave Forces on Offshore Structures* (Van Nostrand Reinhold, New York, 1981).
4. Y.K. Lin, *Probabilistic Theory of Structural Dynamics* (Robert E. Krieger Publishing Company, Huntington, N.Y., 1976).
5. J.B. Roberts and P.D. Spanos, *Random Vibration and Statistical Linearization* (John Wiley & Sons, New York, 1990).
6. T.T. Soong and M. Grigoriu, *Random Vibration of Mechanical and Structural Systems* (Prentice-Hall, Inc., Englewood Cliffs, N.J., 1993).
7. L.E. Borgman, *Journal of Waterways and Harbors Division, ASCE* **93** (1967) 129-56.
8. R. Eatock-Taylor and A. Rajagopalan, *Earthquake Engineering and Structural Dynamics* **11** (1983) 831-42.
9. H. Tuah and R.T. Hudspeth, *Applied Ocean Research* **5** (1983) 63-68.
10. M. Grigoriu, *Journal of Engineering Mechanics, ASCE* **110** (1984) 1731-42.
11. S.K. Thampi and J.M. Niedzwecki, *Journal of Engineering Mechanics, ASCE* **118** (1992) 942-60.
12. Y. Yuan and C.C. Tung, *Ocean Engineering* **11** (1984) 593-607.
13. A. Kareem and Y. Li, *Technical Report No. UHCE88-18* (Dept. of Civil Engrg., University of Houston, Tex., 1988).
14. Y. Li and A. Kareem, *Journal of Wind Engineering and Industrial Aerodynam-*

ics **36** (1990) 915-20.

15. M.G. Donley and P.D. Spanos, *Dynamic Analysis of Non-linear Structures by the Method of Statistical Quadratization, Lecture Notes in Engineering* (Springer-Verlag, New York, 1990).
16. P.D. Spanos and M.G. Donley, *Proceedings of the 9th OMAE Conference, ASME* (1990).
17. P.D. Spanos and M.G. Donley, *Journal of Engineering Mechanics, ASCE* **117** (1991) 1289-1309.
18. A. Kareem and C.C. Hsieh, *Technical Report CEND 91-1* (Department of Civil Engineering, University of Notre Dame, 1991).
19. A. Kareem and J. Zhao, *Proceedings of the OMAE '94 Conference, Vol. I, ASME* (1994).
20. A. Kareem and J. Zhao, *Technical Report No. NDCE93-002* (Dept. of CE/ GEOS, University of Notre Dame, 1993).
21. A. Kareem and J. Zhao, *Journal of Offshore Mechanics and Arctic Engineering* **116** (1994) 137-44.
22. J. Hu and L.D. Lutes, *Journal of Engineering Mechanics, ASCE* **113** 196-209.
23. L.E. Borgman, *Proceedings of the Ocean Structural Dynamics Symposium '82* (1982).
24. M. Olagnon, M. Prevosto and P. Joubert, *Journal of Offshore Mechanics and Arctic Engineering, ASME* **110** 278-81.
25. M. Grigoriu and S.T. Ariaratnam, *Journal of Applied Mechanics, ASME* **55** 905-10.
26. S. Krenk and H. Glover, *Stochastic Structural Dynamics: Progress in Theory and Application* (1988).
27. K. Sobczyk, *Stochastic Differential Equations with Applications to Physics and Engineering* (Kluwer Press, 1990).
28. J.S. Bendat, *Nonlinear System Analysis and Identification from Random Data,* (John Wiley and Sons, New York, N.Y., 1990).
29. M.A. Tognarelli, A. Kareem, J. Zhao and K.B. Rao, *Proceedings of the Tenth ASCE Engineering Mechanics Specialty Conference* (1995).
30. E. Bedrosian and S. Rice, *Proceedings of IEEE* **59** (1971) 1688-707.
31. M. Kac and A. Siegert, *Journal of Applied Physics* **18** (1947) 1688-707.
32. S. Winterstein, *Journal of Engineering Mechanics* **111** (1985) 1291-95.
33. L.R. Mead and N. Papanicolaou, *Journal of Math. Phys.* **25** (1984) 2404-17.
34. K. Sobczyk and J. Trebicki, *Probabilistic Engineering Mechanics* **5** (1990) 102-10.
35. N. Wiener, *Report No. 129, Radiation Laboratory* (MIT, Cambridge, Mass., 1942).
36. J.F. Barrett, *Journal of Electronics and Control* **15** (1963) 567-615.

37. M. Schetzen, *The Volterra and Wiener Theories of Nonlinear Systems* (John Wiley & Sons, New York, 1980).
38. A. Kareem. *Offshore Technology Conference* (1980).
39. A. Kareem and C. Dalton, *Proceedings of the Ocean Structural Dynamics Symposium '82* (1982).
40. N. Salvesen, et al., *Offshore Technology Conference* (1982).
41. E. Simiu and S.D. Leigh, *Journal of Structural Engineering, ASCE* **110** (1984) 785-802.
42. A. Kareem, *Journal of Wind Engineering and Industrial Aerodynamics* **14** (1983) 345-56.
43. B.J. Vickery, *Journal of Wind Engineering and Industrial Aerodynamics* **36** (1990) 905-14.
44. A. Kareem, *Journal of Structural Engineering*, **111** (1985) 37-55.
45. P. Krée and C. Soize, *Mathematics of Random Phenomena* (D. Reidel Publishing Company, Holland, 1983).
46. A. Kareem, *Journal of Engineering Mechanics, ASCE* **110** (1984) 1573-78.
47. K.R. Gurley and A. Kareem, *Applied Ocean Research* **15** (1993) 137-54.
48. J.N. Newman, *Proc. Int. Symp. on Dynamics of Marine Vehicles and Offshore Structures in Waves* (1974) 182-86.
49. J.A. Pinkster, *Ocean Engineering* **6** (1979) 593-615.
50. Y. Li and A. Kareem, *Journal of Offshore Mechanics and Arctic Engineering, ASME* **114** (1992) 175-84.
51. Y. Li and A. Kareem, *Applied Ocean Research* **15** (1993a) 63-83.
52. Y. Li and A. Kareem, *Journal of Engineering Mechanics, ASCE* **119** (1993b) 91-112.
53. Y. Li and A. Kareem, *Journal of Engineering Mechanics, ASCE* **119** (1993c) 161-83.
54. A. Kareem and Y. Li, *Probabilistic Engineering Mechanics* **9** (1994) 1-14.
55. S.R. Winterstein, T. C. Ude and T. Marthinsen, *Journal of Engineering Mechanics, ASCE* **120** (1994) 1369-85.
56. A. Kareem, A.N. Williams and C.C. Hsieh, *Ocean Engineering* **21** (1994) 129-54.
57. H.O. Madsen, S. Krenk and N.C. Lind, *Methods of Structural Safety* (Prentice-Hall, New York, 1986).
58. A. Papoulis, *Probability, Random Variables, and Stochastic Processes* (McGraw-Hill, Tokyo, 1965).
59. J.A. Cochran, *The Analysis of Linear Integral Equations* (McGraw-Hill, New York, 1972).
60. E. Neal, *Applied Ocean Research* **8** (1972).
61. T. Vinje, *International Shipbuilding Progress* **30** (1983) 58-68.

62. A. Naess, *Journal of Ship Research* **29** (1985) 270-84.
63. A. Naess and J.M. Johnsen, *Journal of Offshore Mechanics and Arctic Engineering, ASME* **114** (1992) 195-204.
64. R.S. Langley and S.A. McWilliams, *Applied Ocean Research* **15** (1993) 13-23.
65. V.S. Pugachev, *Probability Theory and Mathematical Statistics for Engineers* (Pergamon Press, 1984).
66. M.K. Ochi, *Probabilistic Engineering Mechanics* **1** (1986) 28-39.
67. M. Grigoriu, *Journal of Engineering Mechanics, ASCE* **110** (1984) 610-20.
68. E.T. Jaynes, *Physical Review* **106** (1957) 620-30.
69. A. Tagliani, *Probabilistic Engineering Mechanics* **5** (1991) 167-70.
70. M. Shinozuka and C-M. Jan, *Journal of Sound and Vibration* **25** (1972) 111-28.
71. Y. Li and A. Kareem, *Journal of Engineering Mechanics, ASCE* **119** (1993) 1078-98.

Uncertainty Modeling in Vibration and Fuzzy Analysis of Structural Systems, pp. 147–165
edited by B. M. Ayyub, A. Guran and A. Haldar
Series on Stability, Vibration and Control of Systems Series B: Vol. 10

ELEMENTS OF FUZZY STRUCTURAL CONTROL

LUCIA FARAVELLI
Department of Structural Mechanics, University of Pavia, via Abbiategrasso 211
27100 Pavia, Italy

and

TIMOTHY YAO
Visiting Researcher, Department of Structural Mechanics, University of Pavia

ABSTRACT

The motivations for a non-standard form of structural control in civil engineering are illustrated with reference to environmental actions like wind and earthquake loading. Fuzzy control is robust, efficient and can be implemented as an adaptive strategy that can react to changes in system parameters. The theory underlying the fuzzy inference scheme is explained in light of the implementation of a control device using displacement and velocity feedback and producing a crisp control force. This paper summarizes the results of ongoing research aimed at the implementation of Fuzzy Logic Controllers (FLC) in active structural control. The numerical applications cover linear and nonlinear systems with one or several degrees of freedom.

1. Introduction

A central problem of modern civil engineering is the retro-fitting of structures that must be used beyond their original design lifetime or that have been exposed to severe environmental loads. This includes historically-valuable structures that require strengthening to improve the overall system reliability. The use of active control through tendons or actuators is a viable alternative to conventional techniques that might require the use of more unsightly bracing to achieve the same level of strengthening. Active structural control can also be used to counteract potentially severe environmental loads acting upon civil engineering structures of new design.

The area of structural control has matured considerably since being formalized by Yao [27]. In particular, a large amount of literature has been produced to testify the ongoing research effort into active control theory as applied to civil engineering systems [21,25,12,13]. Active structural control consists of three main steps:

1. identification of changes to the structure's current state due to modifications of systems parameters and to external disturbances such as earthquakes or winds;

2. compensation for internal and external disturbances;

3. learning and reasoning; i.e., the reaction strategy must be updated on the basis of past successes/failures.

For this purpose, given a structural system subject to external forces, a number of sensors must be incorporated into the structure to monitor the deflection and acceleration at joints or the strain at internal points. Actuators, tendons or other control devices can be used to actively modify the stiffness and damping characteristics of the structure. A control unit must eventually use the sensor readings to manipulate the actuators to modify in real time the behavior of the structure subjected to a dynamic action.

Indeed, control literature in civil engineering is very rich in algorithms [12 21 25] to solve the equations that govern the system behavior. Many of these algorithms, however, require an exact knowledge of the system and do not apply in the presence of an imprecise description. Moreover, within the area of active control strategies, most research has focused upon linear control [22]. In particular, the popular LQ control strategy (linear optimal control with quadratic cost function) has been shown to be deficient when systems move significantly into the nonlinear range of behavior. The unpredictability of natural hazards such as earthquakes and hurricanes makes it necessary to examine how active control strategies react within the nonlinear range of behavior [18]. In addition, control systems should be able to accommodate noisy input measurements, uncertainty in system parameter values and possible changes to the system. Few active control strategies, however, can deal adequately with a lack of exact knowledge of system parameter values or with nonlinear behavior. One promising strategy is the use of fuzzy control [3,4,5,7,8,28,1,11] for its inherent robustness and its ability to deal with linear and nonlinear behaviour of the structure. Imprecise linguistic descriptions of system conditions (e.g., the velocity is slightly negative and the displacement is somewhat positive, so apply a small force in the negative direction) can be used as the basis for activating control forces through the mathematical rules created by Zadeh [29]. Fuzzy set theory provides a mathematical structure for resolving imprecise or uncertain information that can be presented in fuzzy terms. This information is usually provided by an expert. The tuning of fuzzy controllers (generally accomplished through adjustments to the membership functions) is not always intuitive, however, and would likely benefit from automatic processing by neural networks [14] or genetic algorithms [26]. In addition fuzzy control does not require a mathematical model of the process being controlled and provides a structural control that is capable of adapting to changing environmental conditions on the basis of sensor data.

Another problem that should be considered in structural control is that of time delay, which is usually due in large part to the need to generate large control forces. The effect of time delay between the computed control action at time t and the actual time at which it becomes operative generally affects the optimality of a linear controller adversely due to the introduced nonlinearity.

The use of fuzzy controllers is not restricted to linear systems and optimality criteria are not applied. Nevertheless, the design of a fuzzy controller can be improved by using a neural network, which can be trained to compensate for time delay. The network is asked to predict a future control force based upon current response information: this generally requires longer learning times but is effective in compensating for the time delay.

There are no standard schemes for comparing the effectiveness of structural control strategies for nonlinear systems quantitatively. Qualitatively, each strategy has some known strengths and weaknesses. For example, neural networks are known to be computationally fast in execution but slow in training. Fuzzy control can make use of engineering judgement and human decision-making paradigms but has no built-in and systematic method of tuning. Development of a new standard for quantitative comparison of strategies was planned but not yet conducted at the time of the writing of this chapter.

2. Fuzzy control

Fuzzy control converts a linguistic control strategy based upon expert knowledge into an automatic control strategy [16]. The operations of a fuzzy controller comprise four parts:

1. *fuzzification interface*: the state variables to be monitored during the process must be measured. These values are fuzzified using fuzzy linguistic terms defined by the membership functions of the fuzzy sets, which are defined on an appropriate universe of discourse;
2. *knowledge base*: these linguistic terms are then used in the evaluation of the fuzzy control rules
3. *decision making logic*: the result of the application of these rules is a fuzzy set defined on the universe of possible control actions
4. *defuzzification interface*: the crisp control action is produced.

To design a fuzzy controller, one must specify the fuzzy sets and define their membership functions for each of the input and output variables. One can then compile the heuristic control and determine a method for selecting a crisp output action. In the research work presented here, Larsen's product rule is used for the combination of the fuzzy values and the center of gravity method (COG) is used for the defuzzification. These steps are discussed briefly in the following sub-sections.

2.1 Fuzzification interface

The first step is to assign admissible linguistic values to each interested variable by a membership function $\mu(.)$. There are two methods depending on whether the universe of discourse is discrete or continuous:

numerical: a crisp value is translated into a fuzzy singleton

functional: the membership function can be expressed in a functional form as a triangle-shaped function, trapezoid, shaped function etc.

2.2 Knowledge base and decision making logic

Fuzzy control rules are evaluated by means of the compositional rules of inference. This is a fuzzy generalization of the modus ponens (GMP) rule for which:

premise 1: x is A'

premise 2: if x is A then y is B

consequence: y is B'

where A, A', B, B' are fuzzy sets defined on universal sets X, of inputs, and Y of outputs.

Various implication functions in which the antecedents and consequences contain fuzzy variables have been proposed; they can be classified into three categories: fuzzy conjunction, fuzzy disjunction and fuzzy implication. For each statement S, the fuzzy conjunction, which is more commonly used, is defined by $A \to B = A \times B$ specified by:

$$\mu_S(x, y) = \min(\mu_A(x), \mu_B(y)) \tag{1}$$

where

$$x \in X \text{ and } y \in Y$$

The relations S are then aggregated to form the rule-base relation R defined on $X \times Y$; the grade of membership of (x, y) in R is the maximum of its membership grades in any of the fuzzy relations S.

Then the resulting output fuzzy set B' is defined by

$$B' = A' \bullet R \tag{2}$$

where

A' is an actual input fuzzy set;

$\bullet$ is the sup-min compositional operator

The membership function $\mu_{B'}(y)$ is defined by

$$\mu_{B'}(y) = \sup_{x \in X} [\min(\mu_{A'}(x), \mu_{B'}(y))] \tag{3}$$

The fuzzy values of the output variable generated by the whole set of rules are combined by a union operator to derive the final fuzzy value of the output.

2.3 Defuzzification

Many techniques have been proposed to perform the defuzzification, including:

- the max criterion method, which selects the point where the possibility distribution reaches its maximum;
- the mean of maximum method, which selects the mean value of all local control actions with maxima in the membership functions;
- the center of area method (COA), which computes the crisp value of the output as the center of area of its fuzzy distribution; and
- the center of gravity method (COG), which combines weighted centers of gravity to achieve the output crisp value from the component output membership functions.

COA is one of the most widely used techniques; COG is computationally simpler and more efficient.

3. Implementing a fuzzy controller

Routines from the MathWorks Control Toolbox [9] and the Delta Toolbox can be used to solve for the system response. Moreover, the Fuzzy Inference Systems Toolbox [17] provides the basic routines for performing the fuzzy calculations. Although computations were too slow, with this setup, to be used in an actual applications, there exist hardware implementations of fuzzy controllers that make fuzzy control feasible for real-time control.

The fuzzy control covers two input quantities (displacement and velocity) and one output (the control force). The fuzzy control can be considered *closed loop* control in that only the measured response of the system (i.e., the displacement and velocity) are used to determine the control force. Three descriptive fuzzy subsets for each variable are introduced: NE (negative), ZE (zero), and PO (positive). Trapezoidal membership functions map these fuzzy subsets to the appropriate universe of discourse. The combinations of input membership functions then yield nine fuzzy rules in the simple case of a single degree of freedom dynamic system. Figure 1 shows the combinations of membership functions corresponding to the nine rules.

Several types of inference operations upon fuzzy if-then rules (*fuzzy reasoning*) have been proposed to combine the membership value of the premise part to obtain the weight for each rule. Two types of fuzzy inference scheme have been implemented. The first uses Larsen's product rule to combine the membership values for each rule and the center-of-gravity (COG) approach to obtain the output crisp value. The COG defuzzification scheme can be described for discrete universes by the equations [20]

$$\begin{aligned} A_i &= \sum \mu_i(f_j) \\ z_i &= \frac{\sum \mu_i(f_j) f_j}{\sum \mu_i(f_j)} \end{aligned} \tag{4}$$

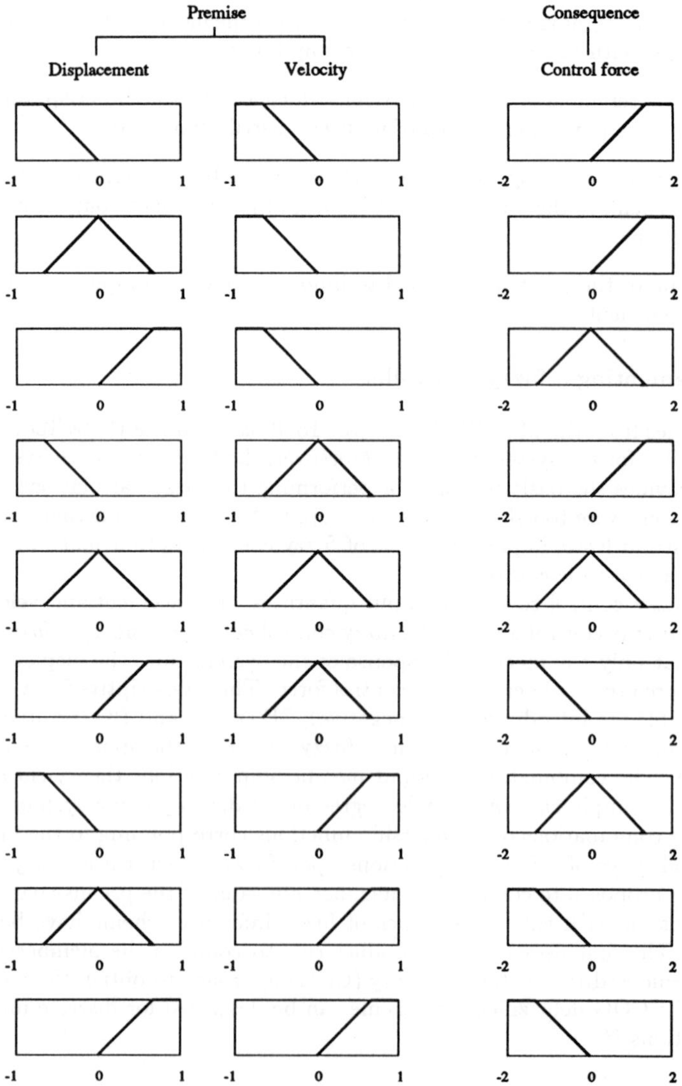

Fig. 1. Membership functions of the SDOF nine rules

$$\bar{z} = \frac{\sum z_i A_i}{\sum A_i}$$

where f denotes the consequent variable and $\mu_i(f_j)$ is the consequent membership function value at f_j arising from the i-th rule. Then, A_i is the area of the consequent membership function, z_i is the center of gravity of each consequent membership function and $\bar{z}$ is the crisp output value.

The second is the Takagi and Sugeno [23] inference system. This computes the fuzzy output for each rule as a linear combination of input variable membership values plus a constant term. For a system with two input variables and one output variable, a general Takagi and Sugeno rule can be expressed as

$$\text{Rule } i: \text{If } x \text{ is } A_i \text{ and } y \text{ is } B_i, \text{ then } f_i = p_i x + q_i y + r_i \tag{5}$$

The final crisp output is achieved using a weighted average, with the weights being obtained through multiplication and normalization of the membership values of the input variables for each rule.

4. Hybrid fuzzy-neural control system

One thorny issue that remains in the implementation of fuzzy control is the formation of appropriate fuzzy if-then rules and membership functions. In addition, static fuzzy rules and membership functions are vulnerable to changes in system parameters such as those occurring when accumulating structural damage [10]. A promising approach for the optimization of fuzzy controllers is the combination of fuzzy systems and artificial neural networks [19]. One possible combination is called hybrid fuzzy-neural control: it creates or modifies the fuzzy rulebase by using the structure of a neural network. This approach is capable of learning, resulting in changes of the initial design through creation or deletion of connections or adaptation of the weights to optimize the system. An example of the hybrid fuzzy-neural approach is the Adaptive-Network-Based Fuzzy Inference System (ANFIS), developed by Jang [14]. This is a fuzzy inference system that uses the structure of an artificial neural network (ANN) to improve the fuzzy inference computational efficiency and for system identification.

4.1. Network architecture

An example of ANFIS is shown in figure 2. It is a network structure consisting of nodes and relational links where some of the nodes are adaptive. The architecture of ANFIS can be fitted to different fuzzy inference schemes, for instance to the Takagi and Sugeno type. The membership function $\mu_{A_i}(x)$ is bell-shaped with maximum equal to 1 and minimum equal to 0. A possible choice is:

$$\mu_{A_i}(x) = \exp\{-[(\frac{x - c_i}{a_i})^2]^{b_i}\} \tag{6}$$

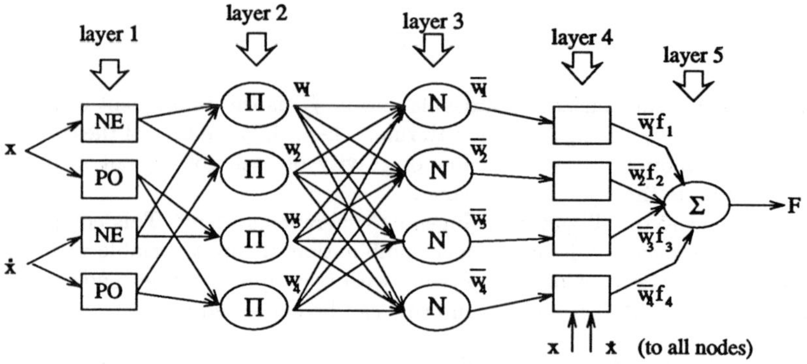

Fig. 2. ANFIS network architecture. Square nodes are adaptive; circular nodes are not.

where a_i, b_i and c_i represent the set of premise parameters.

For layer 1, every node is associated with a node function that is the input membership function

$$\hat{F}_i^1 = \mu_{A_i}(x) \tag{7}$$

where x is the input to node i, μ_{A_i} is the membership function, A_i is a linguistic label (e.g., NE=negative) and $\hat{F}_i^1$ is the output from this layer 1 node.
In layer 2, each node's output is the firing strength of the rule

$$w_i = \mu_{A_i}(x) \times \mu_{B_i}(y) \qquad i = 1, 2 \tag{8}$$

The outputs of layer 3 are the normalized firing strengths

$$\bar{w}_i = \frac{w_i}{w_1 + w_2} \qquad i = 1, 2 \tag{9}$$

In layer 4, one computes the output of each rule using the linear combination as given by Takagi and Sugeno:

$$\hat{F}_i^4 = \bar{w}_i f_i = \bar{w}_i(p_i x + q_i y + r_i) \tag{10}$$

where p_i, q_i and r_i are the set of consequent parameters. The final, crisp output is the sum of all inputs to this node

$$\hat{F}_1^5 = \sum_i \bar{w}_i f_i = \frac{\sum_i w_i f_i}{\sum_i w_i} \tag{11}$$

4.2. Learning from training data

The hybrid learning procedure is composed of a forward pass and a backward pass. In the forward pass the premise parameters are fixed and the consequent parameters (layer 4) are identified by a least-squares estimates. In the backward pass the error rates propagate from the output end toward the input end. For a training set of size P, the error measure for the p-th entry (with $1 \leq p \leq P$) is $E_p = (f_p - \hat{f}_p)^2$ where f_p is the target output and $\hat{f}$ is the network estimation; the overall error will be $E_{tot} = \sum E_p$. Minimization is performed using gradient descent over the parameter space and, in this way, the premise parameters are updated.

5. Numerical examples

The loading in all the following numerical examples is a seismic-like simulated time history record consisting of a filtered Gaussian white noise without time modulation (filter parameters: frequency 15.6 rad/sec; damping 0.6).

For simplicity, where it is not specified otherwise, there are three membership functions to describe each input and output variable (negative, zero and positive). The nine rules are formed through the combination of the two input variables.

All computations were performed using MATLAB 4.1 with the Control Toolbox [9] and the Fuzzy Toolbox [17]. Because of the nature of fuzzy control as implemented in the Fuzzy Toolbox, the control force is computed at each time step. Because the shape of the membership functions (e.g., trapezoidal, triangular or bell) does not significantly affect the effectiveness of the fuzzy controller, trapezoidal membership functions are used in all examples except for the case where membership functions are modified by ANFIS; in those cases, bell shaped membership functions yield a better result.

5.1. Basic Procedure

In general, the following steps are followed in writing the m-files for the fuzzy controller:

1. The universes of discourse are set for the fuzzy variables. This places limits on the domain used in the defuzzification to improve computational speed.

2. The membership functions are defined using TRAP.M or BELL.M from the fuzzy toolbox.

3. A table of rules is built up from the membership functions with each input and output variable represented by a matrix of fuzzy subsets.

4. The structural properties are defined.

5. The random loading is generated. In these numerical examples, this is a simulated seismic acceleration.

6. For each time step, the state vector is computed. In the case of the nonlinear system, ODE23.M is used.

7. State variables are fuzzified and a fuzzy decision is found using LARSEN.M from the fuzzy toolbox.
8. DELFZFIR.M is used to defuzzify the control forces which are then applied to the system.

5.2. Linear SDOF

The first numerical example comprises a linear, single-degree-of-freedom (SDOF) system subjected to simulated seismic activity. The emphasis of this example is upon implementing the fuzzy controller. The dynamic behavior of the plane frame is defined by its natural frequency and damping ratio and influenced by the intensity of the simulated seismic excitation. The damping ratio is kept constant at 0.0355. Two natural frequencies (4.14 and 2.14 radians/s) are considered. In addition, the peak acceleration a_p of the underlying white noise excitation is set to 0.45 and 0.90 m/s^2 to see how the fuzzy control reacts to changes in the load intensity. The universes of discourse are defined to be from -1 to 1 m for the displacement, from -1 to 1 m/s for the velocity and from -2 to 2 m/s^2 ($\times$ the mass of the plane frame) for the control force.

Figure 3 [4] shows a comparison of the controlled displacement with the uncontrolled displacement for combinations of the structural natural frequency and the peak acceleration of the filtered white noise excitation.

5.3. Nonlinear SDOF

In this example, the restoring force follows the smooth differential model developed by Wen [24] and modified by Casciati [2] to introduce hysteretic behavior:

$$\dot{z} = (1/\eta)[A\dot{x} - \nu(\beta|\dot{x}||z|^{n-1}z - \gamma\dot{x}|z|^n)] \tag{12}$$

where the equation of motion for the structure is given by:

$$\ddot{x} + 2\zeta\omega_n\dot{x} + \alpha{\omega_n}^2 x + (1-\alpha){\omega_n}^2 z = -\ddot{x}_g + F \tag{13}$$

F is the control force, and

$$\eta = A = n = 1; \alpha = 0.05; \text{ and } \beta = \gamma = 7$$

Figure 4 [3] shows a comparison of the controlled displacement with the uncontrolled displacement for combinations of the structural natural frequency and the peak acceleration of the white noise excitation. Figure 5 shows the hysteresis plots for the uncontrolled (left column) and controlled (right column) cases.

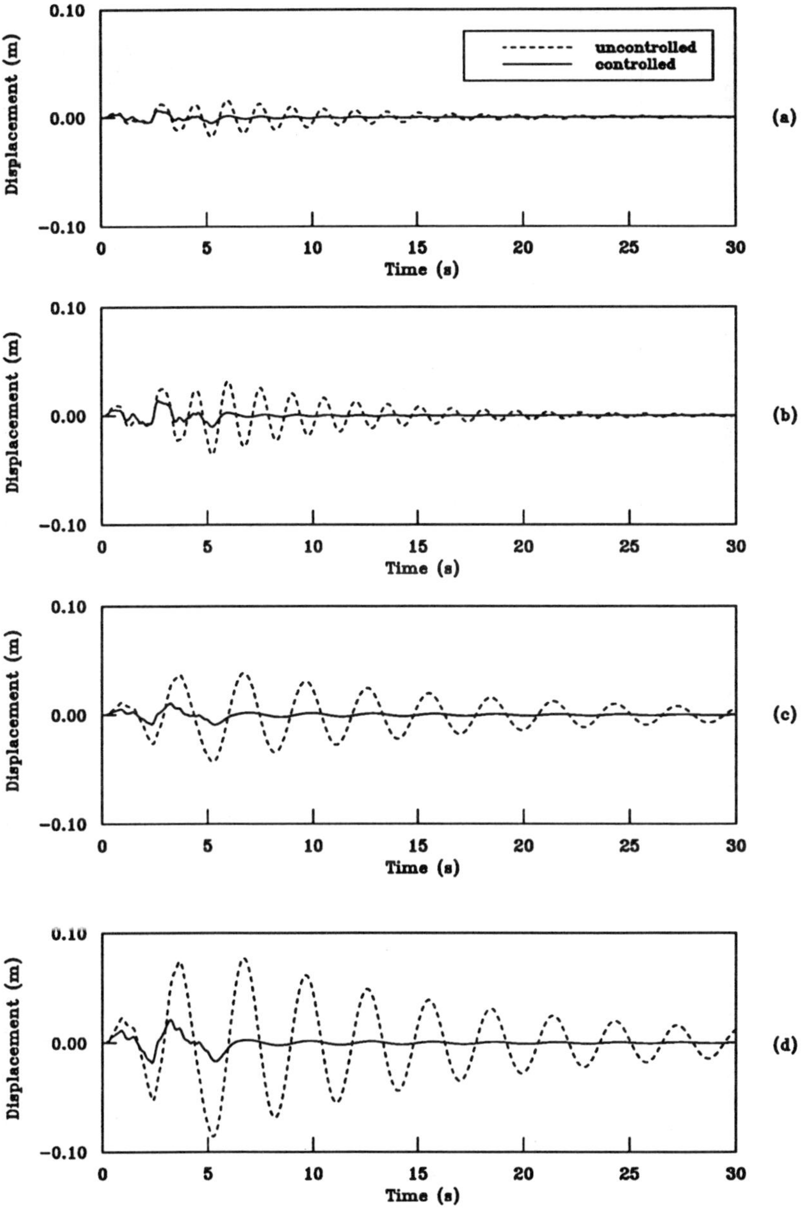

Fig. 3. Comparison of controlled and uncontrolled displacements, SDOF linear example: (a) $\omega_n = 4.14$ rad/s and $a_p = 0.45$ m/s^2; (b) $\omega_n = 4.14$ rad/s and $a_p = 0.90$ m/s^2; (c) $\omega_n = 2.14$ rad/s and $a_p = 0.45$ m/s^2; and (d) $\omega_n = 2.14$ rad/s and $a_p = 0.90$ m/s^2.

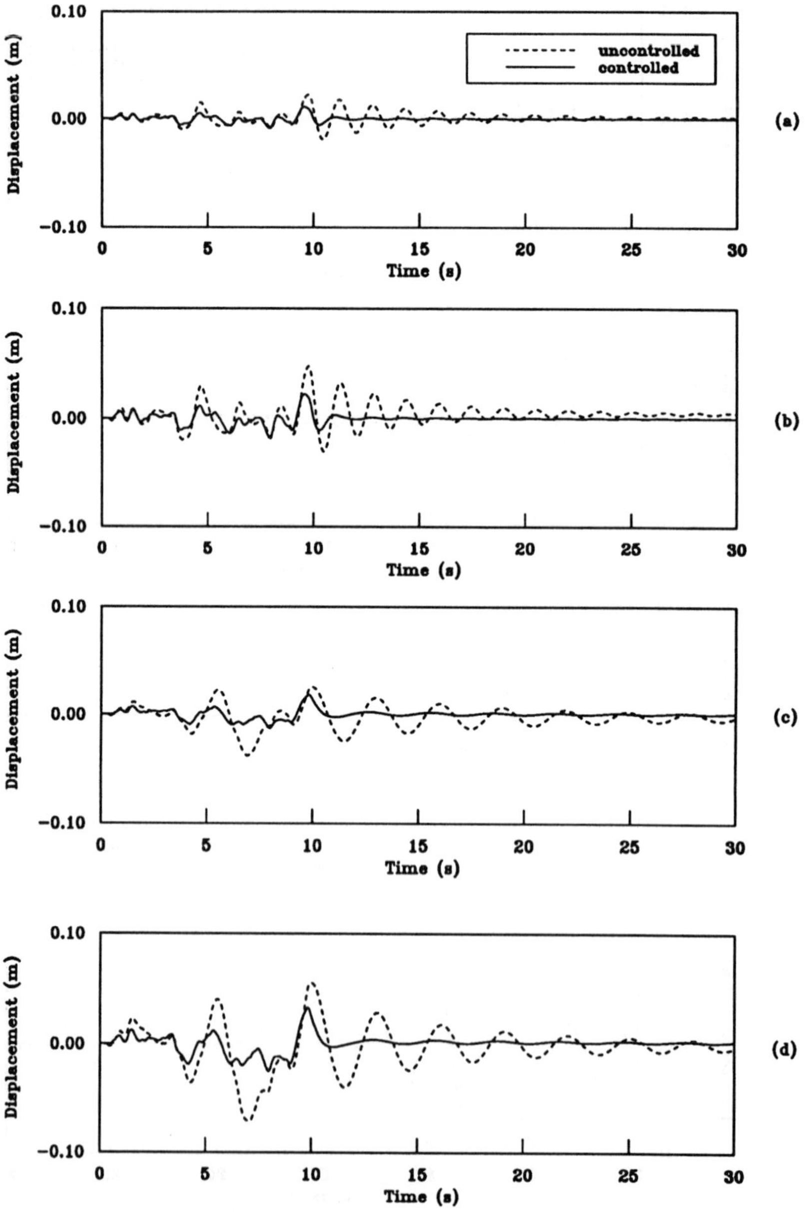

Fig. 4. Comparison of controlled and uncontrolled displacements, SDOF nonlinear example: (a) $\omega_n = 4.14$ rad/s and $a_p = 0.45$ m/s^2; (b) $\omega_n = 4.14$ rad/s and $a_p = 0.90$ m/s^2; (c) $\omega_n = 2.14$ rad/s and $a_p = 0.45$ m/s^2; and (d) $\omega_n = 2.14$ rad/s and $a_p = 0.90$ m/s^2.

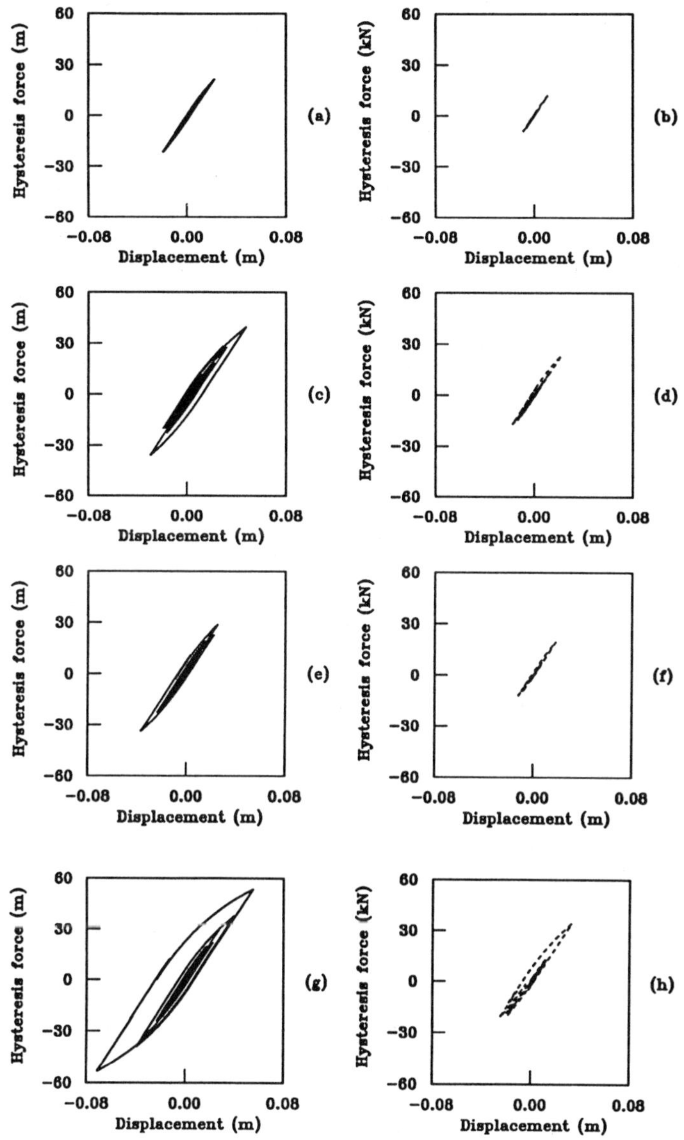

Fig. 5. Force-displacement plots for controlled and uncontrolled cases, SDOF nonlinear example: (a) $\omega_n = 4.14$ rad/s and $a_p = 0.45$ m/s^2, uncontrolled; (b) with control; (c) $\omega_n = 4.14$ rad/s and $a_p = 0.90$ m/s^2, uncontrolled; (d) with control; (e) $\omega_n = 2.14$ rad/s and $a_p = 0.45$ m/s^2, uncontrolled; (f) with control; (g) $\omega_n = 2.14$ rad/s and $a_p = 0.90$ m/s^2, uncontrolled; and (h) with control.

5.4. Linear 2DOF

The numerical example is a linear, two-degree-of-freedom (2DOF) system subjected to simulated seismic activity. The emphasis of this example is upon successful implementation of the fuzzy controller for a multi-degree-of-freedom (MDOF) system.

The fuzzy control covers two input quantities (displacement and velocity) and one output (the control force) for each degree of freedom for a total of six fuzzy variables. We consider three descriptive fuzzy subsets for each of these: NE (negative), ZE (zero), and PO (positive). NL (negative large) and PL (positive large) are added to the control force variable. Trapezoidal membership functions map these fuzzy subsets to the appropriate universe of discourse. The combinations of input membership functions then yield eighteen fuzzy rules. This approach, while cost effective, assumes that there is no explicit interaction between the actuators acting upon each degree of freedom (DOF); i.e., we assume that the state of one DOF does not directly influence the computation of the fuzzy control force of the other DOF. For this example problem, the other approach would require 81 rules to be complete. It is not easy to tune the membership functions or to determine the extent of coverage of each membership function within the domain of the universe of discourse. The dynamic behavior of the two-story plane frame is defined by the mass ratio of its stories (1.0), natural frequencies (2.6 rad/s and 6.7 rad/s) and damping ratio (0.0355), and it is influenced by the intensity of the simulated seismic excitation (0.9 m/s^2 peak acceleration of the underlying white noise). The load duration is 5 seconds. The universes of discourse are defined to be from -0.06 to 0.06 m for the displacement and from -0.1 to 0.1 m/s for the velocity for both degrees of freedom. Both story's universes of discourse range from -2.0 to 2.0 m/s^2 ($\times$ the mass of the appropriate story of the plane frame) for the control force.

Figure 6 [5] shows a comparison of the controlled and uncontrolled displacement time histories for the two degrees of freedom.

5.5. Nonlinear 2DOF

The numerical example is a two-degree-of-freedom (2DOF) system with hysteretic behavior subjected to simulated seismic activity. The dynamic behavior of the two-story plane frame is defined by the mass ratio of its stories (assumed to be equal), natural frequencies (2.6 and 6.7 rad/s) and damping ratios (both set to 0.0355), and the parameters of its hysteretic behavior. The Wen-Bouc hysteresis model [24] was used with the following values of the parameters:

$$\eta_1 = \eta_2 = A_1 = A_2 = n_1 = n_2 = 1; \alpha_1 = \alpha_2 = 0.05; \text{ and } \beta_1 = \beta_2 = \gamma_1 = \gamma_2 = 7$$

The simulated ground motion is defined by its duration (5 seconds) and the peak acceleration (0.9 m/s^2) of the white noise excitation. The universes of discourse are defined to be from -0.06 to 0.06 m for the displacement and from -0.1 to 0.1 m/s for the velocity for both degrees of freedom. Both story's universes of discourse go

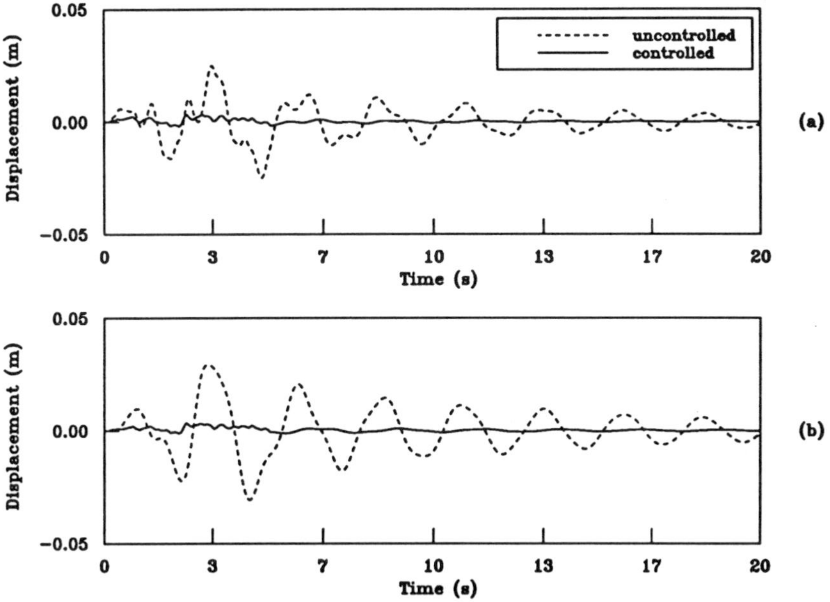

Fig. 6. Comparison of controlled and uncontrolled displacement and velocity time histories, 2DOF linear example: (a) first story displacements; (b) second story displacements.

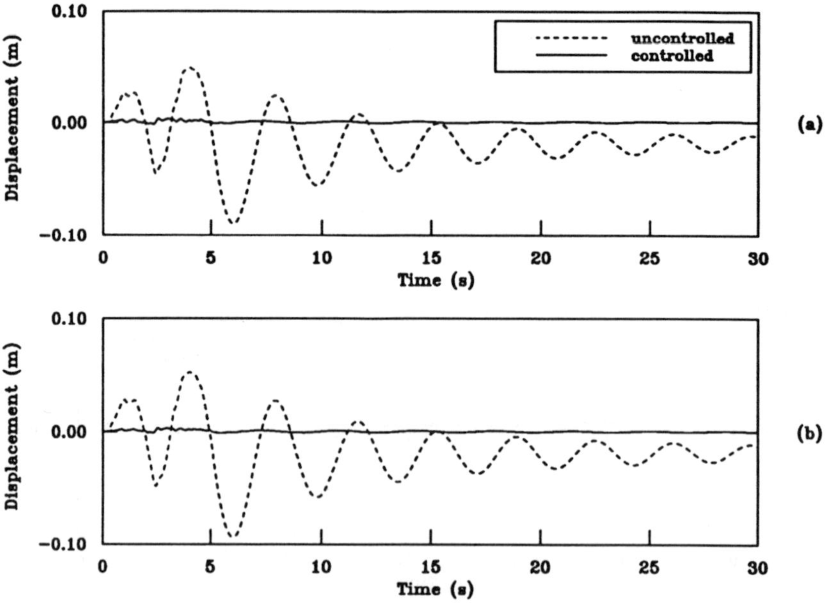

Fig. 7. Comparison of controlled and uncontrolled displacement and velocity time histories, 2DOF nonlinear example: (a) first story displacements; (b) second story displacements.

from -2.0 to 2.0 m/s^2 ($\times$ the mass of the appropriate story of the plane frame) for the control force.

Figure 7 [8] shows a comparison of the controlled and uncontrolled displacement time histories for the two degrees of freedom.

5.6. *Adaptive control*

With reference to a nonlinear single degree of freedom system, the initial membership functions are shown in figure 8. The final input membership functions of the Takagi and Sugeno inference system are shown in figure 9. Although the ANFIS-modified membership functions are not symmetrical, this can be easily remedied by adding to the training data the inverse of the state-variables/control-force records. The computational effort required by the standard controller for this example result 35 times the one of the Takagi-Sugeno controller fitted by ANFIS. required only 1.2 megaflops.

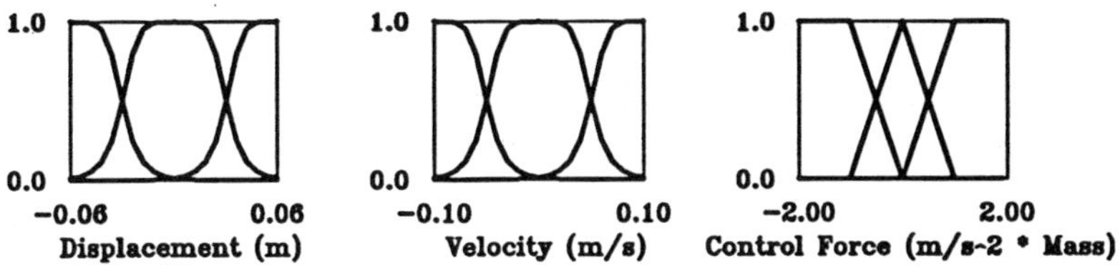

Fig. 8. Initial membership functions for SDOF system

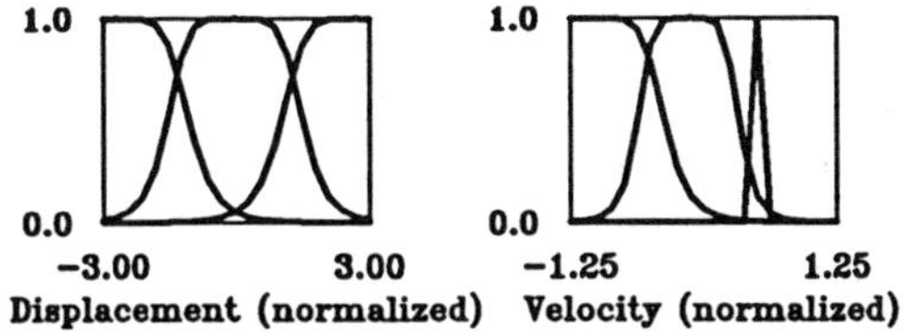

Fig. 9. Modified membership functions for SDOF system

6. Conclusions

Qualitatively, these examples demonstrate some of the potential power of fuzzy control as applied to a civil engineering problem. Note that only three fuzzy subsets were used to describe each input and five fuzzy subsets to define the output variable in the control. Further refinement would likely improve the results, though it would increase the computational cost.

Tuning the existing membership functions and possibly coupling the controllers for each story through additional rules would probably also improve the performance. The tuning of fuzzy controllers (generally accomplished through adjustments to the membership functions) is not always intuitive and would likely benefit from automatic processing by neural networks [14] or genetic algorithms [26]. Although computations were too slow with this setup to be used in an actual applications, there exist hardware implementations of fuzzy controllers that make fuzzy control feasible for real-time control. Future research will focus on finding a systematic approach to tuning the fuzzy controllers.

7. Acknowledgement

This work has been supported by grants from the Italian Ministry of University and Scientific and Technological Research (MURST).

8. References

1. Ayyub B.M. and Hassan M.H.M., Control of Construction Activities: III. A Fuzzy-Based Controller, *Civil Engineering Systems*, Vol 9, (1992) 123-146.
2. F. Casciati and L. Faravelli, *Fragility Analysis of Complex Structural Systems* (Research Studies Press, Taunton, 1991).
3. F. Casciati, L. Faravelli and T. Yao, The Effects of Nonlinearities Upon Fuzzy Structural Control, submitted for publication in *Nonlinear Dynamics* (1994).
4. F. Casciati, L. Faravelli and T. Yao, Fuzzy Control of Civil Structures Subjected to Earthquake Loading, *Proceedings EUFIT'94 (2nd European Congress on Intelligent Techniques and Soft Computing*, Aachen, Germany (1994), Vol. 2, 1050-1054.
5. F. Casciati, L. Faravelli and T. Yao, Application of Fuzzy Logic to Active Structural Control, *Proceedings 2nd European Conference on Smart Structures and Materials*, eds. A. McDonach, P.T. Gardiner, R.S. McEwen and B. Culshaw, Glasgow (1994), 206-209.
6. F. Casciati and T. Yao, Comparison of Strategies for the Active Control of Civil Structures, *Proceedings 1st WCSC (World Conference on Structural Control)*, Pasadena (1994).
7. F. Casciati, L. Faravelli and T. Yao, Fuzzy Logic in Active Structural Control, *NAFIPS/IFIS/NASA '94: Proceedings of the First International Joint Conference of The North American Fuzzy Information Processing Society Biannual Conference, The Industrial Fuzzy Control and Intelligent Systems Conference, and The NASA Joint Technology Workshop on Neural Networks and Fuzzy Logic*, eds. Larry Hall, Hao Ying, Reza Langari, and John Yen, (1994), 268-272.
8. L. Faravelli and T. Yao, Application of an Adaptive-Network-Based Fuzzy Inference System (ANFIS) to Active Structural Control, *Proceeding 1st WCSC (World Conference on Structural Control)*, Pasadena (1994).
9. A. Grace, A.J. Laub, J.N. Little and C.M. Thompson, *Control System Toolbox for use with MATLAB*, (The Mathworks, Inc., Cochituate Place, 24 Prime Park Way, Natick, MA 01760, 1992).
10. S.K. Halgamuge and M. Glesner, Neural networks in designing fuzzy systems for real world applications, *International Journal for Fuzzy Sets and Systems* (1994).
11. Hassan M.H.M. and Ayyub B.M., A Fuzzy Controller for Construction Activities, *Fuzzy Sets and Systems*, Vol. 5, No. 3, (1993), 253-271.

12. G.W. Housner, S.F. Masri, F. Casciati and H. Kameda, *Proceedings of the U.S.-Italy-Japan Workshop/Symposium on Structural Control and Intelligent Systems* (University of Southern California, Los Angeles, 1992).
13. G.W. Housner, S.F. Masri, Proceedings of the International Workshop on Structural Control, Honolulu, eds. G.W. Housner and S.F. Masri, University of Southern California, CE 9311, (1993), 483-490
14. J-S.R. Jang, ANFIS: Adaptive-Network-Based Fuzzy Inference System, *IEEE Transactions on Systems, Man and Cybernetics* (1992).
15. J-S.R. Jang, Self-Learning Fuzzy Controllers Based on Temporal Back Propagation, *IEEE Transactions on Neural Networks* (1992).
16. C.C. Lee, Fuzzy Logic in Control Systems: Fuzzy Logic Controller, Parts 1 and 2, *IEEE Transactions on Systems, Man and Cybernetics*, **20** 2 (1992), 404-435.
17. A. Lotfi, Fuzzy Inference Systems toolbox for MATLAB (FISMAT), Department of Electrical and Computer Engineering, University of Queensland (1994).
18. R.. Miller, S.F. Masri, T. Denghanyar and T. Caughey, Active Vibration Control of Large Civil Structures, *Journal of Engineering Mechanics, ASCE*, **114** (1988), 1542-1570
19. D. Nauck and R. Kruse, Choosing Appropriate Neural Fuzzy Models, *Proceedings EUFIT'94 (2nd European Congress on Intelligent Techniques and Soft Computing*, Aachen, Germany (1994), Vol. 1, 552-557.
20. T.A. Runkler and M. Glesner, DECADE - Fast Centroid Approximation Defuzzification for Real Time Fuzzy Control Applications, *ACM Symposium on Applied Computing (SAC'94), Phoenix, AZ* (1994).
21. T.T. Soong, *Active Structural Control: Theory and Practice*, (Longman Scientific and Technical, 1990).
22. T.T. Soong and R.H. Hanson, Recent Development in Active and Hybrid Control Research in the US, Proceedings of the International Workshop on Structural Control, Honolulu, eds. G.W. Housner and S.F. Masri, University of Southern California, CE 9311, (1993), 483-490
23. T. Takagi and M. Sugeno, Derivation of fuzzy control rules from human operator's control actions, *Proc. of the IFAC Symp. on Fuzzy Information, Knowledge Representation and Decision Analysis*, (1983), 55-60.
24. Y.K. Wen, Equivalent Linearization for Hysteretic Systems Under Random Excitations, *Journal of Applied Mechanics*, **47** 1 (1980), 150-154.
25. Y.K. Wen, (ed.), *Intelligent Structures 2: Monitoring and Control* (Elsevier, 1992).
26. T. Wolf, Optimization of Fuzzy Systems using Neural Networks and Genetic Algorithms, *Proceedings EUFIT'94 (2nd European Congress on Intelligent Techniques and Soft Computing*, Aachen, Germany (1994), Vol. 1, 544-551.
27. J.T-P. Yao, Concept of structural control, *Journal of the Structural Division, ASCE*, **98** 7 (1972).

Uncertainty Modeling in Vibration and Fuzzy Analysis of Structural Systems, pp. 167–178
edited by B. M. Ayyub, A. Guran and A. Haldar
Series on Stability, Vibration and Control of Systems Series B: Vol. 10

UNCERTAINTIES IN STRUCTURAL CONTROL

James T.-P. Yao
Dept. of Civil Engineering
Texas A & M University, College Station, TX 77843-3136, USA

Timothy Yao
Impact Forecasting, L.L.C.
230 West Monroe St., 9th Floor, Chicago, IL 60606, USA

ABSTRACT

The effect and handling of uncertainties is an important topic in the control of civil engineering structures. There exist uncertainties in (1) evaluating the condition and resistance of the structure and (2) predicting future loads. Because the consequence of any unsuitable control strategy for serviceability is not critical, applications of structural control in buildings have been implemented mainly for comfort purposes to date. On the other hand, active control of structures for safety purposes would involve much larger forces; the effect of uncertainties in such cases requires further investigation. The authors describe the results of some research that has been published in this area and outline open questions that need to be examined.

1. Introduction

In his science fiction novel *The R Masters*, Dickson[1] predicts that the lateral motion of very tall buildings will be automatically controlled through the judicious use of large jet engines. The buildings themselves would be supported by anti-gravity devices. Yao[2] formalized the problem of structural control in 1972 through application of automatic control theory. At present, while practical applications of anti-gravity devices (e.g., mag-lev for future railroads) are still being developed, structural control has become a reality since 1989[3-4]. Ideally, a structure can be made completely safe and/or serviceable through the use of a perfect control system. In practice, however, neither the structure nor the control system is perfectly made. In addition, the precise environmental and loading conditions the structure may encounter in the future (e.g., the location, magnitude, and timing of strong earthquakes and/or severe wind storms during the next twenty years) is not well known.

There are two main factors in deciding strategies in structural control, namely the present condition of the structure and the expected loads in the future. Whenever the structure behaves in a linear manner, no damage is accumulated. Consequently, the same control strategy may continue to be used. On the other hand, whenever the structural condition is changed substantially, the damage should be evaluated and the control strategy or usage of the structure should be modified accordingly.

In this chapter, various types of uncertainties associated with structural control will be reviewed and discussed, along with proposed methods for dealing with them. In addition, future research work is also suggested.

2. The Problem of Uncertainty in Structural Control

In the classical theory of structural reliability, random variables are classified into those related to load effects, S, and those related to structural strengths[5], R. The probability of failure refers to the probability of the event that R is less than or equal to S. Later, the concept of limit states was used[6-8]. The problem of calculating the probability of failure becomes more complicated when dynamic effects are included even without elements of structural control[9-11].

In structural control, the following items are used:

1. instruments to sense forcing functions as well as structural response to them;
2. an algorithm to identify the structural system (e.g., linear or nonlinear behavior);
3. an algorithm to activate the control device in case of need;
4. instruments to measure the effectiveness of control devices; and
5. a decision logic to decide if the control devices should be turned off whenever they are not working properly.

All instruments (measuring and recording devices) inherently suffer from noise. Although filters are available to reduce/eliminate this noise, not all acceptable filters produce the same results[12]. In making numerical calculations with idealized mathematical models, the exact timing and magnitude of control forces can be summoned instantaneously. In practice, however, there are time delays in obtaining these control forces (in terms of both the timing and magnitude). Some of these factors have been studied and results of a literature review are summarized as follows.

3. Literature Review

In 1972, Goldberg, Tang, and Yao[13] examined the probability of satisfactory performance of a structure with an active control system. The success of the control system requires the successful operation of sensors, decision logic devices, and control force devices. Assuming that there is no chance for the survival of the structural system given the condition that the control system fails, they obtained a lower bound of the probability for the survival of the combined system (of the structure and control system). Results of numerical examples indicated that

- the important factors for the overall structural reliability include that reliability of each component of the control system and the level of excitation used in the design, and
- the concept of a structural system with control is feasible provided that highly reliable components of control system are used.

In 1984, Basharkhah and Yao[14] used simulated earthquakes to study the reliability aspects of structural control. Prior to that time, almost all the investigators assumed that control forces are available immediately with precise magnitude as required. It was realized that there always exist a "time lag" between the instant when a control force is needed and the instant when the required control force is actually generated. They also called the time required to generate the desired magnitude of the control force the "time constant." The effects of various levels of time lag and time constant were studied numerically. Some six values of time lag and 13 values of time constants were used. For each case, 20 simulations were made. A Markov process as developed by Bogdanoff and Kozin[15] was used to find the reliability of the structure. Results indicated that

- the probability of failure of a structure with a perfect/ideal control system is reduced as expected, and
- the probability of failure of realistic systems (with time lag and/or time constant) may be higher than that of idealized cases.

Later, Yang et al.[16] carefully examined the effect of time delay on structural control. They concluded that a time delay always degraded the efficiency of the control system. This conclusion was confirmed by experimental studies[17]. Therefore, it is necessary to consider reliability aspects of structural control before such systems can be used in civil engineering practice.

In 1987, Yao[18] discussed effects of various types of uncertainties on the reliability of structures with active control and suggested several approaches to deal with these uncertainties. These uncertainties include those associated with

- prediction of future loading conditions with limited number of available records (e.g., only thirty years of good records of strong earthquakes in the States are available),
- incomplete information or inexact data concerning loading conditions,
- lack of understanding of the behavior of complex systems (e.g., the exact damage/failure paths of most existing structures), and
- reliability of control systems.

An interrelationship between structural control and identification was suggested. As an example, several levels of limiting conditions may be established (e.g., warning limit state, elastic limit state, tolerable damage limit state, repairable damage limit state, and near-collapse limit state). If and when the warning limit state is exceeded, the control system is activated. Whenever the structural response exceeds the tolerable nonlinear limit state, the structure should be inspected carefully in order to determine whether the structure should be repaired or demolished. Also, it was suggested to use

- small control forces frequently for comfort control,
- moderate control forces occasionally for controlling tolerable damage, and
- large control forces only for the near-collapse range of the structural response.

To stimulate further discussion and possible implementation, the fuzzy control logic is recommended along with a computer program incorporating data processing and decision logic for the purposes of structural control. The theory of fuzzy sets was started by Zadeh[19] in 1965. It is an effective way of dealing with uncertainty resulting from ambiguity and vagueness. During these past couple of decades, many practical applications have been made (e.g., Klir and Yuan[20], Terano et al.[21], and Zimmerman[22]).

In 1991, Ayyub[23] reviewed 100 papers on civil engineering applications of fuzzy sets. Yao and Natke[24-26] discussed the use of fuzzy sets in dealing with uncertainties in assessing the current condition of the structure as well as in predicting future loads. As an example, good acceleration records of strong-motion earthquakes in the United States have been collected only since the 1964 Alaska Earthquake. It is likely to encounter future ground motions which possess different dynamic characteristics from those recorded to date. Moreover, it is still a challenging task to assess the condition of complex structures. Consequently, there are various types of uncertainties in the equations of motion of existing structures.

In automatic control theory, the effects of uncertainty in the system are quantified in the robustness of the control system. Robustness can be defined as the sensitivity of a system's performance to a large range of uncertain parameters[27]. Reliability, on the other hand, is a probabilistic measure of a system's ability to perform adequately over a specified time period, possibly subject to uncertain parameters. The two concepts appear to be closely related but not equivalent; i.e., the robustness of control systems affects their reliability. Spencer et al.[28] presented a method for computing probabilistic measures of stability and performance for assessing and comparing the robustness of controlled structures subject to the uncertainty inherent to civil engineering structures. Smith and Chase[29] developed a variation of the robust H_∞ optimal control algorithm to account for specified ranges of uncertainty in structural properties.

Li and Ang[30] examined the effect of failure of an actuator on the seismic reliability of buildings. By matching the measured response to the simulated response of the failed system, they identified the failed actuator and then modified the original control algorithm accordingly. A Monte-Carlo simulation method was used to evaluate the response statistics and the system failure probability by prescribing a limiting value of the relative displacement. Results of two numerical examples indicate that the proposed "fault tolerant control" was effective in maintaining the structural reliability under strong earthquakes.

4. Discussion

Because of the conservative nature of engineering design and the many advances of engineering analyses, it is often said that a structure is safe unless two or more major errors are committed during the process of analysis/design/construction. Consequently, almost all the "engineered" structures are safe with or without application of active control. On the other hand, we have had a relatively short period of time (in comparison with the lifetime of the earth) for the scientific measurement of forces resulting from strong earthquakes and other natural disasters. There is always a possibility that, regardless of how well we build our structures, one or more limit states of a given structure may be exceeded someday. At least initially, the concept of structural control was conceived for considerations of safety. To date, practical applications of structural control seem to be mostly for comfort/serviceability considerations.

Since 1989, several structures, mostly buildings, have been built with various active control systems. Their performance is highly satisfactory under wind and earthquake loading conditions. It is understood that engineers who designed these structures are still reluctant to leave the control devices operating in case safety of the structure is involved.

One school of thought is that if and when (a) a destructive force (either due to some unforeseen extremely strong earthquake or wind) occurs and (b) the structure is near collapse, the use of an active control device may save the structure[31]. Then, even if the reliability of the control device is not high, it may still be preferable to doing nothing. If we follow this school of thought, the installation of control devices may be considered as an additional insurance against undesirable outcomes. With the idea in mind that such a device may never be used, its presence would have the effect of giving peace of mind to owners and users of this structure. To illustrate this point, consider the following case.

Let

F_o = event that the structure fails without control devices

F_c = event that the structure fails with control devices

E = occurrence of an extreme event during the lifetime of the structure

E^* = occurrence of no extreme event during the lifetime of the structure

C = the control devices perform as designed

C^* = the control devices do not perform as designed

then,

$$P(F_o) = P(F_o|E^*)P(E^*) + P(F_o|E)P(E) \tag{1}$$

$$\begin{aligned} P(F_c) = {} & [P(F_c|C \cap E^*)P(C|E^*) + P(F_c|C^* \cap E^*)P(C^*|E^*)]P(E^*) \\ & + [P(F_c|C \cap E)P(C|E) + P(F_c|C^* \cap E)P(C^*|E)]P(E) \end{aligned} \tag{2}$$

There exist several conditions under which it may be found that

$$P(F_c) \le P(F_o) \tag{3}$$

As an example, we can design the system such that

$$P(F_c|C \cap E^*)P(C|E^*) + P(F_c|C^* \cap E^*)P(C^*|E) \le P(F_o|E^*) \tag{4}$$

and

$$P(F_c|C \cap E)P(C|E) + P(F_c|C^* \cap E)P(C^*|E) \le P(F_o|E) \tag{5}$$

Furthermore, if the control device can be turned off whenever it is found to be faulty, then

$$P(F_c|C^* \cap E) = P(F_o|E) \tag{6}$$

and

$$P(F_c|C^* \cap E^*) = P(F_o|E^*) \tag{7}$$

It is reasonable to assume that

$$P(C|E^*) \approx 1 \tag{8}$$

thus

$$P(C^*|E^*) << 1 \tag{9}$$

and

$$P(F_c | C \cap E) < P(F_o | E^*) \tag{10}$$

It is also reasonable to assume that

$$P(F_c | C \cap E) << 1 \tag{11}$$

which satisfies the inequalities in Eq. 4.

For purposes of illustration, assume that

$$P(F_o | E) = 10^{-1} \qquad P(F_c | C^* \cap E^*) = 10^{-10}$$
$$P(F_o | E^*) = 10^{-10} \qquad P(F_c | C \cap E) = 10^{-8}$$
$$P(E) = 10^{-5} \qquad P(F_c | C^* \cap E) = 10^{-1}$$
$$P(E^*) = 0.99999 \qquad P(C | E^*) = 0.999999$$
$$P(F_c | C \cap E^*) = 10^{-12} \qquad P(C^* | E^*) = 10^{-6}$$

Then, approximately,

$$P(F_o) \approx 10^{-6} \tag{12}$$

and

$$P(F_c) \approx 10^{-6} \times P(C^* | E) \leq P(F_o) \tag{13}$$

This simple exercise demonstrates that, as long as the faulty control devices can be shut down, the reliability of the structure with active control is at least as good as that of the structure without it. The actual and realistic evaluation or estimation of these quantities is, of course, complicated.

Willsky[32] summarized several methods for failure detection including failure-sensitive filters, voting systems, and multiple hypothesis filter-detectors. Meanwhile, Clark et al.[33] proposed a method using only one set of instruments. The required redundancy in a voting system is provided with a multiple observer system in the computer. The existence of these methods indicates that it is possible to detect and remedy faulty control devices.

From the literature review, much consideration has been given to reliability of structures with active control. However, many investigators looked at only parts of the problem. Few, if any, studies are concerned with the reliability of the entire system with consideration of all components and all possible loading conditions. Results of numerical simulations are indeed valuable for illustrating the effectiveness of certain methodologies and/or the reliability of structural control. Nevertheless, not knowing (a) what types of strong earthquakes or extreme wind may occur in the future nor (b) what failure paths a complex structure may follow under such loads, it is not possible to make any simulations of situations that we do not yet know.

One of the difficulties in studying structural reliability has been the lack of actual failure statistics. Because civil engineering structures are

- mostly individually designed and constructed,
- complex and expensive, and
- usually overdesigned,

few of them ever fail by total collapse. Frequently, the few sets of failure data that exist are proprietary and thus not available for general studies. Consequently, there have not been sufficient failure data that could be used to verify reliability analyses.

Although it is possible to discuss all kinds of uncertainty related to reliability of structures with active control, there is still no comprehensive study of it to date. It is believed that the following problems need to be studied in order to answer the question of the reliability of structural control.

- At the policy level, it is desirable to know whether structural control for safety is needed. If it is not required for safety considerations, reliability of structural control for serviceability is relatively insignificant because the consequence of its failure does not necessarily involve either bodily harm or loss of property.
- In case structural control is needed for safety considerations, a comprehensive study of various types of uncertainties seems to be desirable. Whether the uncertainty is represented with probabilistic or fuzzy measures, it is useful to enumerate these quantities on the basis of available data and/or expert (subjective) evaluations.
- The reliability of the entire system, including the structural system and control system, should be studied. The interrelationship of structural behavior and effect of each and every component of the control system on the overall safety should be studied.
- There exists a need for collecting failure data of structures and components of control systems. It is a time consuming and difficult task because (a) few structures ever fail and (b) people usually do not want to disclose incidences of failure. Nevertheless, it is important to have such data so that results of theoretical studies may be verified. In

addition, by analyzing actual data, insight may be gained to further the progress of theoretical studies.

5. Concluding Remarks

We consider active control as an additional degree of redundancy in ensuring structural safety. Ideally, the safety of the structure is enhanced if

- the structure is carefully designed and well-built,
- future environmental and loading conditions can be predicted with sufficient confidence, and
- the control devices are reliable.

Continued development of more effective and reliable sensing hardware and active force devices is needed. Likewise, advanced methodologies and software for monitoring the structural condition and making condition assessments of the structure are desirable.

We need to learn more about the behavior of the structure under various possible loading conditions. As we collect more data on wind storms and strong earthquakes, we will learn to predict future loads with better accuracy. Considering the relatively short period of our scientific records, it is indeed a slow process to gain more confidence in these predictive techniques.

Refinements are needed in the calculations of failure probabilities. In addition, we should develop better monitoring systems and fail-safes for active control systems. These can ensure that malfunctions of the control system will not cause any adverse effects; i.e., in Eqs.1-2,

$$P(F_c | C^* \cap E) \leq P(F_o | E) \tag{14}$$

and

$$P(F_c | C^* \cap E^*) \leq P(F_o | E^*) \tag{15}$$

Much progress has been made in practical applications of structural control in the last several years. Results of all available studies indicate that structural control can be effective in reducing structural response to environmental loads as designed. Many questions on uncertainties associated with structural control, however, still need to be studied.

The most important uncertainty is about our knowledge on extreme loads. We simply do not know with any certainty what possible extreme events will occur during the lifetime of the structure. To use active control devices as an additional degree of redundancy against failure seems to be a reasonable approach to structural safety.

6. References

1. G. R. Dickson, The R-Master, J.B. Lippincott Company, Philadelphia and New York, 1973, 46.
2. J. T. P. Yao, Concept of structural control, Journal of the Structural Division, ASCE, 98 (7), 1972.
3. T. T. Soong, Active Structural Control: Theory and Practice, Longman Scientific and Technical, 1990, 194 pages.
4. T. T. Soong and R. H. Hanson, Recent Development in Active and Hybrid Control Research in the US, Proceedings of the International Workshop on Structural Control, Honolulu, eds. G. W. Housner and S. F. Masri, University of Southern California, CE 9311, 1993, 483-490.
5. A. M. Freudenthal, J. M. Garrelts, and M. Shinozuka, The Analysis of Structural Safety, *Journal of the Structural Division*, ASCE, 92 (ST1), 1966, 267-325.
6. A. H-S. Ang and W. H. Tang, *Probability Concepts in Engineering Planning and Design, Vol. II*, John Wiley and Sons, Inc., 1984.
7. H. O. Madsen, S. Krenk, and N. C. Lind, *Methods of Structural Safety*, Prentice-Hall, Inc., Englewood Cliff, NJ, 1986.
8. S-H. Dai and M-O. Wang, *Reliability Analysis in Engineering Applications*, Van Nostrand Reinhold, New York, NY, 1982.
9. Y. K. Lin, Probabilistic Methods in the Theory of Structures, McGraw-Hill, New York, NY, 1967.
10. Y. K. Wen and H-C. Chen, System reliability under time-varying loads, *Journal of Engineering Mechanics*, ASCE, 115 (4) 1989.
11. T. T. Soong, and M. Grigoriu, *Random Vibration of Mechanical and Structural Systems,* PTR Prentice-Hall, Inc., Englewood Cliffs, NJ, 1993.
12. J. E. Stephens and J. T. P. Yao, *Data Processing of Earthquake Acceleration Records*, Structural Engineering Report No. CE-STR-85-5, Purdue University, West Lafayette, Indiana, 1985, 152 pages.
13. J. E. Goldberg, J. P. Tang, and J. T. P. Yao, Reliability of Structures with Control Systems, *Proceedings, International Symposium on Systems Engineering and Analysis, Vol. II-Contributed Papers,* Purdue University, W. Lafayette, Indiana, 1972, 153-155.
14. M. A. Basharkhah, and J. T. P. Yao, Reliability Aspects of Structural Control, *Civil Engineering Systems*, 1, 1984, 224-229.

15. J. L. Bogdanoff and F. Kozin, A New Cumulative Damage Model: Part 4, Journal of Applied Mechanics, 47 (1), 1980, 240-257.
16. J. N. Yang, A. Akbarpour, and G. Askar, Effect of Time Delay on Control of Seismic-Excited Buildings, Journal of Structural Engineering, ASCE, 116 (10), 1990, 2801-2814.
17. R. C. Lin, T. T. Soong, and A. M. Reinhorn, *Experimental Evaluations of Instantaneous Optimal Algorithm for Structural Control*, Tech. Rep.#NCEER-87-0002, NCEER, Buffalo, NY, 1987.
18. J. T. P. Yao, Uncertainties in Structural Control, *Vibration Control and Active Vibration Suppression-DE-Vol.4*, Editors: D. J. Inman, and J. C. Simonis, ASME Book No. H00404, 1987.
19. L. A. Zadeh, Fuzzy Sets, *Information and Control*, 8, 1965, 338-353.
20. G. Klir, and B. Yuan, *Fuzzy Sets and Fuzzy Logic: Theory and Applications*, Prentice Hall PTR, Upper Saddle River, NJ 1995.
21. T. Terano, K. Assai, and M. Sugeno, *Fuzzy Systems Theory and Its Applications*, Academic Press, Boston, MA, 1992.
22. H. J. Zimmerman, *Fuzzy Set Theory and Its Applications*, Second Edition, Kluwer Academic Publishers, Boston, MA, 199
23. B. Ayyub, Systems Framework for Fuzzy Sets in Civil Engineering, *Fuzzy Sets and Systems*, 40, 1991, 491-508.
24. J. T. P. Yao, and H. G. Natke, Uncertainties in Structural Identification and Control, *Proceedings, International Fuzzy Engineering Symposium*, Yokohama, Japan, 2, 1991, 844-849.
25. J. T. P. Yao and H. G. Natke, Effect of Active Control to Structural Reliability, *Proceedings, ASCE Specialty Conference on Probabilistic Mechanics, Structural and Geotechnical Reliability*, Denver, Colorado, 1992.
26. J. T. P. Yao and H. G. Natke, Reliability of Structures with Active Control, *Proceedings, IEEE International Conference on Fuzzy Systems (FUZZ-IEEE*), San Diego, 1992.
27. R. C. Dorf, *Modern Control Systems: Sixth Edition,* Addison-Wesley, 1992, 578-581.
28. B. F. Spencer, M. K. Sain, C. H. Won, and D. C. Kaspari, Reliability based measures of structural control robustness, *Structural Safety*, 15 (1-2), August, 1994, 111-129.
29. H. A. Smith and J. G. Chase, Robust disturbance rejection using H_∞ control for civil structures, *Proceedings 1st WCSC (World Conference on Structural Control),* Pasadena 1994.
30. P. Li and A. H-S. Ang, Seismic Reliability of Buildings with Fault Tolerant Active Control, *Proceedings of the First World Conference on Structural Control*, 1, U.S. Panel on Structural Control, c/o Department of Civil Engineering, University of Southern California, Los Angeles, CA 90089-2531, 1994, WA3-3 through WA3-12.
31. F. Casciati and T. Yao, Comparison of Strategies for the Active Control of Civil Structures, Proceedings 1st WCSC (World Conference on Structural Control), Pasadena, 1994.

32. A. S. Willsky, A survey of Design Methods for Failure Detection, *Automatica*, 12, 1976, 601-611.
33. R. N. Clark, D. C. Fosth, and V. M. Walton, Detecting Instrument Malfunctions in Control Systems, *IEEE Transactions on Aerospace and Electrical Systems*, AES-11 (4), 1975, 465-473.

Uncertainty Modeling in Vibration and Fuzzy Analysis of Structural Systems, pp. 179–232
edited by B. M. Ayyub, A. Guran and A. Haldar
Series on Stability, Vibration and Control of Systems Series B: Vol. 10

STRUCTURAL FUZZY CONTROL

MAGUID H.M. HASSAN
Department of Civil Engineering, Higher Technological Institute
P.O.Box 228, 10th Of Ramadan City, Egypt

and

BILAL M. AYYUB
Department of Civil Engineering, University of Maryland
College Park, Maryland 20742, USA

ABSTRACT

A general framework for structural fuzzy control is proposed. The proposed framework is organized in three consecutive stages system definition, state evaluation and fuzzy control. The first stage involves the development of a generic framework for system modeling and identification which should serve as a guide for structural system identification. The framework is capable of reflecting the same structural system in several forms based on its nature, the control scheme objectives and its behavior. The second stage defines a generic scheme for state evaluation. The proposed scheme provides a guide for several state evaluation techniques based on the system model and its properties. Because of the complexity and uncertainty involved in structural systems, fuzzy-based control is considered a potential candidate for structural control. The proposed system is capable of monitoring and controlling several control attributes. The main objective is to identify and improve variables responsible for any unsatisfactory state condition. The system also includes a self learning unit that is capable of defining and extracting new rules in order to improve the system's performance. The proposed concepts are applied to several structural systems with different control objectives in order to demonstrate the generality of the framework.

1. Introduction

The overall objective of structural control is to provide optimum mechanisms necessary to react to the dynamics of any structural system. This is considered an alternative approach to ensure a safe performance against all expected loading conditions during the structure's lifetime[30,31,42]. In a broader sense, structural control should result in safe and/or satisfactory performances under any uncertain conditions during all of the system's life stages. Any structural system could be idealized in several forms that satisfy different control objectives during several stages of its lifetime. For example, any structural system goes through several changes during its construction period. The safety of the system needs to be ensured and controlled during that period. The same system, when being designed, goes through several elimination and comparison stages among other candidate systems, proposed for the same purpose. At that stage, the reliability of the selected system might be a viable control objective. Yet, the nature of the two objectives are totally different. For the same system, if the objective was to monitor and control its dynamic response under earthquake excitation, yet another view of the system needs to be identified and other properties should be emphasized. Thus, the need for a general system identification framework is warranted. The needed framework should be able to adapt and emphasize different properties and components of structural systems based on the control objective and the nature of the behavior of the system. In other words, it should be capable of reflecting different views of the same system within different control environments.

Because of the complexity and uncertainty involved in structural systems, fuzzy-based control is considered a potential candidate approach for structural control. In spite its usefulness in such applications, very little work has been reported in reference to structural fuzzy control[2,3,5,9,20]. In this chapter, a generic structural fuzzy control framework is outlined. The proposed framework has a wide range of practical applications some of which are presented in the following discussion. The first step in the development of a fuzzy-based controller is to build a model for the controlled system. System analysis provide methods and techniques for identifying general systems. In addition, a performance function that relates the inputs to outputs, i.e., loads to responses, needs to be defined. The performance function represents a general concept that could be translated into different forms depending on the control objectives and the way the system is defined. For example, the equation of motion of a structural system under earthquake excitation represents a form of a performance function. In a different context, a suitable limit state equation could also be considered a performance function for the same system. Considering the construction stage of a structural system, a behavior function that relates the likelihood of occurrence of several variables could be viewed as a viable performance function. The performance function together with a system model are utilized in a state evaluation scheme that results in the states of the control attributes at different points in time.

Traditionally, structural control was performed by applying the automatic control theory to civil engineering structures. The control scheme relies simply on the feedback-control concept. In this chapter, a fuzzy-based structural control strategy is proposed. The proposed controller is intended to act as a central brain-like unit for structural systems. A simple analogy could be drawn between a human being trying to balance himself on a shaking ground and a structural system that needs to restrain its vibration caused by an earthquake. The human brain guides the person to balance himself very efficiently. No mathematical models are built and no time is wasted in evaluating the persons dynamic properties. It seems rather intriguing to apply the same concepts, if possible, to similar structural systems. Structural control schemes should be simple, direct and as accurate as possible[30,31,42]. Fuzzy-based control relies on simple IF THEN rules that reflects a specific strategy for controlling the undesired state of the structural system[3,5,8]. The development of such rules need very close and comprehensive study. The proposed scheme could be modified and integrated to several intelligent new technologies such as neural networks, to suit a specific application. Each structure would have a different rule-base, i.e. a collection of rules, that could handle its control. These controllers should be installed and connected to a set of sensors and actuators. The controller should be activated as soon as any unacceptable state and/or condition is initiated. The use of fuzzy control eliminates the need to develop and update an accurate mathematical model of the controlled system. These models either might not be easily available for some historic buildings and / or monuments or might be difficult and time consuming to develop. The development of such a control system could be performed using several software shells that are available in the market, *fuzzy* Tech, by Intel, is an example for such shells. In the following discussion, a general system identification framework is presented together with a general state evaluation framework. The basic components and performance of a fuzzy-based structural controller are, then, outlined. At some locations some additional upgrades for the basic unit are discussed and viewed to suit specific applications.

2. System Definition

2.1. Introduction

The definition of a system is commonly considered the first step in an overall methodology formulated for achieving a set of objectives[22,44,45,48,50]. In this study, a general system definition framework is developed. A system can be generally defined as an arrangement of elements, with some important properties and inter-relations among them. Systems can be classified based on the available knowledge into several main levels[22,42]. Some researchers classified systems based on their nature[26]. Four main types were defined, namely, natural, designed, human activity, and social and cultural. Others realized the hierarchical formation of systems that is based on the available degree of details and information[15,45,48,50]. The set approach for the system definition problem was also introduced[22,42]. This approach was criticized because of its inability to express the properties of the overall system, knowing the qualities of its elements[13]. However, for some control objectives, the set approach might be suitable for representing the variables of the problem. The ability to infer information about the overall system, knowing the behavior of its components, can be dealt with using special techniques[22]. Once a system is defined, the next step is to define its environment[15,22,26,44,45,48,50]. The environment is defined as everything within a certain universe that is not included in the system. An interesting notion within systems thinking was also introduced[15], which allows the change in boundaries between a defined system and its environment. Systems could be classified, according to the degree of knowledge, into a structure of successive levels. Each level includes all information present at the lower levels, in addition to, more detailed information and relations.

2.2. System Classification

At the first level of knowledge, which is usually referred to as level (0), the system is known as a source system[22,44]. Source systems comprise three different components, namely object systems, specific image systems and general image systems[22,44]. The object system constitutes a model of the original object. It is composed of an object, attributes and a backdrop. The object represents the specific problem under consideration. The attributes are the important and critical properties selected to be measured or observed as a model of the original object. The backdrop is the domain within which the attributes are observed. The specific image system is developed based on the object. This image is built through observation channels which measure the attribute variation within the backdrop. The attributes when measured by these channels correspond to the variables in the specific image system. The attributes are measured within the support set which corresponds to the backdrop. The support can either be time, space or population.

The second level of hierarchical system classification is the data system. The data system includes a source system together with actual data introduced in the form of states of variables for each attribute. The actual states of the variables at the different support instances, yield the overall states of the attributes. Special functions and techniques are used to infer information regarding an attribute. Based on the states of the variables representing it and the nature of problem, i.e., dynamic response, safety evaluation or reliability assessment. A formal definition of a data system could be expressed as follows:

$$D = \{ S, a \} \tag{1}$$

where D is the data system, S is the corresponding source system, and (a) is the observed data that specifies the actual states of the variables at different support instances.

At the next knowledge level, support independent relations are defined to describe the constraints among the variables. These relations could be utilized in generating states of the basic variables for a prescribed initial or boundary condition. In other words, these relations describe the performance of the modeled system. The set of basic variables includes those defined by the source system and possibly some additional variables which are defined in terms of the basic variables. There are two main approaches for expressing these constraints. The first approach utilizes a support independent function that describes the behavior of the system. A function defined as such is known as a behavior function, a probability distribution that relates the likelihood of occurrence of all possible state combinations of the control variables is an example of such a function. Such an approach has been adopted in the safety control of structural systems during their construction period[6]. The second approach depends on relating successive states of the different variables. In other words, this approach describes a relationship between the current overall state of the basic variables and the next overall states of the same variables. A function defined as such is known as a state-transition function. For example, an equation of motion of a structural system under earthquake excitation could be viewed as a state-transition function. Regardless of the type of function used to describe the input/output relation, it could be generally classified as a performance function. The performance function together with the data system comprise what is defined as a generative system.

At the highest knowledge level, structure systems are defined as sets of smaller systems or subsystems. The subsystems could be source, data or generative systems. These subsystems may be coupled due to sharing common variables or due to interaction in some other form. A formal definition of a structure system could be expressed as follows:

$$SE_B = \{ (V_i, E_B{}^i), \text{for all } i \in N_e \} \tag{2}$$

where SE_B is the structure system whose elements are behavior systems, V_i is the set of sampling variables for the element, i.e., the behavior system, $E_B{}^i$ is the $i^{\underline{th}}$ behavior system, i.e., the element or subsystem, and N_e is the total number of elements or subsystems in the structure system. Based on this basic system classification, a general system identification framework is defined in the following section which include all components discussed above.

2.3. General System Definition Framework

As mentioned above, the first step in the development of a fuzzy-based controller is the identification and modeling of the system. In system analysis, the structure of the system, the relevant state variables and their effect on the system performance need to be characterized and idealized. Any structural system could be modeled in several forms each reflecting some relevant set of properties, components and their interconnections. Each model is developed within the context of a specific control problem, i.e., control objectives and nature. Thus, the need for a general system identification framework is warranted. The proposed framework should be able to adapt and emphasize different properties and components of structural systems based on the control objective and the nature of the behavior of the system. In other words, it should be capable of reflecting different views of the same system within different control environments. In this section, the major components that need to be defined in any system identification model are discussed and integrated into a general framework. The proposed framework should serve as a guide for the development of structural system models for structural control applications. In this chapter, the notion of structural control is presented in a broad sense. The objective of structural control is not only to suppress or reduce the amount of vibration of the structural system caused by an earthquake excitation or any other type of dynamic load. In fact, it does represent the control of any undesired condition, being the vibration, safety level, or reliability or simply the performance of the structural system. Based on that definition, structural control would require a rather general system identification scheme that emphasizes several properties and components for different control objectives and environments.

The proposed model should emphasize the hierarchical nature of any structural system. In general, a system can be defined as an arrangement of elements, with some important properties and inter-relations among them. The set of elements being connected to comprise the required system may not be all at the same level of importance, and/or performance. Therefore, the hierarchical structure of these elements needs to be emphasized. Such hierarchical structure should be considered in evaluating the states of the elements and thus the state of the overall system. Thus, any structural system should represent the top level of identification in order to include all information and knowledge regarding variables, components and inter-relations. Referring to the previous system classification, such top level is referred to as a structure system. Underneath that system is a set of subsystems, i.e., components or elements. Such subsystems could be further broken down into smaller subsystems if that is warranted in the problem under consideration. One of the main benefits of breaking down systems into such smaller systems is the accurate state evaluation and accurate feedback control implementation which is discussed in the following sections. Once the overall system is decomposed into a set of subsystems, the importance levels of the individual components need to be specified. In addition, the inter-relations among any two individual components / subsystems should be specified. In other words, if the state variation of one component affects the state of any other component, this effect needs to be quantified and incorporated in the state evaluation scheme.

At this stage, the general structure of the system is defined. Figure 1, shows the framework at this level. The figure shows the hierarchical structure of the overall system and its components. It is obvious that any subsystem / component could be further broken down into smaller subsystems / components in order to reach an acceptable level of detail and accuracy, as mentioned earlier. The figure also identifies the inter-relations among the individual subsystems / components at each hierarchical level. The C/C type of inter-relation is a component-to-component relationship. While the S/S type of inter-relation is a subsystem-to-subsystem relationship. It should be realized that the overall structural system could be a subsystem itself of a larger structural system. Such decomposition techniques are utilized in structural dynamic analysis in order to decompose multi-degree-of-freedom systems into a set of smaller degree-of-freedom systems.

The next step is to identify the attribute(s) of interest. The control attribute is defined as a property of the system that is required to be monitored and its value should be controlled at any point in time. The safety of the system is an example of a control attribute. The displacement amplitude of a structural system under earthquake excitation is another example of a control attribute. It should be realized that the attribute value should be evaluated at all hierarchical levels. In other words, the safety of all underlying subsystems and components should be evaluated and aggregated to result in the safety of the overall system. The aggregation procedure depends on the nature of the problem and the inter-relations among the subsystems and components. The aggregation procedure and the evaluation of the control attribute level are discussed in the following section. However, such evaluation procedures depend on real time measurements of specific control variables. Such control variables are the ones that are known to influence the values of the control attributes. Control variables should be defined at this level of system identification. The system definition should be able to adapt to all types of variables. For example, some

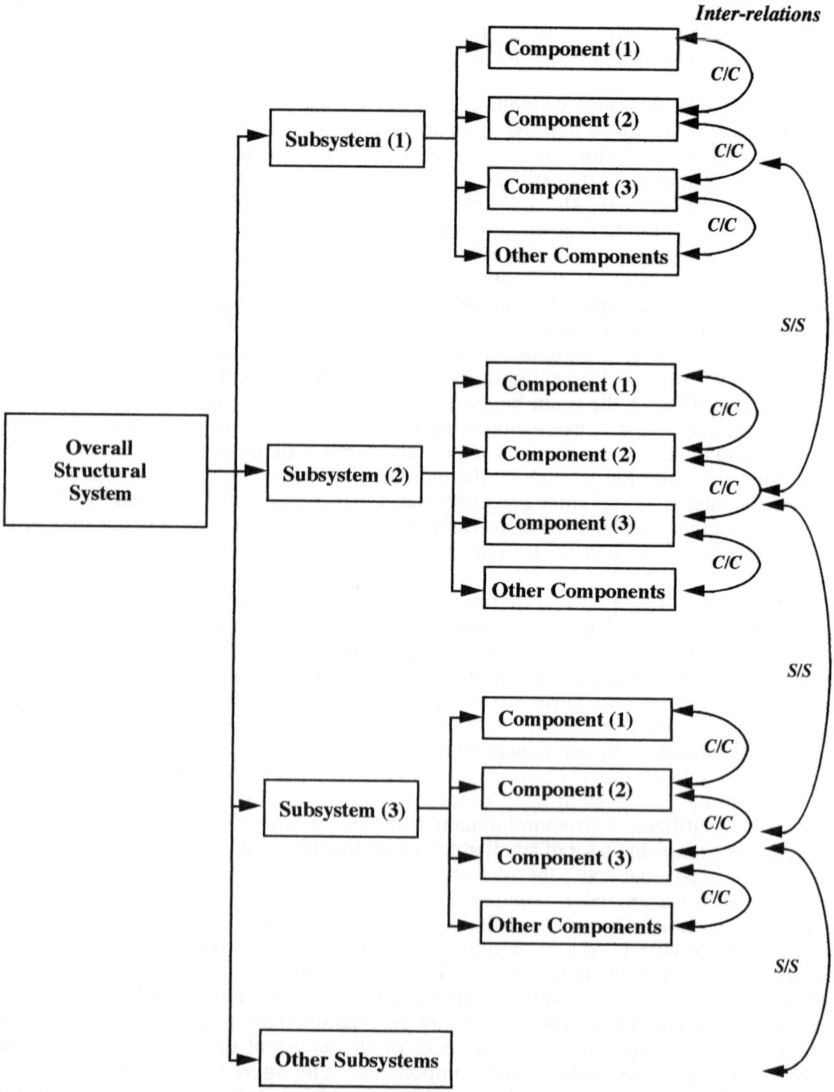

Figure 1. Hierarchical Structure and Inter-relations

variables may not be quantitative in nature. The developed framework should be able to incorporate such types of variables and develop equivalent quantitative measures for such qualitative variables. In addition, variables may be deterministic or random in nature. The system, when defined should be able to adapt to all types of variables and incorporate such variables in the state evaluation scheme.

Figure 2, shows the framework with the control attributes and the control variables defined to represent each attribute. It should be realized that multiple control attributes might be necessary to control a specific problem[4]. For example, for a system under earthquake excitation it might be necessary to control both displacements and accelerations rather than any of these individually. Each controlled quantity, i.e., displacement or acceleration, is considered a single control attribute. The figure shows the hierarchical formation of the system and its impact on the attribute evaluation. The variables are defined and observed at the lowest hierarchical level as shown in figure. The variables affecting any single attribute, are in general different from one component to the other. This might not always be the case, however, the most general case is for different variables. The same attribute might have different variables affecting its value, when observed for several components. For example, the reliability of a beam element being defined as its attribute of interest is defined based on the limit state equation representing the failure mode of that element which is mostly bending failure. However, for a column element, the limit state equation would represent a generally buckling failure mode. Thus, a set of random variables would be required to evaluate the reliability of the beam, i.e., yield strength and section modulus. While a different set would be required to evaluate the reliability of the column element, i.e., modulus of elasticity, radius of gyration and member length. The number of variables necessary to define a given attribute is not limited and it is not necessary for all attributes to have the same number of state variables. The shown diagram only outlines the structure and shows a number of variables just for the sake of explanation. However, the number of attributes must be the same in all hierarchical levels and the same attributes should be observed at all levels. The structure simply reflects an important rule which states that the attribute level of any system, is a reflection of the aggregated attribute levels of its components. The aggregation procedure and the actual evaluation of the attribute level are discussed in the following section.

The final element that is essential in the general system identification framework is the performance function. The performance function, together with a state evaluation scheme, estimates the attribute state values at all respective hierarchical levels and consequently evaluates the attribute level of the overall structural system. The performance function depends on the control objective and the nature of the control problem. For example, the dynamic equation of motion of a structural system is considered its performance function, if its dynamic response under dynamic excitation is the control objective. However, if the reliability of the system is considered as a control objective, the limit state equation for each component would be considered its performance function. Figure 3, shows the proposed general system identification framework outlined in the previous discussion. It should be realized that the figure does not set a limit on the number of subsystems / components, or even the level of hierarchical decomposition. There is no limit also on the number of attributes or their state variables. The numbers shown in the figure are only limited by the available space and physical representation. It should be noted that the lowest hierarchical level, i.e., the component, is the level that includes all the details considering the state evaluation of the attributes. In other words, it includes the state variables, the performance function and the observation channels collecting the data regarding the states of the variables. In the following subsections, three applications for the general system identification framework are developed. The applications relate to a structural system that is modeled for different control objectives at several stages of its expected life. The

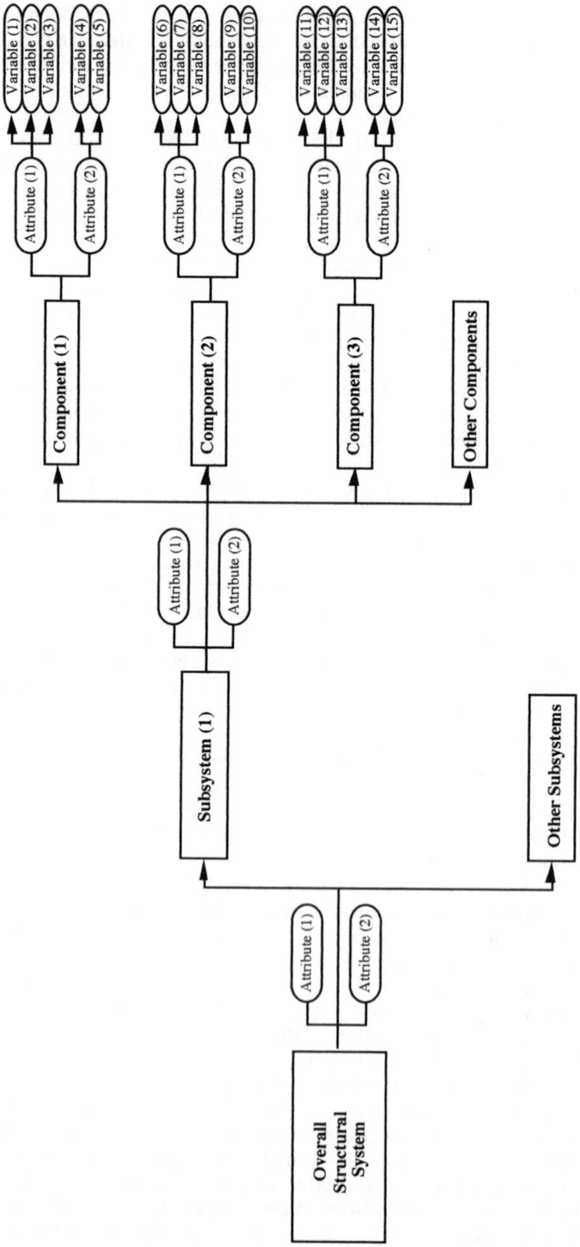

Figure 2. Hierarchical Structure with Control Attributes and Control Variables

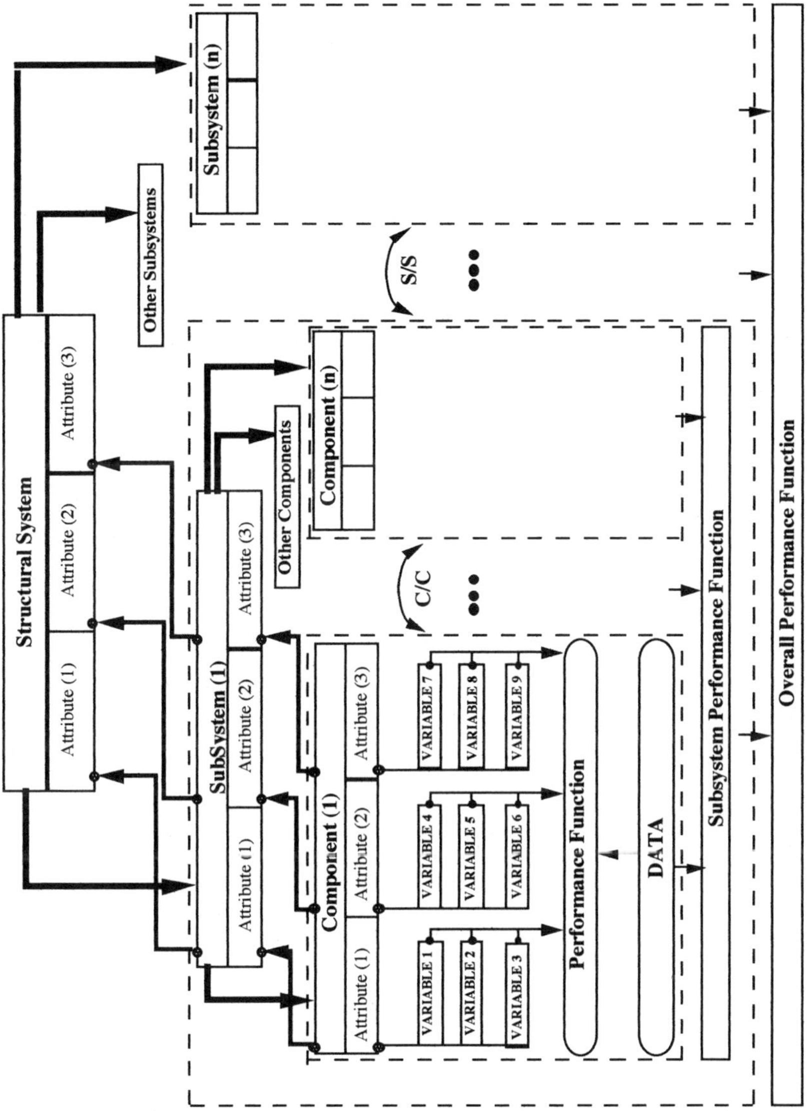

Figure 3. General System Identification Framework

developed applications should demonstrate the use and benefits of the developed framework and how to apply these concepts to practical problems.

2.4. Definition of a Structural System for Safety Control During Construction

In this example, a model is defined for a structural system during its construction period. During construction, several uncertainties are involved in the behavior of the system. Figure 4, shows some of these uncertainties on the developed model and their classification. Because of the uncertainties and complexities incorporated during the construction period, traditional control theory was not a suitable choice for the control of structural systems during that period.

The control objective in this example is the safety of the structural system during the construction period. During that period, several variables affect the safety of the system. The safety of the system at this stage is represented by the safety of the construction activities / processes underway. This representation is due to the fact that the structural system as modeled for analysis and design purposes, is not constructed and operative as a whole yet. Thus, the safety of the construction operation itself is the real measure of the safety of the system at this stage. Figure 5, shows the developed model for the construction activity, i.e., any construction operation underway for which the safety is being monitored and controlled. According to the model, a construction activity is defined as a structure system. For a construction activity, the subsystems represent the different underlying processes. For example, for a concrete placement activity, the processes include falsework construction, rebar placement and concrete pouring[6]. The resulting system definition forms an image of the object of interest, i.e., construction activity, which emphasizes certain important properties. These properties represent the control attributes used in the control system. For the concrete placement example, safety is considered the only attribute of interest. Each subsystem has the same attribute of the structure system. The behavior of the whole activity, i.e., the structure system, is expressed in terms of the behavior of its components, i.e., the processes.

As shown in Figure 5, each process has several potential behavior functions. An algorithm, defined as the replacement procedure is needed in order to determine which behavior function represents the performance of the process at the current support instant. The behavior function at this application represents the performance function that defines the input / output relationship. An overall state represents a given combination of all relevant variables. Probability distribution functions, are a suitable choice for a behavior function for such an application. Any candidate behavior function should be able to take into account all possible combinations of the individual states of the variables which results in all potential overall states. Accordingly, any overall state at any support instant has a probability measure assigned by the behavior function. This is the only suitable approach to express the behavior of such systems. This is due to the uncertainties related to the state of any given variable at any given support instant. Referring to Figure 5, three variables are used to model the safety attribute at any point in time. The states of these three variables at any point in time represents an overall state. In that sense, the behavior function defines the behavior of the system.

At any support instant, the states of these variables provide an image of the state of the control attribute at the process level. Accordingly, an image of the corresponding attribute at the activity level can be formed. This is discussed in detail in the state evaluation stage. If any given state of an individual variable could only be expressed with a certain probability measure, any combination of such states should also be assigned a suitable probability measure. The probability measure of an overall state, assigned by the behavior function, represents the frequency of occurrence of that individual overall state. For crisply defined variables, the frequency of occurrence can be directly translated to the actual

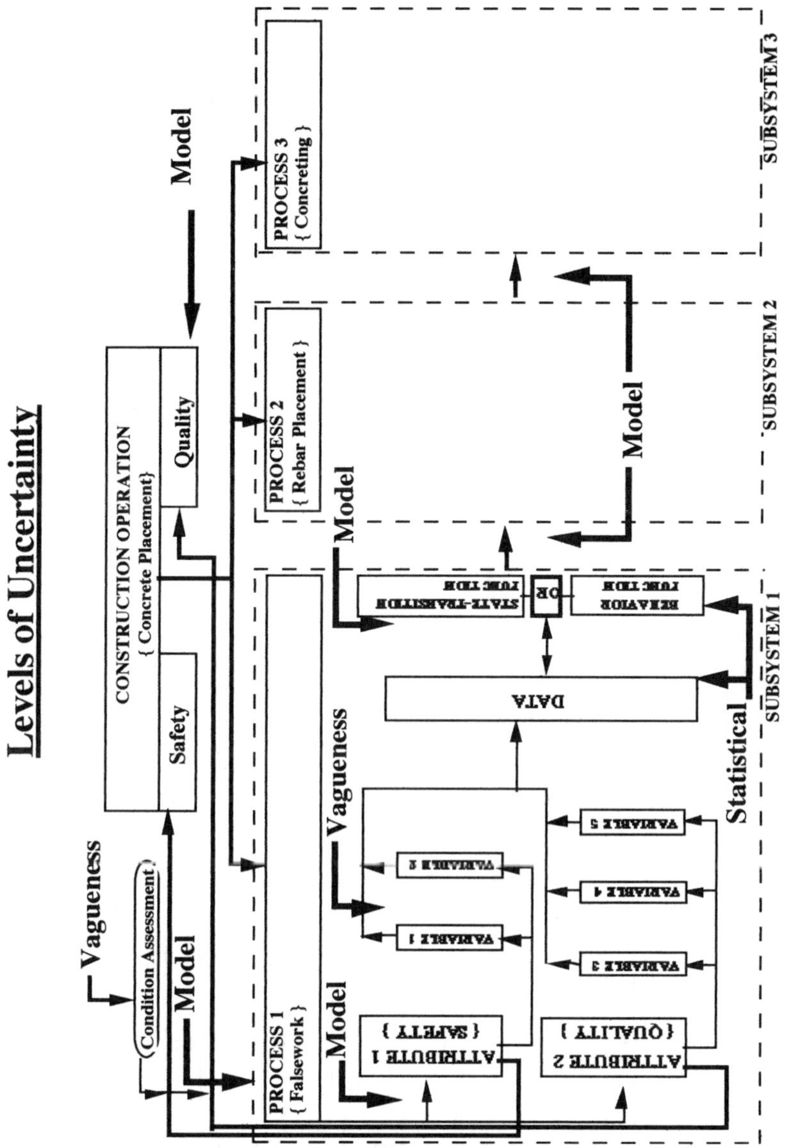

Figure 4. Sources of Uncertainty in Model Building

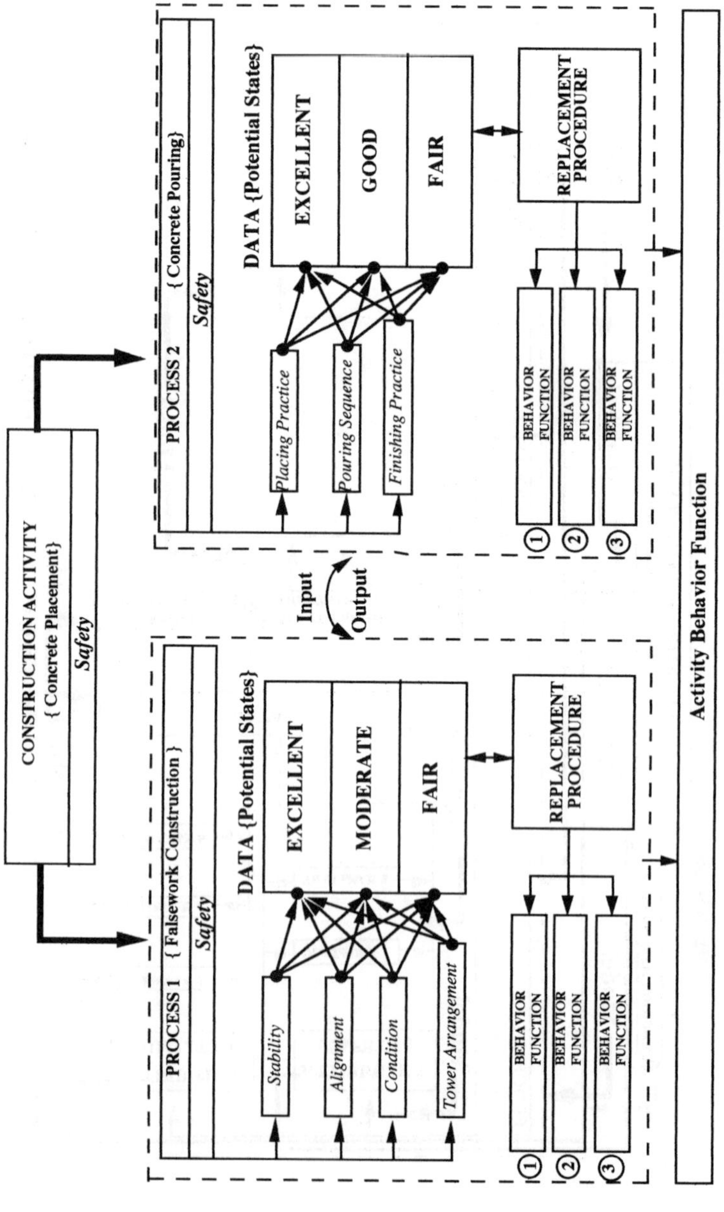

Figure 5. A Model of a Structural System During Construction

number of occurrences of such state. However, for fuzzy variables, an appropriate aggregation function should be defined in order to evaluate the required frequency of occurrence. The introduction of fuzzy variables at this specific application was meant to deal with several qualitative variables that were identified as potential control variables. The inclusion of such an approach serves the generality of the developed model. For qualitative variables, fuzzy set theory is used in defining the potential states, together with a suitable observation channel that yields a quantitative equivalent for each state[19,21,22]. As an example for this type of variable, labor experience (v_l) is considered. This variable is assumed to have four potential states, namely, fair, good, moderate and excellent. These linguistic measures are defined using fuzzy sets[19,21,40,49]. Using a scale of 0 to 10 for the level of experience, these measures are defined as follows:

$$\text{Fair} = \{\ 0|1.0,\ 1|0.8,\ 2|0.6\ \} \tag{3-a}$$

$$\text{Good} = \{\ 1|0.3,\ 2|0.5,\ 3|1.0,\ 4|0.8\ \} \tag{3-b}$$

$$\text{Moderate} = \{\ 4|0.8,\ 5|1.0,\ 6|0.9,\ 7|0.3\ \} \tag{3-c}$$

$$\text{Excellent} = \{\ 8|0.8,\ 9|0.9,\ 10|1.0\ \} \tag{3-d}$$

where Fair, Good, Moderate and Excellent are linguistic measures on a scale from 0 to 10 in which 0 is the lowest level and 10 is the highest level, and 0.1, 0.2, ..., 1.0 are degrees of belief that the corresponding elements belong to the measures. In other words, the element 0|1.0 in the Fair measure means that the degree of belief, i.e., value of membership function, that the level 0 belongs to the Fair measure is 1.0. An aggregation function should be applied to the degrees of belief of the individual states of the different variables, which yield an overall degree of belief for the overall state at each support instant. The maximum operator was utilized in this study as an aggregation function. The sum of the degrees of belief of the overall state over all the support instances represents a measure of the likelihood of occurrence of that state. Table 1, shows the evaluation of the behavior function value for a combined state that involves a fuzzy, as well as, a crisp variable. The corresponding probability of occurrence of each overall state is calculated as the ratio of the likelihood of occurrence of such a state to the sum of all likelihood of occurrence of all potential states. The likelihood of occurrence of each overall state can be expressed as:

$$N_s = \sum_{\text{all } t} d_{s,t} \tag{4}$$

where N_s is the likelihood of occurrence; $d_{s,t}$ is the aggregated degree of belief of overall state (s) at support instant (t), and the summation was performed over all support instances. The corresponding probability of occurrence of overall state (s) was then calculated using the following formula:

$$F_I(s) = \frac{N_s}{\sum_{\text{all } s} N_s} \tag{5}$$

Table 1. Behavior function values for overall states

Overall State (C_i)	Variable (v_1)		Variable (v_2)		Degree of Belief of Overall State (C)	Likelihood of Occurrence (N_c)	Behavior Function (f_B)
	State	Degree of Belief	State	Degree of Belief			
C_1 (1,10)	1	0.8	10	1	0.8		
	1	0.7	10	1	0.7		
	1	0.5	10	1	0.5		
	1	0.3	10	1	0.3	2.3	0.354
C_2 (3,10)	3	0.4	10	1	0.4		
	3	0.7	10	1	0.6		
	3	0.6	10	1	0.7	1.7	0.262
C_3 (2,10)	2	0.5	10	1	0.5		
	2	0.8	10	1	0.8		
	2	0.9	10	1	0.9	2.2	0.338
C_4 (0,10)	0	0.2	10	1	0.2		
	0	0.1	10	1	0.1	0.3	0.046

where $F_I(s)$ is the probability of having state (s) which corresponds to the value of the behavior function for that state, N_s is the Likelihood of occurrence of state (s), and the summation was performed over all the overall states.

Any construction activity consists of a number of processes that should be performed in order to declare that activity as complete[6]. These processes depend on each other in some manner. Considering concrete placement as a construction activity, the underlying processes include falsework construction, rebar placement, concrete pouring and concrete finishing. These processes represent inter-related subsystems within the structure system. Each process is defined as a generative system and all processes should be accomplished in order to declare the concrete placement activity as complete. The inter-relation among the subsystems is represented by the dependence of each process on the preceding one. Another form of inter-relationship is the input / output relation between the successive processes. The behavior of the activity should now be defined using a similar behavior function. However, such behavior function needs to be defined, knowing the behavior functions of the underlying processes. This function is shown in Figure 5 where it relates states of individual processes. The definition of such a function involves a linear optimization problem and the specifics of its definition is not in the scope of this study. However, the overall behavior function was developed for this application by the authors[6].

2.5. Definition of a Structural System for Reliability Control

In this application, a model is defined for a structural system with the objective of reliability assessment and control. In the design stage of any structure, it is a common practice to study several candidate systems and select the best candidate, based on several criteria, e.g., cost and constructability. It is also a common practice to assume a maximum expected live load, wind load and other dynamic loads and perform the design resulting in a structure that is capable of supporting these loads. However, these loads are random in nature, which make it very difficult to ensure the safety of the designed structure during its life time. Thus, the design engineer is left with several sources of uncertainty in the design problem. None of these uncertainties are completely considered in the design process. Structural reliability concepts and techniques, however, provide an answer to these uncertainty problems. Therefore, when selecting a structural system out of several candidate systems, the reliability of the system should be a major criterion in the decision making process. A viable control objective might be to provide and maintain a minimum predefined reliability level to any structural system within its design stage. The reliability of a given structural system could be evaluated knowing the states of the relevant variables. Therefore, for the objective of monitoring and controlling the reliability level, a control system could be developed for highly redundant systems. The structural system should include stand-by elements which could be selectively activated by the control system in order to improve its reliability. For the above mentioned application, the model should be built to reflect all potential failure modes at the component, as well as, the overall system level.

The structural system is decomposed into a set of potential overall failure modes. Each failure mode is defined in terms of several components. At the component level, the reliability is defined based on several failure modes. Each mode has its own limit state equation defined in terms of the related control variables. A limit state equation is expressed mathematically as:

$$CF_i = f_i(V_{1i}, V_{2i}, V_{3i}, \ldots, V_{ni}) \qquad (6)$$

where CF_i is the ith failure mode safety margin, $f_i(.)$ is the performance function of the ith failure mode, and $V_{1i}, V_{2i}, V_{3i}, \ldots, V_{ni}$ are state variables affecting the ith failure mode and (n) is the number of variables. Each state variable is, in general, a random variable. Knowing the probability distribution of each state variable, or the joint probability distribution, a complete definition of the system could be developed. However, it is very difficult to define the probability distribution of the individual state variables, as well as, the joint probability distribution. Thus, statistical properties of the variables could be utilized in the definition of the system and later used in the evaluation of the reliability level of the component. The structural system, in this example, is broken down into four hierarchical levels. At this application, this level of detail is needed for the accurate evaluation of reliability measures. Figure 6, shows the system definition, where the components represent several types of structural elements, i.e., beams, columns or slabs. For each type of structural element, several models are usually present in any given structural system. For each component, a suitable limit state equation is defined for each individual failure mode. At the state evaluation stage, each limit state equation is evaluated to result in the probability of failure of its corresponding failure mode. An overall component reliability level is evaluated based on all potential failure modes, i.e., bending, shear and deflection.

At the subsystem level, a similar structure is defined. However, at this level each overall failure mode is defined in terms of the individual components rather than the state variables. These failure modes are considered as subsystems, and the relevant components at each failure mode should be identified. This model does not require the same components to be present at all potential failure modes. Each component should be assigned a relative importance factor that reflects its impact on all individual failure modes. For a component that is not related to any given failure mode, a relative importance factor of zero shall be assigned. Such importance factors are necessary to define the relevance of the reliability of a given component to the reliability of the structural system as a whole. Each failure mode should also have a suitable limit state equation that relates all relevant components. This limit state equation is considered the performance function of the subsystem, i.e., the failure mode. Knowing the state of each component, i.e., reliability level, at a certain point in time, a reliability assessment for each individual failure mode could be developed. The performance function of the ith failure mode at the system's level is defined as:

$$SF_i = sf_i(C_{1i}, C_{2i}, C_{3i}, \ldots, C_{mi}) \tag{7}$$

where SF_i is the ith failure mode safety margin at the system level, $sf_i(.)$ is the performance function of the ith failure mode at the system level, and $C_{1i}, C_{2i}, C_{3i}, \ldots, C_{mi}$ are components affecting the ith failure mode and (m) is the number of components. Reliability assessment techniques could then be employed to evaluate a reliability level at the overall system's level. Such evaluation techniques are further discussed in the state evaluation stage.

2.6. *Definition of a Structural System for Structural Dynamic Control*

The major structural control objective has always been to restrain the dynamic response of structural systems under random dynamic excitation. Earthquake and wind loads are examples of such dynamic random loads. The evaluation of such loads and the assurance of safety were always problems because of the uncertain and random nature of such loads. Thus, the notion of structural control has emerged as an alternative for ensuring safety of such structural systems under the effect of highly uncertain and random loads[30,31,48]. In traditional structural control, the equation of motion that represents an instantaneous

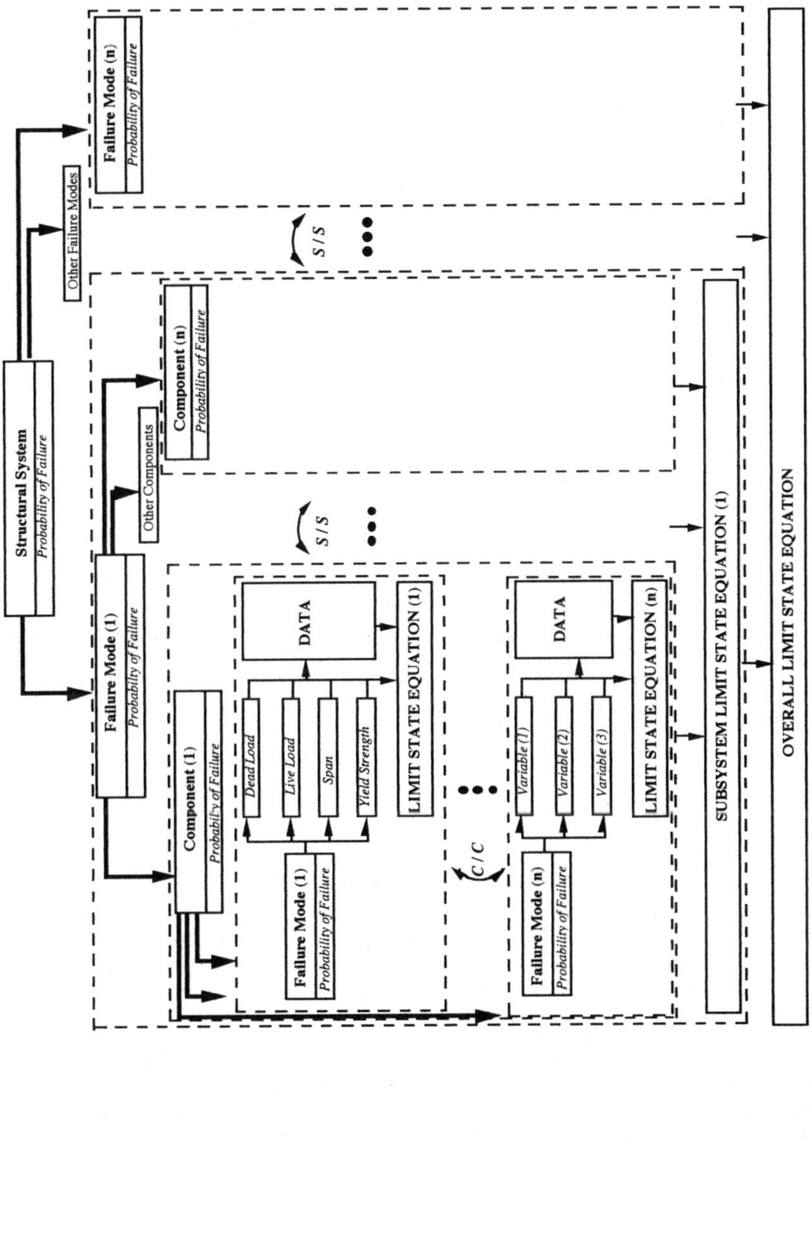

Figure 6. A Model of a Structural System for Reliability Assessment

dynamic equilibrium, is usually used in evaluating the system's response[20]. Since the control objective is to limit the response of the structural system, the evaluated response is continuously compared to a predefined acceptable threshold, beyond which the control system is activated in order to suppress the system's dynamic response.

Structural systems, in practice, are highly complex and involve a huge number of degrees of freedom. The equation of motion of a SDOF system is written as[37]:

$$m\ddot{x}(t) + c\dot{x}(t) + kx(t) = p(t) \tag{8}$$

where m is the mass, c is the damping coefficient, k is the stiffness, p(t) is the forcing function, x(t) is the displacement, $\dot{x}(t)$ is the velocity, and $\ddot{x}(t)$ is the acceleration. For a NDOF system, N equations could be written similar to the SDOF system in Eq. (8), written for each degree of freedom as[37]:

$$m_i\ddot{x}_i(t) + c_i\dot{x}_i(t) + k_{i\,j}x_i(t) + k_{j\,i}x_j(t) = p_i(t) \tag{9}$$

where each mass m_i represents a degree of freedom. Thus, the equation of motion of the NDOF system is given as[37]:

$$M\ddot{X}(t) + C\dot{X}(t) + KX(t) = P(t) \tag{10}$$

where M is the diagonal mass matrix, C is the diagonal damping matrix, K is the stiffness matrix, X(t), $\dot{X}(t)$, $\ddot{X}(t)$ are displacement, velocity and acceleration vectors respectively, and P(t) is the forcing function vector. This equation represents a set of N coupled equations. With the increase of degrees of freedom of structural systems, this set of equations gets more and more complex.

A major step in building such a model is to identify the level of detail and the components that define the overall modeled system. It is well known that the response of any linear structural system could be expressed as the linear combination of its free undamped mode shapes. This linear representation is defined as:

$$X = \Phi Y = \phi_1 y_1 + \phi_2 y_2 + \cdots + \phi_n y_n \tag{11}$$

where ϕ_i is the eigenvector of mode shape i, y_i is the modal amplitude of mode shape i, and X is the displacement vector. Keeping in mind that these mode shapes do not change with time, thus, one might consider these mode shapes as the components of any structural system. Each mode shape has a SDOF equation of motion that is solved independently in order to result in its modal amplitude. The response of the whole structure is then evaluated using linear combination of all mode shapes as shown in Eq. (11). Figure 7, shows a model for the system being defined and outlined herein. The figure reflects three hierarchical levels, the highest of which is the overall structural system. The system is then broken down into subsystems, i.e., substructures. This procedure is often used when dealing with complex and very large number of degrees of freedom[37]. Methods of component mode synthesis are utilized in breaking down such structural systems into

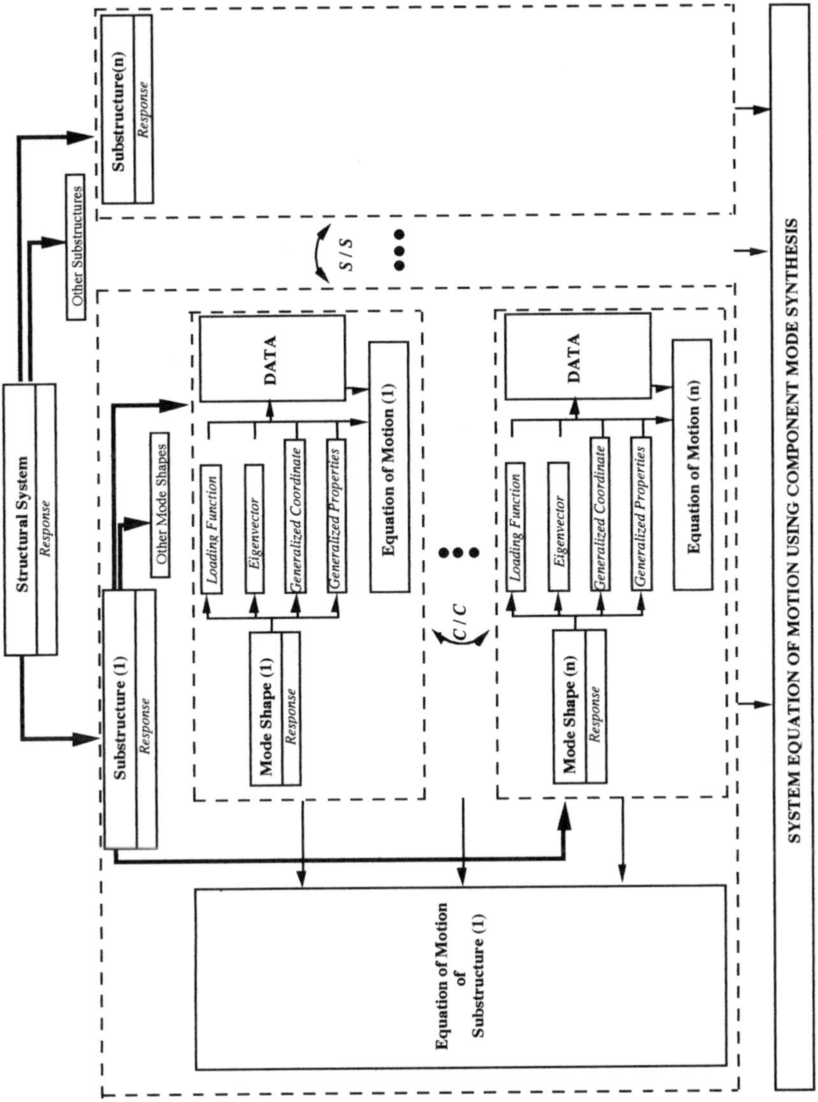

Figure 7. A Model of a Structural System for Active Structural Control

smaller manageable structural systems. Then, the response of the overall structural system is evaluated once the responses of its components, i.e., substructures, are known. Thus, the use of such an approach serves the generality of the developed model. If for a specific example, the size of the problem was manageable, one of the shown hierarchical levels could be easily deleted. This does not affect the performance and integrity of the model in any way. This integral nature is one of the major advantages of the hierarchical system identification framework. Therefore, the first step is to breakdown the structural system into smaller interconnected substructures. The interconnections in this application represent the degrees of freedom at the juncture between any two connecting substructures. The equation of motion of each individual substructure is written as[35,38]:

$$\begin{bmatrix} m_{ii} & m_{ij} \\ m_{ji} & m_{jj} \end{bmatrix} \begin{Bmatrix} \ddot{x}_i \\ \ddot{x}_j \end{Bmatrix} + \begin{bmatrix} c_{ii} & c_{ij} \\ c_{ji} & c_{jj} \end{bmatrix} \begin{Bmatrix} \dot{x}_i \\ \dot{x}_j \end{Bmatrix} + \begin{bmatrix} k_{ii} & k_{ij} \\ k_{ji} & k_{jj} \end{bmatrix} \begin{Bmatrix} x_i \\ x_j \end{Bmatrix} = \begin{Bmatrix} p_i \\ p_j \end{Bmatrix} \tag{12}$$

where i is an internal degree of freedom, j is a juncture degree of freedom where an interconnection with another substructure exists, m is a mass matrix, k is a stiffness matrix, p is a force vector, and $x, \dot{x}, \ddot{x}$ are displacement, velocity and acceleration vectors respectively. The developed model, at this stage, breaks down the NDOF system into a set of MDOF systems where M is a smaller number of degrees of freedom. The model shown in Figure 7, then, defines each of the substructures by further breaking them down into their components, i.e., mode shapes. The model also shows the input variables, i.e., the forcing functions, the output variables, i.e., the response, which is the control attribute in this example. The use of the response as an expression for the control attribute is meant as a generic term that includes one or all of the response quantities, i.e., displacement, velocity and acceleration. All related and required properties are also defined in the developed model. The equation of motion at each level defines the performance of that specific hierarchical level. At the lowest hierarchical level, i.e., the mode shape, a SDOF equation of motion is defined for each mode, based on its properties. Such an equation could be written as:

$$M_n \ddot{Y}_n + C_n \dot{Y}_n + K_n Y_n = P_n(t) \tag{13}$$

where Y_n is the modal amplitude of the nth mode, M_n is the generalized mass matrix defined as $\phi_n^T M \phi_n$, C_n is the generalized damping matrix defined as $\phi_n^T C \phi_n$, K_n is the generalized mass matrix defined as $\phi_n^T K \phi_n$, and $P_n(t)$ is the generalized forcing vector defined as $\phi_n^T P(t)$. The equation of motion of the next hierarchical level, i.e., the substructure, is written as shown in Eq. (12), where the displacement vector is expressed in terms of the modal amplitudes as shown in Eq. (11). The equation of motion for the overall structural system is then developed using the component mode synthesis approach[35,38]. The details of such an approach are beyond the scope of this chapter.

3. State Evaluation

3.1. Introduction

In traditional control theory, a basic mathematical model known as the transfer function defines the behavior of the system. The use of fuzzy control algorithms is meant to result in simple control schemes. Such schemes are easily tailored and applied to several types of applications. Therefore, general system identification was considered a major step in developing such a general fuzzy control scheme. The mathematical transfer function used in traditional control theory is represented in this general control algorithm by two major units. The first unit is the system identification framework, outlined in the previous section. The second unit acts, side-by-side, with the system identification framework in order to translate the input values into output values. The performance function is a general form for expressing the input / output relation for the system. However, such function, by itself, might not be enough to express such a relation. In general, a state evaluation mechanism should be defined, to be applied together with such a performance function, such that an input / output relation is completely defined. In this section, the basic building blocks necessary for the development of any state evaluation scheme are discussed. Such scheme should be dependent on the nature of the system and the form of its performance function. Therefore, a general state evaluation scheme is very difficult to develop because of the wide range of potential performance functions. However, any state evaluation scheme shall include several essential building blocks. Although, these building blocks should be present in every state evaluation scheme, they usually have several physical representations, Yet, they perform the same functions in all schemes. In the following discussion, the main building blocks of any state evaluation scheme are outlined. Several practical applications then follow to demonstrate the implementation and use of the outlined building blocks in the development of the state evaluation schemes.

3.2. State Evaluation Building Blocks

In spite of the wide variety of state evaluation schemes, employed to evaluate output states of different applications, such schemes share essential building blocks. Such building blocks perform important tasks that serve the integrity of the developed scheme. The objective of this section is to outline such essential building blocks and identify their intended role in the whole scheme. The presence of such common building blocks is due to the fact that such state evaluation schemes are intended to operate together with the system identification framework outlined earlier. The definition of the system in such a universal framework would require specific components to be present in any state evaluation scheme. Such basic building blocks are summarized and discussed in the following sections.

Relative Importance Factors. These factors are defined as real numbers less than or equal to one. Such factors should represent the impact of any given variable on the state of the component it is modeling. Similar factors are defined for all hierarchical levels within the same context. This process should carry on to the top hierarchical level. Thus, the impact of any given variable, which is defined at the lowest hierarchical level, could be carried over to the top hierarchical level through such factors. This would result in the impact of that specific variable on the state of the top overall system. This could be viewed as the most practical procedure that would evaluate such impacts, because of the nature of the system model. These factors might have several physical representations, however, the same task is performed regardless of the form of that factor. The need for such factors is justified because of the hierarchical system identification framework outlined earlier. The presence of several hierarchical levels, each comprises one or more sub-hierarchical levels, dictates the need for some factor relating the impacts of such multiple component levels to their superior levels.

Inter-connections. Inter-connections were mentioned in the system identification framework as an important component of such definition. These inter-connections are fully defined and utilized at the state evaluation scheme. Inter-connections represent mutual impacts among components / subsystems at the same hierarchical level. For any given application, all potential inter-connections shall be outlined and considered in the state evaluation scheme. Such inter-connections include the effect of a given component state on all other components at the same hierarchical level. When evaluating the states of all components, the impact of all neighboring components should be considered. These inter-connections might be considered as relative importance between members of the same level. However, such inter-connections take different forms, i.e., they need not be always fractions less than one. These inter-connections might be simply an input / output relationship or correlation of some form between the two components / subsystems.

Aggregation Procedures. These procedures are meant to evaluate the attribute level within a given hierarchical level due to several potential states of its components. These procedures are only necessary whenever several component / subsystem potential states affect the state of its superior subsystem/system. For example, several failure modes might be considered responsible for the failure of a given structural component. The failure likelihood of such a component shall be evaluated taking into consideration all potential failure modes. This is only possible through a suitable aggregation procedure that combines the effect of all potential failure modes into a representative component failure likelihood. The nature of such procedure depends on the nature of the problem and the nature of the control variables. Such procedure might be a union or an intersection function, depending on the nature and type of failure modes. In some applications, this task might be accomplished by the performance function of that level, while other applications might require the introduction of a specific procedure for such task. The development of such a procedure is a major component of the state evaluation scheme. It should be emphasized that for any given example, several aggregation procedures might be used at consecutive steps within the state evaluation scheme. These procedures may or may not be of the same type and form.

Performance Functions. Such functions are defined as input/output relations, which are considered the corner stone of the state evaluation scheme. Performance functions are dependent on the application under consideration. These functions shall also be developed in a manner that accommodates all potential types of control variables that might be related. Such a definition is essential for the generality of the developed scheme. Performance functions should be defined at each hierarchical level. At the lowest level, they should define the relation between the input state values and the output attribute level. At higher levels, they should relate states of components / subsystems, as inputs, to the superior hierarchical level attribute value as outputs. As mentioned earlier, some of these functions might be supplemented by aggregation procedures that aid in the evaluation of attribute values. Performance functions might be mathematical expressions relating inputs and outputs in a manner that defines the behavior of the modeled system. It might be logical expressions relating potential states to expected attribute levels. It might also be defined as probability distributions which relate all potential states of components / variables to expected attribute levels. In some applications, higher performance functions are not always easily defined. In such cases, special techniques are available in order to define a performance function of a subsystem / system knowing the performance functions of its components / subsystems. Such an approach has been adopted and developed, by the authors for a specific application as discussed below[7]. In the following sections, three applications are outlined in order to demonstrate the development of a suitable state evaluation scheme. In the developed examples, all building blocks are outlined and discussed in order to show how such concepts could be implemented for different practical applications.

3.3. State Evaluation for Safety Control During Construction

Referring to the system definition of the same application, the safety of the construction operation was considered a reasonable measure for the safety of the structural system during that stage. State evaluation is essential in two phases of any construction project, the planning and the control phases. In the planning phase, all possible combinations of states of variables should be considered. This could be done by applying the behavior function of the process, which assigns a frequency of occurrence for each potential combination of states[7]. However, in the control phase, real-time observations and measurements provide crisp information about the states of the variables. Thus, an adjusted behavior function which deals with a reduced number of potential combinations, should be considered.

State Evaluation at Process level. The objective at this stage is to evaluate the failure likelihood of each process, due to the occurrence of certain state combinations of control variables. These variables were assumed to be qualitatively assessed using linguistic measures that can be defined using fuzzy set theory as explained in Eq. (3). In this application, the likelihood of failure during construction was chosen as a safety measure for construction activities. The failure likelihood attribute was assumed to be significantly affected by two variables where v_1 = labor experience and v_2 = equipment condition. The probability of failure was used as the likelihood measure. The failure likelihood level was subjectively assessed to be in the range from 10^{-1} to 10^{-6}. Accordingly, a fuzzy set with six elements (10^{-1}, 10^{-2}, 10^{-3} ,10^{-4}, 10^{-5}, 10^{-6}) was assumed to represent the failure likelihood level. Each state is assigned a corresponding expected failure likelihood. In fuzzy set theory, such a mapping results in a fuzzy relation. Thus the relation between the state of the variable and the failure likelihood level, defined as a linguistic measure, is constructed. This relation represents a component of the performance function. For example, an excellent state of a variable was assumed to result in a low failure likelihood. For the four main states defined in Eq. (3), the corresponding failure likelihood measures were assumed as follows:

$$\text{High} = \{10^{-3}|0.16,\ 10^{-2}|0.64,\ 10^{-1}|1.0\} \tag{14-a}$$
$$\text{Moderate-High} = \{10^{-5}|0.16,\ 10^{-4}|0.64,\ 10^{-3}|1.0,\ 10^{-2}|0.64\} \tag{14-b}$$
$$\text{Moderate-Low} = \{10^{-5}|0.64,\ 10^{-4}|1.0,\ 10^{-3}|0.64,\ 10^{-2}|0.16\} \tag{14-c}$$
$$\text{Low} = \{10^{-6}|1.0,\ 10^{-5}|0.64,\ 10^{-4}|0.16\} \tag{14-d}$$

where High, Moderate-High, Moderate-Low and Low are linguistic measures on a scale from 10^{-6} to 10^{-1} in which 10^{-6} is the lowest level and 10^{-1} is the highest level, and 0.1, 0.2, 0.3, ...,1.0 are degrees of belief that the corresponding elements belong to the measures. Such failure likelihood is dependent on a specific state levels rather than a variable by itself. Thus, an importance factor is introduced at this stage to influence the resulting fuzzy relation in a manner that reflects the impact of the variable under consideration. The importance factor is applied as a multiplier to the state membership function. In other words, the whole variable membership function is scaled up or down based on the value of the importance factor. It is important to point out that this effect does not emphasize any specific element in the universe of discourse, and thus retains the uncertainty associated with the fuzzy set definition of each state. The adjusted state fuzzy set was computed as

$$\mu_A{}^a(z) = \omega\ \mu_A(z) \tag{15}$$

where $\mu_A{}^a(z)$ is the adjusted membership function of element z in fuzzy set A, ω is the importance factor, and $\mu_A(z)$ is the original membership function of element z in fuzzy set A. Therefore, the importance factor magnifies the effect of the variable on the failure likelihood, at the process level. However, the magnified variable retains the original shape of the fuzzy set.

In this example, all the involved variables and failure likelihood measures were defined using linguistic measures. Then, fuzzy relations were defined on the Cartesian product space of the corresponding fuzzy sets[19,21,40,49], where the Cartesian product of two fuzzy sets A and B is defined as

$$\mu_{AXB}(w,z) = \mathrm{MIN}\{\mu_A(w), \mu_B(z)\} \qquad (16)$$

where $\mu_{AXB}(w,z)$ is the membership function of the Cartesian product, MIN is a minimum operator, $\mu_A(w)$ is the membership function of element w in fuzzy set A, and $\mu_B(z)$ is the membership function of element z in fuzzy set B. For example, The relation between variable (v_1) with state (1), i.e., Excellent, and the resulting Low failure likelihood is defined using the Cartesian product shown in Table 2. Table 3, shows the relation between variable (v_2) with state (20), i.e., Good, and the resulting Moderate-High failure likelihood. Then, it is necessary to consider the effect of the overall state $C_I(1,20)$, i.e., the combination of variable (v_1) with state (1) and variable (v_2) with state (20), on the failure likelihood level. This relation was defined using the union function for fuzzy relations which is defined as follows:

$$\mu_{R_1 \cup R_2}(w) = \mathrm{MAX}\{\mu_{R_1}(w), \mu_{R_2}(w)\} \qquad (17)$$

where $\mu_{R_1 \cup R_2}(w)$ is the membership function for the union, MAX is a maximum operator, $\mu_{R_1}(w)$ is the membership function of element (w) in fuzzy relation R_1, and $\mu_{R_2}(w)$ is the membership function of the same element (w) in fuzzy relation R_2. The resulting relation matrix is referred to hereafter as the combined relation matrix. Table 4, shows the combined relation matrix for the overall state $C_I(1,20)$ of process I. Accordingly, all relation matrices collectively, together with the behavior function of the process, define its performance. In this application, an aggregation procedure was found necessary to supplement the performance function in order to evaluate the failure likelihood in a fuzzy set format. Applying the maximum operator, as an aggregation tool, to each column in the combined relation matrix defined in Table 4, a degree of belief for each element in the failure likelihood fuzzy set was evaluated.

It is always desirable to have a single measure that represents the failure likelihood level of the process under consideration. Knowing the fuzzy-set estimate of the failure likelihood level, a defuzzification procedure was utilized to evaluate a single point estimate for this fuzzy set. Based on this approach, the probability content (or average probability) P_{f_j} for a fuzzy failure likelihood measure due to the jth combination of states can be determined as[11]

Table 2. Relation between the State of Variable v_1 and Failure Likelihood

Relation Between State and Failure Likelihood		Low Failure Likelihood					
		10^{-6}	10^{-5}	10^{-4}	10^{-3}	10^{-2}	10^{-1}
Excellent State of Variable	0	0.00	0.00	0.00	0.00	0.00	0.00
	1	0.00	0.00	0.00	0.00	0.00	0.00
	2	0.00	0.00	0.00	0.00	0.00	0.00
	3	0.00	0.00	0.00	0.00	0.00	0.00
	4	0.00	0.00	0.00	0.00	0.00	0.00
	5	0.00	0.00	0.00	0.00	0.00	0.00
	6	0.00	0.00	0.00	0.00	0.00	0.00
	7	0.00	0.00	0.00	0.00	0.00	0.00
	8	0.64	0.64	0.16	0.00	0.00	0.00
	9	0.72	0.64	0.16	0.00	0.00	0.00
	10	0.80	0.64	0.16	0.00	0.00	0.00

Table 3. Relation between the State of Variable v_2 and Failure Likelihood

Relation Between State and Failure Likelihood		Moderate-High Failure Likelihood					
		10^{-6}	10^{-5}	10^{-4}	10^{-3}	10^{-2}	10^{-1}
Good State of Variable	0	0.00	0.00	0.00	0.00	0.00	0.00
	1	0.00	0.15	0.15	0.15	0.15	0.00
	2	0.00	0.16	0.25	0.25	0.25	0.00
	3	0.00	0.16	0.50	0.50	0.50	0.00
	4	0.00	0.16	0.40	0.40	0.40	0.00
	5	0.00	0.00	0.00	0.00	0.00	0.00
	6	0.00	0.00	0.00	0.00	0.00	0.00
	7	0.00	0.00	0.00	0.00	0.00	0.00
	8	0.00	0.00	0.00	0.00	0.00	0.00
	9	0.00	0.00	0.00	0.00	0.00	0.00
	10	0.00	0.00	0.00	0.00	0.00	0.00

Table 4. Combined Relation Matrix for Overall State $C_I(1,20)$ of Process I

Combined Relation Matrix		Failure Likelihood					
		10^{-6}	10^{-5}	10^{-4}	10^{-3}	10^{-2}	10^{-1}
$C_I(1,20)$	0	0.00	0.00	0.00	0.00	0.00	0.00
	1	0.00	0.15	0.15	0.15	0.15	0.00
	2	0.00	0.16	0.25	0.25	0.25	0.00
	3	0.00	0.16	0.50	0.50	0.50	0.00
	4	0.00	0.16	0.40	0.40	0.40	0.00
	5	0.00	0.00	0.00	0.00	0.00	0.00
	6	0.00	0.00	0.00	0.00	0.00	0.00
	7	0.00	0.00	0.00	0.00	0.00	0.00
	8	0.64	0.64	0.16	0.00	0.00	0.00
	9	0.72	0.64	0.16	0.00	0.00	0.00
	10	0.80	0.64	0.16	0.00	0.00	0.00

$$\log_{10}\left(P_{f_j}\right) = \frac{\sum_{i=1}^{N_p} \mu\left(P_{f_i}\right)\log_{10}\left(P_{f_i}\right)}{\sum_{i=1}^{N_p} \mu\left(P_{f_i}\right)} \quad \text{for } j = 1,2,\ldots,9 \tag{18}$$

where P_{f_i} is the $i^{\underline{th}}$ element in the failure likelihood fuzzy set, and N_p is the number of elements in the failure likelihood fuzzy set. The logarithm to the base 10 of P_{f_j} and P_{f_i} was used in Eq. (18) in order to obtain a weighted average of the power order of the P_{f_i} and P_{f_j} values. The previous procedure could be considered as a defuzzification procedure by which an expected value of the failure likelihood level P_{f_j} results from its corresponding fuzzy set.

As mentioned earlier, state evaluation is essential in two phases of any construction project. At the planning phase, all possible combinations of states, with associated frequencies, should be considered. This procedure accounts for the uncertainty resulting from the lack of knowledge about the occurrence of any of these combinations. This could be achieved by using the behavior function of the process. This function assigns a probability measure based on the frequency of occurrence of each possible combination of states of the variables, i.e., overall states. Based on the previous discussion, a failure likelihood level was calculated for each potential combination of states. However, the resulting failure likelihood level has a frequency of occurrence which is equal to that of the combined state causing it. In other words, the resulting failure likelihood at the process level is distributed over all possible combinations of states. It is however important to determine a single point estimate of the process failure likelihood level, taking into account all possible combinations of states. This could be accomplished by applying the mathematical expectation which is defined as

$$E(\mathbf{P_f}) = \sum_{j=1}^{9} P_{f_j} F_I\left(C_{Ij}\right) \tag{19}$$

where $E(\mathbf{P_f})$ is the expected value of the probability of failure, P_{f_j} is the probability of the $j^{\underline{th}}$ overall state as defined in Eq. (18), and $F_I(C_{Ij})$ is the frequency of occurrence of overall state C_{Ij} given by the behavior function of process I. Applying Eq. (19) to the probabilities of failure of all possible combinations, an expected value of the probability of failure at the process level was determined. This is another form of an aggregation procedure. This means that at the process level, two aggregation procedures are necessary in order to evaluate the failure likelihood. It should be emphasized, though, that these two procedures are completely different in their form and shape, however they perform the same task. The outlined procedure was applied to the example under consideration. However, in a real-time control operation mode, fewer potential states for all state variables should be defined. Such an improved definition results from the fact that more information should be available at that stage than in the planning stage. Thus, a modified behavior function is defined and a more realistic failure likelihood could be evaluated.

State evaluation at the Activity Level. The objective at this stage is to combine the resulting failure likelihood for all the associated processes, into a failure likelihood assessment of the construction activity as a whole. Two issues need to be addressed at this stage, the first is the impact of the global overall states, i.e., combinations of overall states, on the failure likelihood of the whole activity. The second issue is the overall behavior function that represents the entire activity.

In the previous section, combinations of states of variables, i.e., overall states, and their effect on a specific process were studied. The same rationale could be applied at the next hierarchical level. However, combinations of overall states, i.e., global overall states, and their effect on the whole activity should be studied. It is quite clear that the impact of the overall state $C_I(1,20)$ on the failure likelihood of process I is different from its impact on the construction activity as a whole. This difference results from the fact that the impact of the overall state $C_I(1,20)$ on the construction activity should include an importance factor of process I, that reflects the significance of its condition on the failure likelihood of the construction activity. As discussed earlier, the importance factor is a suitable approach for carrying the impact of a given state over to the top hierarchical level. This is done through multiple importance factors each of which is defined at a specific level. At the process level, the first importance factor defined the impact of a given variable on the process. At the next level, an additional importance factor is needed to carry that impact over to the top hierarchical level, i.e., overall system. This importance factor should be applied as a multiplier to the corresponding combined relation matrix at the process level. For example, the overall state $C_I(1,20)$ of process I as defined in Table 4 was adjusted using an importance factor of 0.8. The adjusted combined relation matrix was therefore calculated for the overall state $C_I(1,20)$ as shown in Table 5. The combined effects of all the involved processes on the activity failure likelihood, were then determined using the union function of fuzzy relations as defined in Eq. (17). The union function was applied to the adjusted combined relation matrices of each potential global overall state. At this stage, all possible combinations should be considered at the process and activity levels. For the example under consideration, two variables were considered for each process with three potential states per variable and two processes were considered for the activity, resulting in nine different combinations considered at the process level and eighty one potential combinations, i.e., global overall states, at the activity level. The resulting matrix is defined hereafter as the combined process matrix. Applying the same aggregation and defuzzification procedures, which were developed at the process level, a failure likelihood level was determined for each combined relation matrix. Table 6, shows the resulting failure likelihood level matrix which contains the failure likelihood levels of the corresponding combinations as its entries. The final computational step is to determine a single point estimate of the activity failure likelihood level taking into account all potential combinations of the overall states. This could be achieved by defining the overall behavior function of the whole activity. The development of such a function is beyond the scope of this chapter. However, a brief discussion follows in order to outline the adopted procedure which could be applied in similar situations. The interested reader is referred to previous publications by the authors[7].

Performance Function of the Overall System. The performance function, at the activity level, should assign a probability of occurrence for each potential global overall state. This means that an overall behavior function of the construction activity should be defined, based on known behavior functions of its components, i.e., the underlying processes. Referring to Tables 7 and 8, in order to define the behavior function of the overall system, all potential combinations of the overall states of both processes should be considered. Table 9, shows these potential combinations, together with the associated unknown frequencies of the combinations, i.e., $F_A(C_{Ii},C_{IIj})$. In order to solve for the unknown frequencies, two conditions should be satisfied. The first condition requires that the

Table 5. Adjusted Combined Relation Matrix for Overall State $C_I(1,20)$ of Process I

Adjusted Combined Relation Matrix		Failure Likelihood					
		10^{-6}	10^{-5}	10^{-4}	10^{-3}	10^{-2}	10^{-1}
$C_I(1,20)$ Overall State	0	0.00	0.00	0.00	0.00	0.00	0.00
	1	0.00	0.12	0.12	0.12	0.12	0.00
	2	0.00	0.13	0.20	0.20	0.20	0.00
	3	0.00	0.13	0.40	0.40	0.40	0.00
	4	0.00	0.13	0.32	0.32	0.32	0.00
	5	0.00	0.00	0.00	0.00	0.00	0.00
	6	0.00	0.00	0.00	0.00	0.00	0.00
	7	0.00	0.00	0.00	0.00	0.00	0.00
	8	0.51	0.51	0.13	0.00	0.00	0.00
	9	0.58	0.51	0.13	0.00	0.00	0.00
	10	0.64	0.51	0.13	0.00	0.00	0.00

Table 6. Activity Failure Likelihood Matrix

Failure Likelihood Matrix of the Activity		Process II								
		$C_1(1,10)$	$C_2(1,20)$	$C_3(1,30)$	$C_4(2,10)$	$C_I(2,20)$	$C_6(2,30)$	$C_I(3,10)$	$C_8(3,20)$	$C_9(3,30)$
Process I	$C_1(1,10)$	$3.9x10^{-6}$	$4.0x10^{-5}$	$9.3x10^{-5}$	$2.6x10^{-5}$	$2.6x10^{-5}$	$1.0x10^{-4}$	$5.6x10^{-5}$	$6.0x10^{-5}$	$9.3x10^{-5}$
	$C_2(1,20)$	$5.6x10^{-5}$	$5.6x10^{-5}$	$1.3x10^{-4}$	$5.6x10^{-5}$	$5.6x10^{-5}$	$1.3x10^{-4}$	$1.1x10^{-4}$	$1.1x10^{-4}$	$1.3x10^{-4}$
	$C_3(1,30)$	$1.4x10^{-4}$	$1.6x10^{-4}$	$1.4x10^{-4}$	$1.5x10^{-4}$	$1.5x10^{-4}$	$1.5x10^{-4}$	$1.4x10^{-4}$	$1.4x10^{-4}$	$1.4x10^{-4}$
	$C_4(2,10)$	$1.5x10^{-4}$	$1.5x10^{-4}$	$3.2x10^{-4}$	$1.5x10^{-4}$	$1.5x10^{-4}$	$3.1x10^{-4}$	$2.7x10^{-4}$	$2.7x10^{-4}$	$3.2x10^{-4}$
	$C_5(2,20)$	$2.0x10^{-4}$	$2.7x10^{-4}$	$5.9x10^{-4}$	$2.0x10^{-4}$	$7.2x10^{-4}$	$1.5x10^{-3}$	$3.6x10^{-4}$	$1.2x10^{-3}$	$1.5x10^{-3}$
	$C_6(2,30)$	$5.0x10^{-4}$	$6.9x10^{-4}$	$6.9x10^{-4}$	$5.0x10^{-4}$	$1.7x10^{-3}$	$1.7x10^{-3}$	$5.0x10^{-4}$	$1.7x10^{-3}$	$1.7x10^{-3}$
	$C_7(3,10)$	$7.0x10^{-4}$	$6.2x10^{-4}$	$7.0x10^{-4}$	$6.5x10^{-4}$	$6.5x10^{-4}$	$6.5x10^{-4}$	$7.0x10^{-4}$	$6.8x10^{-4}$	$7.0x10^{-4}$
	$C_8(3,20)$	$1.0x10^{-3}$	$1.3x10^{-3}$	$1.6x10^{-3}$	$9.5x10^{-4}$	$5.4x10^{-3}$	$5.4x10^{-3}$	$1.0x10^{-3}$	$7.2x10^{-3}$	$7.2x10^{-3}$
	$C_9(3,30)$	$1.0x10^{-3}$	$1.3x10^{-3}$	$1.7x10^{-3}$	$9.5x10^{-4}$	$5.4x10^{-3}$	$5.4x10^{-3}$	$1.0x10^{-3}$	$7.2x10^{-3}$	$2.5x10^{-2}$

Table 7. Behavior Function for Process I

Behavior Function Value		Equipment Condition		
		Excellent (10)	Good (20)	Fair (30)
Labor Experience	Excellent (1)	$F_I(C_{I1}) = 0.10$	$F_I(C_{I2}) = 0.05$	$F_I(C_{I3}) = 0.10$
	Good (2)	$F_I(C_{I4}) = 0.10$	$F_I(C_{I5}) = 0.25$	$F_I(C_{I6}) = 0.20$
	Fair (3)	$F_I(C_{I7}) = 0.05$	$F_I(C_{I8}) = 0.05$	$F_I(C_{I9}) = 0.10$

Table 8. Behavior Function for Process II

Behavior Function Value		Equipment Condition		
		Excellent (10)	Good (20)	Fair (30)
Labor Experience		$F_{II}(C_{II1}) = 0.10$	$F_{II}(C_{II2}) = 0.15$	$F_{II}(C_{II3}) = 0.05$
	Good (2)	$F_{II}(C_{II4}) = 0.05$	$F_{II}(C_{II5}) = 0.20$	$F_{II}(C_{II6}) = 0.10$
	Fair (3)	$F_{II}(C_{II7}) = 0.15$	$F_{II}(C_{II8}) = 0.10$	$F_{II}(C_{II9}) = 0.10$

Table 9. Activity Behavior Function

Behavior Function Value for the Activity		Process II								
		$C_1(1,10)$	$C_2(1,20)$	$C_3(1,30)$	$C_4(2,10)$	$C_5(2,20)$	$C_6(2,30)$	$C_7(3,10)$	$C_8(3,20)$	$C_9(3,30)$
Process I	$C_1(1,10)$	f_1	f_2	f_3	f_4	f_5	f_6	f_7	f_8	f_9
	$C_2(1,20)$	f_{10}	f_{11}	f_{12}	f_{13}	f_{14}	f_{15}	f_{16}	f_{17}	f_{18}
	$C_3(1,30)$	f_{19}	f_{20}	f_{21}	f_{22}	f_{23}	f_{24}	f_{25}	f_{26}	f_{27}
	$C_4(2,10)$	f_{28}	f_{29}	f_{30}	f_{31}	f_{32}	f_{33}	f_{34}	f_{35}	f_{36}
	$C_5(2,20)$	f_{37}	f_{38}	f_{39}	f_{40}	f_{41}	f_{42}	f_{43}	f_{44}	f_{45}
	$C_6(2,30)$	f_{46}	f_{47}	f_{48}	f_{49}	f_{50}	f_{51}	f_{52}	f_{53}	f_{54}
	$C_7(3,10)$	f_{55}	f_{56}	f_{57}	f_{58}	f_{59}	f_{60}	f_{61}	f_{62}	f_{63}
	$C_8(3,20)$	f_{64}	f_{65}	f_{66}	f_{67}	f_{68}	f_{69}	f_{70}	f_{71}	f_{72}
	$C_9(3,30)$	f_{73}	f_{74}	f_{75}	f_{76}	f_{77}	f_{78}	f_{79}	f_{80}	f_{81}

frequency of an overall state of a process, calculated from the unknown overall behavior function, i.e., unknown frequencies in Table 9, should be compatible with the frequency of the same overall state, calculated from the behavior function for the process. This condition can be expressed as follows:

$$F_I(C_{Ii}) = \sum_{j=1}^{9} F_A(C_{Ii}, C_{IIj}) \qquad \text{for } i = 1,2, \ldots,9 \qquad (20\text{-a})$$

$$F_{II}(C_{IIj}) = \sum_{i=1}^{9} F_A(C_{Ii}, C_{IIj}) \qquad \text{for } j = 1,2, \ldots,9 \qquad (20\text{-b})$$

where $F_I(C_{Ii})$ is the value of the behavior function for process I for the overall state C_{Ii} as defined in Table 7, $F_{II}(C_{IIj})$ is the value of the behavior function for process II for the overall state C_{IIj} as defined in Table 8, $F_A(C_{Ii},C_{IIj})$ is the value of the overall behavior function for the global overall state (C_{Ii},C_{IIj}). This condition results in a set of equations with the values of $F_A(C_{Ii},C_{IIj})$, i.e., f_1, f_2, ..., f_{81} as unknowns. Referring to Tables 7, 8 and 9, an example constraint equation could be written as:

$$f_1 + f_2 + f_3 + f_4 + f_5 + f_6 + f_7 + f_8 + f_9 = 0.1 \qquad (21)$$

where f_1, f_2, ..., f_9 are values of the overall behavior function for the combinations $C_{I1}(1,10)$ and all potential overall states of process II, and 0.1 is the value of the behavior function of process I for the overall state $C_{I1}(1,10)$. The total number of constraints is equal to the total number of overall states in both processes. In the example under consideration eighteen equations similar to Eq. (21) were developed. The second condition requires that all resulting probabilities should be positive. This condition could be stated as follows:

$$f_i \geq 0 \qquad \text{for } i = 1,2, \ldots,81 \qquad (22)$$

Solving the set of constraints and the inequalities defined in Eqs. (21) and (22), a solution for the overall behavior function results. However, it is obvious that this solution is not unique. In other words, the solution defines a range within which the predefined conditions are satisfied. In order to select the optimal overall behavior of the overall system, the principle of maximum entropy was utilized[19,22]. The principle of maximum entropy states that the selection of a probability distribution, among several candidate distributions, should be based on the maximization of the entropy, subject to all additional constraints required by the available information. This means that an optimization problem should be solved in order to obtain the required solution. However, all additional constraints should be linearly independent. Therefore, the objective is to define the frequency distribution (or probability mass function) that maximizes the Shannon entropy, with linearly independent set of constraints. The specifics of such a procedure are beyond the scope of this chapter. However, the interested reader is referred to the authors work published earlier[7]. By the definition of the overall behavior function, the final aggregation step could be applied. Thus, resulting in a single point estimate of the failure likelihood of the overall system, i.e., construction activity.

3.4 . State Evaluation for Reliability Assessment and Control

The main objective of the state evaluation scheme, in this application, is to evaluate the reliability of the structural system. However, as demonstrated in the previous application, in order to evaluate the attribute level at the overall system, component and/or subsystem attribute levels should be evaluated first. In other words, the reliability of the system is a function of the reliability of its components. This is easily generalized in reference to any given attribute level required to be evaluated. By now, it should be realized that the state evaluation scheme for any application has to be performed at consecutive levels, as demonstrated in the previous application. In the following discussion, a proposed scheme is developed and outlined with the emphasis on the role of the previously defined building blocks.

State Evaluation at the Component Level. Traditional structural reliability techniques could be utilized to evaluate a reliability assessment at both hierarchical levels, i.e., the component and the system[10,11,12,18,24,27,43,46,47]. At the component level, the defined limit state equation for each individual failure mode, defines the cutoff limit between the safe and failure zones. The cutoff limit is known as the failure surface where the loads are equal to the resistances[10,27]. The failure surface, i.e., the performance function, could be expressed as shown in Eq. (6). Based on this definition, the safe zone is defined as $CF_i(.) > 0$ and the failure zone is defined as $CF_i(.) < 0$. The reliability of the component is defined as the probability of its survival during its regular operation[10,27]. This could be expressed mathematically as:

$$P(CF_i(.) > 0) = P_s = 1 - P_f = \text{Reliability} \tag{23}$$

where P_s is the probability of survival, and P_f is the probability of failure. The general form of any performance function could be expressed as :

$$CF_i(.) = R - L \tag{24}$$

where R is a measure of resistance, and L is a measure of load effect. As shown in the system model, each component usually comprises several potential failure modes. Each mode is defined in terms of a set of basic random variables representing the resistance and load effects. For example, for a beam under bending, one potential failure mode is a bending failure where the resistance is defined as :

$$R = F_Y S_X \tag{25}$$

where F_Y is the yield strength and is represented as a random variable with a suitable distribution, and S_X is the section modulus and is also represented as a random variable. The load effect acting on a simply supported beam uniformly loaded, would be defined as:

$$L = \frac{wl^2}{8} \tag{26}$$

where w is the load intensity represented as a random variable, and l is the beam span which may or may not be represented as a random variable. In this example, (R) represents the moment capacity of a given cross section, i.e., the resistance. If such a capacity exceeds the load effect, i.e., the maximum moment acting on the beam (L), the beam is considered safe. The state evaluation in this example should result in the reliability of the beam and ultimately the reliability of the whole system, i.e., overall structure. As in the previous state evaluation example, the evaluation has to start at the lowest hierarchical level, i.e., the beam, and proceed to the next higher level until the reliability of the overall structural system is evaluated. Reliability engineering techniques could be utilized in defining a suitable state evaluation scheme[10,11,12,18,24,27,43,46,47]. The evaluation of the system / component reliability depends on several factors, such as, the performance function being linear or nonlinear, the correlation between the variables and the type of distribution for each variable[10,27]. In general, the reliability problem could be solved by the integration of the joint probability distribution as follows:

$$P_s = \int_0^{\infty}\int_0^{R} f_{R,L}\ dL\ dR \tag{27}$$

where $f_{R,L}$ is the joint probability distribution of load and resistance. A general approach is utilized in order to evaluate an approximate value of the reliability of the system / component because of the difficulty of performing the above integration. In traditional reliability assessment, the minimum distance to the failure surface is considered a measure of safety, i.e., reliability, of the system[24,25,27]. The probability distributions and statistical properties of the involved variables are assumed to be known and constant throughout the whole analysis. In general, knowing the distribution or statistical properties of each variable, the reliability index could be evaluated as follows:

$$\beta_j = \frac{\sum_{i=1}^{nv} v_{ij}^{`*} \left(\frac{\partial f_j}{\partial V_{ij}^{`}}\right)_*}{\sqrt{\sum_{i=1}^{nv}\left(\frac{\partial f_j}{\partial V_{ij}^{`}}\right)_*^2}} \tag{28}$$

where β_j is the reliability index of the $j^{\underline{th}}$ failure mode, and $(v_{1j}^{`*}, v_{2j}^{`*}, \ldots, v_{nv}^{`*})$ is the most probable failure point of the $j^{\underline{th}}$ failure mode in the reduced coordinates, i.e. equivalent standard normal variates. The most probable failure point is evaluated using an optimization scheme where the distance to the failure point from the origin is minimized subject to the constraint that the point should lie on the failure surface[24,25,27]. The most probable failure point is defined as

$$v_{ij}^{`*} = -\alpha_{ij}^{*}\,\beta_j \tag{29}$$

and

$$\alpha_{ij}^{*} = \frac{\left(\frac{\partial f_j}{\partial V_{ij}^{`}}\right)_*}{\sqrt{\sum_{i=1}^{nv}\left(\frac{\partial f_j}{\partial V_{ij}^{`}}\right)_*^2}} \tag{30}$$

where $v_{ij}`^*$ is the most probable failure point for the $j^{th}$ failure mode, $\alpha_{ij}^*$ is the direction cosine along the axes $v_{ij}`$, and β_j is the reliability index of the jth failure mode. In general, the solution of this problem has an iterative nature where the failure point is assumed and the reliability index is evaluated, the failure point is then checked and the reliability index updated until convergence is attained. The probability of failure of the jth failure mode could be evaluated using the following relation:

$$P_{f_j} = 1 - \Phi(\beta_j) \tag{31}$$

where P_{f_j} is the probability of failure of the jth failure mode, $\Phi(.)$ is the cumulative distribution function of a standard normal variate. The previous discussion is based on the basic problem where the variables are considered uncorrelated, normally distributed, with linear performance functions and the failure modes are considered totally independent. Any variations from these conditions could be easily addressed by an orthogonal transformation, by which correlated variables are transferred into an equivalent set of uncorrelated variables. In case of correlated failure modes, by evaluating upper and lower bounds for the probabilities of failure which correspond to totally correlated and totally independent modes[10,11,12,18,24,27,43,46,47]. For non-normally distributed variables, a transformation operation could be utilized in order to transform these variates into equivalent normally distributed variates and then the same procedure could be adopted[24,25,27]. For nonlinear performance functions, a linear approximation of the failure surface is acceptable, using the tangent to the failure surface at the most probable failure point[24,25,27]. The direction cosines introduced at the previous approach represent a form of importance factors in the state evaluation scheme. These factors evaluate the component of every random variable at the failure point. In other words, it is a measure of the contribution of each random variable in evaluating the failure point which is the point with minimum distance from the failure surface to the origin. This distance, as defined in reliability theory, is a measure of the reliability of the system. Therefore, these direction cosines are the importance factors that define the impact of each random variable on the reliability of the component / system.

As outlined in the system model of this example in Fig. 6, every component might have several potential failure modes. Each failure mode has its own performance function which is evaluated as discussed above. The final component reliability should be evaluated such that all potential failure modes are considered. This could only be accomplished by introducing a suitable aggregation function. For example, for the simple beam considered earlier, three potential failure modes are possible. These are bending failure, shear failure or both. Each failure mode could be expressed separately using its own performance function and a set of random variables. These performance functions could be written as:

$$CF_1 = F_Y\, S_x - \frac{wl^2}{8} \tag{32-a}$$

$$CF_2 = \tau_Y\, A_w - \frac{wl}{2} \tag{32-b}$$

$$CF_3 = \frac{\frac{wl^2}{8}}{F_Y\, S_x} + \frac{\frac{wl}{2}}{\tau_Y\, A_w} - 1 \tag{32-c}$$

These performance functions would result in three potential reliability measures for the same component. Now the aggregation function would combine all three measures into a single representative reliability measure, taking into consideration all potential failure modes. A logical aggregation operation could be performed at this example. For any given simple beam, the failure might be due to one or all of the previously defined failure modes. Therefore, if a union operator is considered as an aggregation function, a single reliability measure would result. This could be expressed as:

$$P_{s_{C_i}} = \bigcup_{j=1}^{n} P_{s_{C_{ij}}} \tag{33}$$

where $P_{s_{C_i}}$ is the reliability of the ith component, and $P_{s_{C_{ij}}}$ is the reliability of the ith component based on the jth failure mode and n is the total number of potential failure modes. This approach is only introduced for the sake of demonstration of the mechanics of building a successful state evaluation scheme. However, it might not result in good approximations because of the neglect of the connectivity between the individual failure modes. In other words, the correlation between the failure modes has to be included in the aggregation procedure. The correlation, at this example, may be due to sharing one or more random variables between several failure modes. Correlation between failure modes is a form of inter-connection, which introduces some problems to the evaluation of the reliability of any given component / system. It has been shown[12,18,43,46] that the direction cosines between any two failure modes are equivalent to the correlation coefficient between these two failure modes. Each failure mode is represented by the shortest distance to its failure surface. The line with the shortest distance, intersects the failure surface at the most probable failure point. The tangent to the failure surface at this failure point is considered a linear approximation of the actual failure surface. Thus, the direction cosines of the angles between any two tangents of any two failure surfaces is a measure of the linear correlation between the two failure modes. Knowing such coefficients, suitable bounds for the interval including the actual reliability of the component / system could be evaluated. The process of evaluating upper and lower bounds for such a reliability measure is another form of an aggregation procedure aiming at the fusion of several reliability measures into a single representative measure.

State Evaluation at the System Level. At the system's level, the probability of failure evaluated for each component is utilized, together with the appropriate performance function in order to define each potential failure mode. At this level, a two-staged aggregation operation is applied. The first stage aggregates the probabilities of failure of the individual components interrelated in each potential failure mode, into an overall probability of failure for the failure mode under consideration. The second stage aggregates the probabilities of failure of the individual failure modes, into an overall probability of failure of the overall system. The aggregation functions could be union or intersection functions depending on the nature of the individual failure modes and components under consideration. The correlation between the individual failure modes should also be considered. If the failure modes showed a certain degree of correlation, upper and lower bounds of the probability of failure should be evaluated rather than single point estimates. Several problems arise at the system level. First, as the system gets bigger, a greater number of potential failure modes should be considered. In order to be able to accurately assess the reliability of the system, all potential failure modes should be identified and considered in the state evaluation scheme which, by itself, is a formidable task. Second, this large number of failure modes is bound to result into one or more components being shared by several failure modes. Thus, creating connectivity, i.e., correlation, between the individual failure modes. However, this correlation is very difficult to quantify. Therefore, the state evaluation scheme at this level should be able to identify as much failure modes as possible and practical, in order to evaluate reasonable approximations for the reliability of the system. In addition, it should be able to solve the correlation problem through the utilization of suitable connectivity and aggregation procedures.

Several studies have considered the identification of multiple failure modes[12,18,43,46], where two basic approaches were identified[12]. The first is known as the failure mode approach where all potential failure modes are systematically identified. This approach is suitable for ductile systems where the sequence of component failure is not important, thus reducing the total number of potential failure modes. However, if the system is brittle or, includes both brittle and ductile components, a large number of potential failure modes would result. This is mainly due to the fact that the sequence of component failure would affect the final probability of failure and accordingly the reliability of the system. The second approach, is known as the stable configuration approach[12]. In this method, the structure, in its damaged state due to the failure of a given component, is studied in order to identify an alternate load path. In other words, for a given set of failed components an alternate safe load path is identified in order to render a stable structural system. This means, if one or more safe paths are identified, the structure will survive. Therefore, survival of the system depends on potential stable configurations which in turn might be interrelated. The aggregation procedure, in this case, would involve the union of all potential stable configurations. This aggregation procedure would result in the probability of survival rather than the probability of failure, assuming that individual stable configurations are totally independent. Thus, the problem of failure mode / stable configuration identification could be dealt with based on the type of components being used in the structural system.

In cases where multiple failure modes are involved in the evaluation of the probability of failure of a given system, it usually happens that these modes are somehow correlated. The correlation results at two different levels. First, in evaluating the failure probability of an individual failure mode. This is defined in terms of the failure modes of a set of components, as shown in the system's model in Fig. 6. These components might share one or more basic random variables. Thus, the correlation between the components which is defined as C/C inter-relation, has to be quantified and included in the evaluation of the probability of failure of each individual failure mode. At another level, when evaluating the probability of failure of the system which should depend on the probability of failure of all potential failure modes, the inter-relation, i.e., correlation, between these failure modes should be quantified and included in the procedure. The correlation, at this level, results from the presence of one or more components in several failure modes. It is practically very difficult to accurately evaluate or quantify the degree of correlation between failure modes. A general approach has always been to evaluate upper and lower bounds that define the interval within which the exact probability of failure lies. In general, two intervals could be developed, namely, unimodal bounds and bimodal bounds. The later should result in smaller intervals and, thus, more accurate results. In some cases where these intervals are too large, they loose their significance and the resulting range would be too wide to be useful. Bimodal bounds consider the correlation between couples of modes, thus reducing the intervals and resulting in better approximations. Usually, these bounds correspond to the case of perfect correlation, i.e., $\rho = \pm 1$, or perfect independence, i.e., $\rho = 0$. In practice, actual correlation lies somewhere in between these cases. When evaluating bimodal bounds, the correlation coefficient between each two modes should be evaluated. As an approximation, this coefficient corresponds to the direction cosine between the tangents to the failure surfaces at their failure points as discussed earlier[12,18,43,46].

3.5. State Evaluation for Active Structural Dynamic Control

Active structural control has been introduced as a practical alternative for safety checking of structural systems resisting severe environmental loads[30,31,42]. Since the inception of the notion of active structural control, automatic control theory was the sole approach in developing suitable control algorithms[20]. In these algorithms, the response and mathematical modeling of the system were an essential step in evaluating the state of the system at any given point in time. For the application of fuzzy control strategies, different approach has to be utilized in evaluating the state of any structural system. In the present context, the state of a structural system is defined by its displacement pattern at any given point in time. Once the system has identified such a pattern, a suitable control strategy could be selected in order to suppress its resulting vibration. It should be emphasized, though, that the actual response of the system is not required. A rather conceptual view of the displaced shape is of more significance. The basic problem would then be how would the system be able to identify its displaced pattern. Neural network technology could be utilized in performing such a task[13,16,17]. In this application, no mathematical models for the system are needed, no material properties or structural properties are necessary and no complicated mathematical calculations are required. A simple neural network could be trained to identify potential displacement patterns. It should signal out those degrees of freedom that are highly influencing such a pattern. It also evaluates a contribution level for each contributing mode shape. Such a task is one of perfect nature for neural network applications.

Why neural networks. Neural networks have been conceived as a result of studying the neural system within the human brain[13,16,17]. The type of tasks which a human can perform better than a computer system, depends on the performance of such a neural structure. Such tasks belong to a very special category which is identified as pattern recognition problems. When a human tries to balance himself on a shaking ground, his / her brain, some how, identifies the mode shape of vibration, i.e., the deformed position of the human body, through a set of internal sensors. This little piece of information, when being processed by the human brain, triggers a specific control strategy that aims at balancing the human body. Such strategy depends on the mode shape and previous experience stored within the human brain. The developed strategy is translated into specific commands to a set of actuators, i.e., muscles, that perform the required task. A structural system vibrating under earthquake excitation is now examined. If the structure had a brain-like unit that stores experiences, connects to a set of sensors, identifies modes of vibration, and controls a set of actuators. A similar balancing procedure could be defined. Neural networks have always been closely related to pattern recognition problems[13,16,17]. Therefore, if a neural network could be developed such that it could identify the structure's displacement pattern, this would serve as a first stage for an active structural control scheme.

Neural networks offer a wide range of characteristics that suit the nature of the problem at hand[13,16,17]. The parallel processing nature, could solve the issue of structure reduction adopted in traditional structural analysis of multi-degree-of-freedom systems. The learning property, also solves the problem of estimating structural properties and characteristics. There is no need to identify those characteristics in order to solve for the vibration mode shape. No mathematical models are required and no intensive calculations are necessary.

The proposed network is intended to receive the actual vibration pattern through a set of sensors. The network should try to develop the same mode shape as an output. In order to perform this task, it has to breakdown the displacement pattern into its components based on modal superposition approach[37]. The network then reconstructs the displacement pattern and present it as an output. If this task is successfully performed, i.e., the output pattern coincides with the input pattern, two important pieces of information would be available. First, the contribution level of each mode shape in the final displacement pattern. Second, the main contributing mode shapes in the final displacement pattern. This information would be used later by the control system, in selecting a suitable control strategy and defining the location of control forces to be applied. The displacement pattern of any given structural system, thus, defines its state at any point in time. By identifying such a displacement pattern, the current state of the system is evaluated.

Modal superposition of linear systems. For linear NDOF systems, it is well known that any final displacement pattern could be broken down into a linear combination of its free undamped mode shapes[37]. These mode shapes represent a number of independent displacement patterns that do not change with time and are equal to the number of degrees-of freedom of the system. An analogy could be drawn between the modal superposition approach and the furrier representation of any polynomial function. The main theme of this approach is based on the orthogonality property possessed by the mode shapes with respect to mass, stiffness and damping matrices. According to this approach, any displacement pattern could be written as defined in Eq. (11). It should be realized, though, that not all modes significantly contribute to the overall displacement pattern, and those which do contribute, have variable significance levels. In general, few of the first mode shapes are quite enough to express accurately the displacement shape of a given structural system. However, this number of critical mode shapes grows with the increase of the number of degrees-of-freedom.

The basic concept of modal superposition, at any point in time, is utilized. The actual modal superposition approach is not implemented, since the actual response of the system is not required for the proposed control algorithm. However, the main objective is to use the modal superposition concepts in evaluating the critical mode shapes significantly contributing to a given displacement pattern. In addition, the contribution level of each individual mode is also evaluated.

Neural network model . The proposed neural network is used in the real-time identification of the overall displacement pattern of any structural system[1]. The network accepts the actual displacement pattern as input, decomposes it into a linear combination of the undamped free mode shape patterns. The network, then, reconstructs the displacement pattern and ensures that both patterns coincide. The network comprises five consecutive layers, in addition to the input layer. The structure and performance of each layer is outlined in the following discussion.

The network accepts the input pattern through an input layer which is considered as an inactive layer. As a standard notation, layers are considered active if certain calculations are performed within them[13,16,17]. The input layer has a number of nodes equal to the degrees-of-freedom of the system. Each node is attached through a sensor to a given degree-of-freedom at the structural system. An additional node is attached to the ground in order to provide real-time relative datum. At a given point in time, the sensors transfer an image of the overall displacement pattern, through the input layer, to the first active layer. This layer is defined as the mode shape layer, where each individual neuron represents an individual mode shape. Each node is connected to all sensors through a set of weights defined as

(m_{ij}) where (m) is the connection weight between the ith degree of freedom and the jth mode shape. Figure 8 shows a block diagram of the neural network model. The model is developed for a three degree of freedom system. The sensor locations are identified as an input node (S_i) and its connection weights to all three mode shapes are shown in figure. Each neuron in the mode shape layer, performs a summation of the weighted inputs and results in a net result which is acted upon by the activation function of the neuron[13,16,17]. This operation is defined as:

$$NET_i = \sum_{i=1}^{N} S_i \, m_{ij} \tag{34-a}$$

$$OUT_i = F(NET_i) = \frac{1}{1+e^{-NET_i}} \tag{34-b}$$

where NET_i is the weighted sum of inputs to the ith neuron, OUT_i is the output of the activation function as defined in Eq. (34-b). Such a function is known as a sigmoidal function and is widely used in neural network applications[13,16,17]. According to the mathematical definition of such a function, OUT_i would be in the interval (0,1). The output of the activation function, i.e., OUT_i, represents the contribution level of the ith mode shape in the overall displacement pattern. This layer is fully connected, in contrast to all other succeeding layers. Full connectivity results when each neuron accepts input from all neurons in the previous layer.

The second layer is defined as the threshold layer, which is partially connected. That is, each neuron of the mode shape layer, which delivers the input to the threshold layer, is connected to one neuron only in the threshold layer as shown in figure. This layer performs a threshold operation that only permits the mode shapes with an activation level beyond a specific threshold, to be included in the development of the overall displacement pattern. In other words, a threshold (δ) is defined whereby each neuron performs a test regarding its input OUT_i. This could be expressed mathematically as follows:

$$\begin{aligned} Y_i &= OUT_i \quad \text{IF} \quad OUT_i \geq \delta \\ Y_i &= 0 \quad \text{IF} \quad OUT_i < \delta \end{aligned} \tag{35}$$

where Y_i is the output of the threshold layer that corresponds to the ith mode shape pattern. This layer basically suppresses the mode shapes that are not significantly contributing to the overall displacement pattern. Thus, eliminating unnecessary large number of basic pattern components being included in the linear combination process without any significant change in the final displacement shape. The threshold (δ) should be defined based on previous experience regarding the contribution of basic mode shapes in the overall displacement pattern. However, a value of (0.5) might be a good starting value that could be refined later during the learning process.

The third layer comprises a number of neurons equal to NxN where N is the number of degrees-of-freedom. This layer is defined as the scaling layer and is partially connected. Every N neurons represent a mode shape pattern, Such that, each neuron in the threshold layer sends its output, i.e., the actual activation level for significant mode shapes and zero for non significant mode shapes, to N neurons in the scaling layer. These N neurons represent the components of the ith mode shape, represented by the corresponding neuron

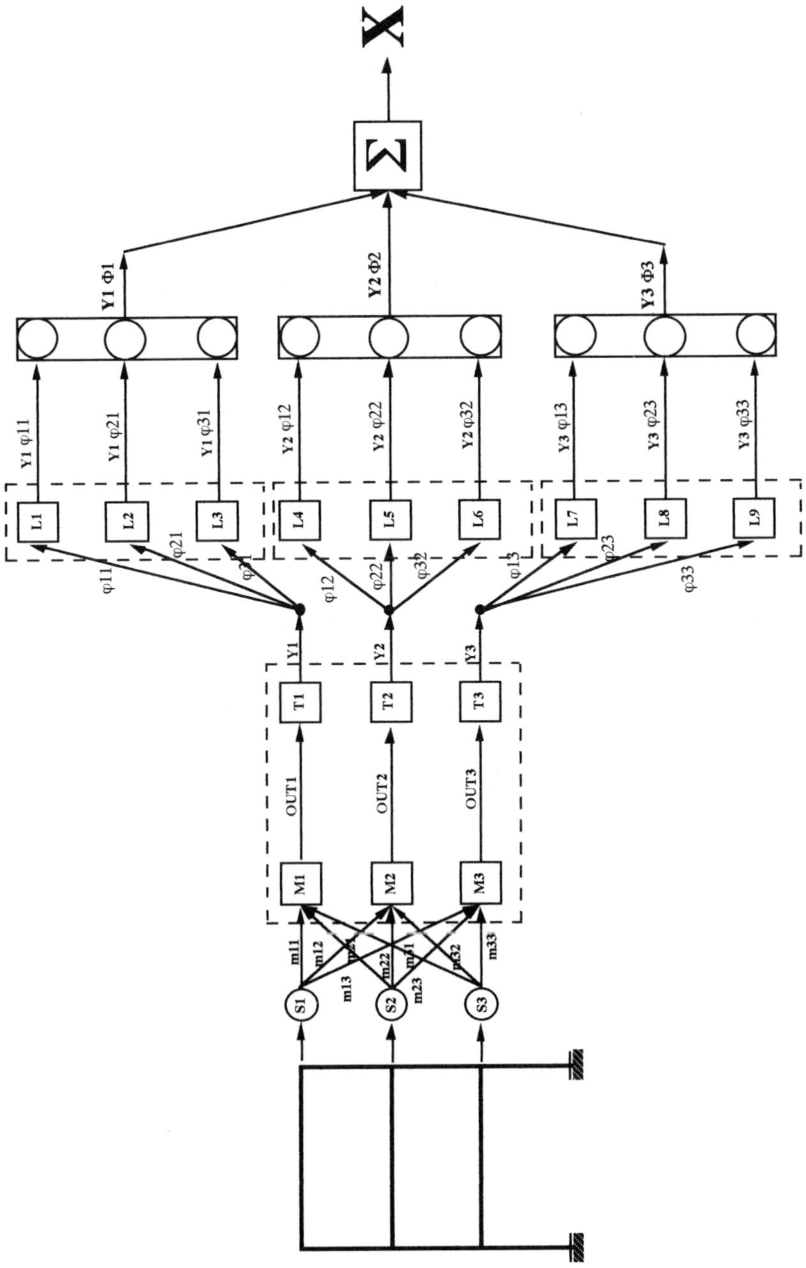

Figure 8. Block Diagram of Neural Network Model

in the mode shape layer, as shown in figure. According to this definition, the first neuron in the threshold layer connects to the first N neurons in the scaling layer. The second neuron connects to the following N neurons in the scaling layer. A simple mathematical rule could be developed to decide the location of the fist neuron, in the scaling layer, that should be connected to any given neuron in the threshold layer. This expression could be defined as:

$$L_j = (j-1)\, N + 1 \tag{36}$$

where L_j is the location of the first neuron in the scaling layer that connects to the jth neuron in the threshold layer, as shown in figure and N is as defined earlier. The connection weights represent the corresponding mode shape components of the jth mode shape. Referring to Fig. 8, T_1 is connected to the first three neurons where the connection weights are defined as:

$$T_1 \xrightarrow{\;Y_1\;} \left| \begin{array}{l} \xrightarrow{\;\phi_{11}\;} L_1 \xrightarrow{\;Y_1\phi_{11}\;} \\ \xrightarrow{\;\phi_{21}\;} L_2 \xrightarrow{\;Y_1\phi_{21}\;} \\ \xrightarrow{\;\phi_{31}\;} L_3 \xrightarrow{\;Y_1\phi_{31}\;} \end{array} \right. \tag{37}$$

where T_1 is the first neuron in the threshold layer with an output Y_1 and connects to the first three neurons in the scaling layer L_1, L_2 and L_3, ϕ_{11}, ϕ_{21} and ϕ_{31} are the connection weights, which corresponds to the components of the first mode shape, as shown in Eq. (37). The output of this layer is, in principle, a scaled mode shape. In other words, the mode shape is scaled by its contributing or activation level as shown in Eq. (37). It should be realized that these connecting weights are not trainable.

The next layer could be considered as a component of the scaling layer. It performs an aggregation process whereby each set of neurons representing a scaled mode shape are aggregated through a set of unit weights into one neuron. This neuron outputs a scaled mode shape rather than its scaled components. This is expressed symbolically as:

$$\left. \begin{array}{l} Y_1\phi_{11} \longrightarrow \\ Y_1\phi_{21} \longrightarrow \\ Y_1\phi_{31} \longrightarrow \end{array} \right| \longrightarrow Y_1\Phi_1 \tag{38}$$

where Φ_1 is the mode shape pattern of the first degree of freedom.

One last step is required in order to reconstruct the original displacement pattern. At this stage, a one neuron layer is utilized. The neuron performs a summation process which is required in the final linear combination of the scaled mode shapes in order to reconstruct the overall displacement pattern. This operation could be expressed mathematically as:

$$Y_1\Phi_1 + Y_2\Phi_2 + Y_3\Phi_3 = X \tag{39}$$

where Y_i is the activation level, i.e., scaling factor or generalized coordinate, of the ith mode shape, Φ_i is the mode shape of the ith mode, and X is the overall displacement pattern fed to the network, in the first place, through the input layer. The developed displacement pattern should be compared to the actual input to the network. Any error value is propagated through the network and the trainable weights changed until the network is trained to identify the displacement patterns correctly. The actual training of the network is beyond the scope of this chapter. If the network could successfully reconstruct the input pattern, then the breakdown performed by the network should result in a set of significant contributing mode shapes and the contribution level of each mode. This information is essential in driving the next stage of this system which is the fuzzy controller.

4. Fuzzy Control

4.1. Introduction

Several practical problems with traditional control algorithms[14] have been identified which limited their practical implementation. Some of these problems are, time delay, prediction of actual behavior of existing structures, mathematical modeling of such behavior and uncertainties incorporated in the lack of exact knowledge[14]. Such practical limitations led the way to the development and introduction of new control technologies and their applications in active structural control. Such technologies include neural networks, fuzzy control and fuzzy neural control. In this section, the state of the art of fuzzy control is outlined together with a comprehensive discussion of fuzzy control strategies, their integration with other control algorithms and their practical applications in active structural control.

Recent work has been cited during the last five years concerning the use of fuzzy control, fuzzy expert control, neural networks and the integration of some of these algorithms with traditional control algorithms[2,3,5,9]. Other cited literature studied the integration of fuzzy control with traditional control algorithms[5,9]. Such integration would result in supplementing traditional control algorithms with better characteristics representative of fuzzy control, while keeping those advantages of traditional control. In most of these studies, it was concluded that fuzzy control algorithms are potential candidates for active structural control. Several key problems were identified, such problems, when addressed, would ultimately improve the performance of such controllers even better. One major drawback for such control algorithms is the lack of generality. However, it is considered a fair price to pay for a simple and effective control algorithm. In addition, such control algorithms might prove very effective in very special situations where mathematical models are impractical to develop, such as historic monumental structures.

4.2. Fuzzy Controller

A fuzzy controller can be considered as an abstracted collection of rules that summarize the experiences of a human controller. This collection of rules form what is known as the rule-base. Each rule is in the form of an IF THEN implication rule where if the antecedent, i.e., the input of the rule, is satisfied then the consequent, i.e., the output of the rule, is implied. In most fuzzy control applications, rules include two antecedents, i.e., rule inputs. Usually such antecedents are some form of a control variable that measures the error or amount of deviation from a perfect well defined state and the rate of change of such a control variable. In current fuzzy control applications, some have used displacement and velocity while others have used velocity and acceleration as the two antecedents[2,3,5,9]. It should be emphasized that usually one control variable and its derivative are selected. The derivative serves as the rate of change of such a control variable. The inclusion of the second input, i.e., the rate of change, helps in stabilizing the performance of the control system. A rule comprising two inputs and one output is defined as a three dimensional rule. The development of a suitable rule-base is not an easy task. Some principal rules could be developed, however, these rules are not, by far, enough to successfully control the vibration of a structural system. Thus, fuzzy controllers should posses the ability to learn from previous experiences and real situations. A property that mimics the performance of a human controller if being responsible for the same task. This requires the inclusion of a self learning mechanism as an integral component of the control system. Thus, simple fuzzy controllers that depend on simple look-up tables would not be good enough for the application at hand. A self-learning fuzzy controller should be considered which has the ability to improve its performance and expand its rule-base.

Rule manipulation which is one of the major tasks performed by a fuzzy controller, should be performed through mathematical operations. Therefore, a mathematical model that represent each individual rule has to be defined[7]. This model is usually referred to as the implication function[23,33,34,36,39,41]. An inferring mechanism should then be developed in order to evaluate a single representative control action for a given set of rules and input values.

Implication Function. Several implication functions were presented in the reviewed literature[28,29,34,36,39]. However, the most widely used function models the rule using the fuzzy Cartesian product[34,36,39]. For example, a rule that relates a fuzzy set A and a fuzzy set E is expressed as

$$\text{IF A THEN E} \tag{40}$$

where A is the fuzzy event that represent the antecedent, and E is the fuzzy event that represent the consequent. Fuzzy sets are defined by a membership function that evaluates the degree to which elements belong to a given set. This degree usually ranges from 0 to 1 which makes the fuzzy set a generalization of crisp sets. This rule is expressed using the fuzzy Cartesian product of both fuzzy events A and E which results in a two dimensional matrix where each one of its entries is defined as follows:

$$\mu_{A\,X\,E}(u,z) = \text{MIN}\{\ \mu_A(u)\ ,\ \mu_E(z)\ \} \tag{41}$$

where $\mu_{A\,X\,E}(u,z)$ is the membership value for the Cartesian product, MIN is a minimum operator, $\mu_A(u)$ is the membership value of element (u) in fuzzy set A, and $\mu_E(z)$ is the membership value of element (z) in fuzzy set E. This function in Eq. (41) has been extensively used in many fuzzy control applications[23,32,33,34,36,39] However, in other studies this implication function was critically analyzed[28,29]. A set of criteria by which an implication function could be tested were presented. Several implication functions were developed in order to satisfy the relevant criteria . The implication function that satisfies these criteria was selected as suitable for structural fuzzy control. For example, if the rule under consideration states the following:

$$\text{IF A is Small THEN E is Small} \tag{42}$$

the following criteria were determined as the necessary conditions for selecting the implication function:

Criterion 1 if A is small Then E should be small
Criterion 2 if A is very small Then E should be very small
Criterion 3 if A is more or less small Then E should be more or less small
Criterion 4 if A is not small Then E should be unknown.

where

$$\mu_{\text{very small}}(u) = \mu_A(u)^2 \tag{43-a}$$

$$\mu_{\text{more or less small}}(u) = \mu_A(u)^{0.5} \tag{43-b}$$

$$\mu_{\text{not small}}(u) = 1 - \mu_A(u) \tag{43-c}$$

where $\mu_A(u)$ is the membership value of element (u) in fuzzy set A, i.e., small, $\mu_{\text{very small}}(u)$ is the membership value of element (u) in the fuzzy set very small, $\mu_{\text{more or less small}}(u)$ is the membership value of element (u) in the fuzzy set more or less small, and $\mu_{\text{not small}}(u)$ is the membership value of element (u) in the fuzzy set not small. The implication function that satisfies these criteria is expressed as[28]:

$$A \times Z \longrightarrow U \times E \tag{44}$$

where A is the fuzzy event that represent the antecedent, E is the fuzzy event that represent the consequent, $\longrightarrow$ is 1 if $\mu_A(u) \leq \mu_E(z)$, and 0 if $\mu_A(u) > \mu_E(z)$, Z is the universe of discourse of fuzzy event E, U is the universe of discourse of fuzzy event A, A X Z is the Cartesian product of fuzzy set A and the universe of discourse Z, and U X E is the Cartesian product of fuzzy set E and the universe of discourse U. The implication function defined in Eq. (44) could be simplified as follows:

$$A \longrightarrow E \tag{45}$$

This function renders a two dimensional matrix that models the rule defined in Eq. (42). However, for the purposes of this application, all the involved rules should have two antecedents and one consequent as mentioned earlier. Thus, the implication function should be expanded in order to be able to handle a three dimensional rule. Three possible options were analyzed, using the same criteria, mentioned earlier, for identifying the appropriate function, a suitable implication function was developed[7]. This function could be written as follows:

$$(A \times B) \longrightarrow E \tag{46}$$

and could be interpreted as

$$\text{IF (A AND B) THEN E} \tag{47}$$

This function results in a three dimensional relation matrix with a membership function defined as:

$$\mu_{3R}(u,w,z) = \mu_{A \times B}(u,w) \longrightarrow \mu_E(z) \tag{48}$$

where $\mu_{3R}(u,w,z)$ is the membership value of error element (u), change in error element (w) and the resulting control action element (z) in the three dimensional implication function defined in Eq. (46), $\mu_{A \times B}(u,w)$ is the membership value of error element (u) and change in error element (w) in the Cartesian product of the error fuzzy set A and change in error fuzzy set B, and $\mu_E(z)$ is the membership value of element (z) in the control action fuzzy set E. The resulting three dimensional relation matrix has entries of zeros and ones as defined in Eq. (44). The relation matrix was tested by performing a composition with some input values of A and B, the resulting output value of E met the specified criteria of Eq. (43)[7].

Inference Mechanism. An inference mechanism is a procedure by which a control action could be inferred given a rule-base and a set of input values. For a single rule, the compositional rule[40,49] was found satisfactory according to the reviewed literature[23,33,34,36,39]. However, if the rule-base contains a number of rules that apply, a procedure needs to be defined such that all the applicable rules contribute to the final control action[23,34,36,39]. The first step in this approach should be to identify all the rules that should be included in a given control situation. An applicability factor that is greater than zero and less or equal one, is defined for each rule. This factor represents how close the input values are to the antecedents of the rule under consideration.

Applicability Factor. It is desirable ,when evaluating the final control action, to include all the available information from all the relevant rules. However, not all the rules are applicable in all cases. Moreover, the applicability levels of such rules are not the equal. Therefore, a non-negative value that is less than or equal to one is evaluated, for each rule, for a given set of input values. This value, which is referred to as the applicability factor, is evaluated based on how close the input values are to the rule antecedents. As mentioned earlier, each rule represents a relation between two antecedents and an implied consequent

For the sake of explanation, a simple rule that has the velocity and acceleration as antecedents and the control force as a consequent is assumed. The applicability factor is then defined as the minimum membership value of both input values, i.e., velocity and acceleration levels, in the corresponding rule antecedents fuzzy sets respectively. Thus, the applicability factor could be expressed mathematically as:

$$AF_i = \text{MIN}\{\ \mu_{(A)_i}(v)\ ,\ \mu_{(B)_i}(a)\ \} \tag{49}$$

where AF_i is the $i^{\underline{th}}$ rule applicability factor and i ranges from 1 to the total number of rules n, MIN is a minimum operator, $\mu_{(A)_i}(v)$ is the membership value of the input velocity value (v), and $\mu_{(B)_i}(a)$ is the membership value of the input acceleration value (a). For example, for the following rule:

IF Velocity is Positive-Small **AND** Acceleration is Positive-Big
THEN Control Force is Positive-Big (50)

These linguistic measures are defined using fuzzy sets[19,21,40,49] where the universe of discourse, i.e., the range within which any given fuzzy event changes, has to be defined based on previous experience. For example the linguistic measures introduced in the previous rule could be defined as follows:

$$\text{Positive-Small} = \{\ 0|0.7,\ 0.1|1.0,\ 0.15|0.7,\ 0.2|0.3\ \} \tag{51-a}$$
$$\text{Positive-Big} = \{\ 0.5|0.3,\ 1.5|0.7,\ 2|1.0\ \} \tag{51-b}$$
$$\text{Positive-Big} = \{\ 1m|0.3,\ 1.5m|0.7,\ 2m|1.0\ \} \tag{51-c}$$

where the velocity ranges on a scale from -0.1 to +0.1 in which -0.1 is the lowest level and +0.1 is the highest level, the acceleration ranges on a scale from -2 to +2 in which -2 is the lowest level and +2 is the highest level, the control force ranges on a scale from -2m to +2m in which -2m is the lowest level and +2m is the highest level, m is the associated mass, and 0.1,0.2,..., 1.0 are degrees of belief that the corresponding elements belong to the linguistic measures. It should be emphasized, though, that these linguistic measures and their universe of discourse are problem dependent and might change from a structural system to the other. Practical experience should be advised when establishing the fuzzy sets expressing the individual variables. The values used in this example are only defined for the sake of demonstration. Thus, if the rule is not applicable, the final action would be zero. However, if the rule is applicable with some degree, its final action is scaled based on its applicability factor. This step could be expressed mathematically as:

$$\mu_{(FA)_i} = AF_i\ \mu_{(IA)_i} \tag{52}$$

where $\mu_{(FA)_i}$ is the membership function of the final i^{th} rule action, AF_i is the ith rule applicability factor and i ranges from 1 to the total number of rules n, and $\mu_{(IA)_i}$ is the membership function of the initial i^{th} rule action, as calculated in the next section. Once an action is evaluated, for every single rule, and scaled based on its applicability factor, an overall action can then be evaluated taking into account all the relevant rules.

Compositional Rule of Inference.. The compositional rule of inference[40,49] was used extensively and successfully in several cited applications of fuzzy control[23,33,34,36,39]. If a fuzzy relation is defined between two universes U and Z, the compositional rule of inference evaluates a fuzzy subset E of the universe Z which is induced by the fuzzy subset A of the universe U. For example, for the rule defined in Eq. (50), using the implication function developed earlier, a three dimensional relation matrix is defined. This relation matrix represents a fuzzy relation between the three universes of the fuzzy variables Velocity, Acceleration and Control Force. Thus, applying the compositional rule of inference, a control force could be evaluated given a set of fuzzy measures for the two antecedents, i.e., Velocity and Acceleration. The Velocity and Acceleration levels are both crisp numbers. Thus, using a singleton fuzzy set with a membership value of 1.0 at the corresponding input level, a fuzzyfied Velocity and Acceleration result. These fuzzyfied input values when composed with the implication function, results in the initial control force. There are several definitions for the compositional rule of inference[19,40,49]. The definition used in this study is the MAX-MIN matrix product. This definition was chosen because of its simplicity and success in similar control applications[19]. The compositional rule of inference was originally defined for two dimensional problems[40,49]. This definition was expanded in order to apply to three dimensional rules[7] which is defined as:

$$\mu_{(IA)_i}(z) = \underset{\text{all } u \in U}{\text{MAX}} \ \text{MIN}\left[\mu_{\text{Error}}(u)\left\{\underset{\text{all } w \in W}{\text{MAX}} \ \text{MIN}\left(\mu_{\text{Change in Error}}(w), \mu_{3R}(u,w,z)\right)\right\}\right] \quad (53)$$

where $\mu_{(IA)_i}(z)$ is the membership value of element (z) in the ith rule initial control force fuzzy set, $\mu_{Velocity}(u)$ is the membership value of element (u) in the fuzzyfied input Velocity level, $\mu_{Acceleration}(w)$ is the membership value of element (w) in the fuzzyfied input Acceleration level, and $\mu_{3R}(u,w,z)$ is the membership function of the three dimensional fuzzy relation defined by the implication function in Eq. (48). Equation (53), results in a fuzzy subset of the Control Fore universe that represents the initial control force for each rule. This initial action is then scaled by the applicability factor as explained in the previous section to get the final rule action. The final step is to evaluate an overall action based on all applicable rules. The overall action is defined using the union function of fuzzy relations which is defined as:

$$\mu_{A \cup B}(z) = \text{MAX}\ \{\ \mu_A(z)\ ,\ \mu_B(z)\ \} \quad (54)$$

where $\mu_{A \cup B}(z)$ is the membership value of element (z) in the union of fuzzy events A and B, $\mu_A(z)$ is the membership value of element (z) in fuzzy event A, and $\mu_B(z)$ is the membership value of element (z) in fuzzy event B. Thus, the overall action is defined as:

$$\mu_{(OA)}(z) = \underset{i=1}{\overset{n}{\text{MAX}}} \ \mu_{(FA)_i}(z) \quad (55)$$

where $\mu_{(OA)}(z)$ is the membership value of element (z) in the overall action fuzzy set, $\mu_{(FA)_i}(z)$ is the membership value of element (z) in the overall action of the ith rule, and MAX is a maximum operator applied over all the applicable rules, where i ranges from 1 to the number of applicable rules (n). The resulting overall action is a fuzzy subset of the universe of discourse of the fuzzy variable Control Force. In order to be able to apply the control force, it should be in the form of a crisp number. A defuzzyfying procedure has been developed based on the center of gravity method[7]. According to this procedure, the crisp control action could be defined as follows:

$$\mathbf{OA} = \frac{\sum_{i=1}^{n} z\ \mu_{(OA)_i}(z)}{\sum_{i=1}^{n} \mu_{(OA)_i}(z)} \quad (56)$$

where **OA** is the crisp overall action, $\mu_{(OA)}(z)$ is the membership value of element (z) in the overall action, and (n) is the total number of applicable rules.

4.3. Self Learning System

A fuzzy controller depends on a rule-base where all previous experiences and information are stored. However, it is difficult to construct a complete rule-base that is capable of handling all potential situations. Thus, it is necessary to include a self-learning unit within the control system that is responsible for expanding and updating the current rule-base. The self learning unit should identify situations or cases that are not covered in the current rule-base. It should also extract necessary information and construct new rules to handle these situations. In general, the self-learning unit monitors the performance of the control system. Two basic approaches could be adopted in the learning stage of a fuzzy controller. The first utilizes a performance matrix while the second utilizes a neural network in the learning stage. In the latter a fuzzy neural controller is developed.

Performance Matrix. The performance matrix is a two dimensional matrix that summarizes the required output correction for the system, knowing the Error and the Change in Error levels[32]. The performance of the control system is measured using the already known Error and Change in Error to look-up the required output correction according to the performance matrix. The performance matrix stores an ideal performance which should be a reflection of how the control system should react to each potential combination of Error and Change in Error. If the control system deviates from this ideal track, the learning system should be able to identify the underlying causes and improve the control system's performance. The performance matrix should be developed based on previous experiences and knowledge of experienced controllers. Unsatisfactory performances may be due to one of two main reasons. The first is due to missing rules, i.e., the current rule-base cannot handle the current situation because none of its rules is applicable. Accordingly, the control system cannot suggest any corrective action. The second is due to untuned rules. In other words, the rules that apply to the current case needs to be adjusted in order to result in the expected ideal performance. Figure (9), summarizes the logic behind the operation of this learning strategy in a flow chart format.

Neural Network Learning. In this form of learning, a learning scheme is used to train the controller to behave in a specific manner. The controller should be modeled in a neural network format. In such a model, each individual rule is mapped to a single neuron of the neural network. The controller comprises multiple active layers where each layer connects to its neighbors through a set of synaptic weights. The learning process could be defined as supervised or unsupervised training. In supervised training, pairs of inputs and suggested outputs are presented to the controller[2,13,16,17]. During the learning process, the synaptic weights are altered, according to s specific learning algorithm, in order to result in the required performance. While in unsupervised training, no output vector is required in advance. However, an input vector is used and the network weights are adjusted according to a defined algorithm such that the network response is consistent. In other words, the controller responds with similar actions to similar inputs. The learning algorithm should also adjust the performance of the network by developing new rules as explained earlier. Therefore, adjusting an already existing rule would corresponds to changing the connection weights to that specific neuron. However expanding the rule-base and developing a new rule requires a separate algorithm that is capable of expanding an existing neural network and adjusting its structure accordingly.

4.4 . Control Action Implementation

Once a control force vector is evaluated, its action need to be implemented. Physical implementation of control forces is beyond the scope of this chapter. Further research work is needed in this area in order to develop suitable methods and mechanisms for the employment of the evaluated control forces to the system. However, hydraulic systems and active bracing members could still be utilized in applying such forces as in traditional control. An optimization problem has to be solved in order to decide the optimum number

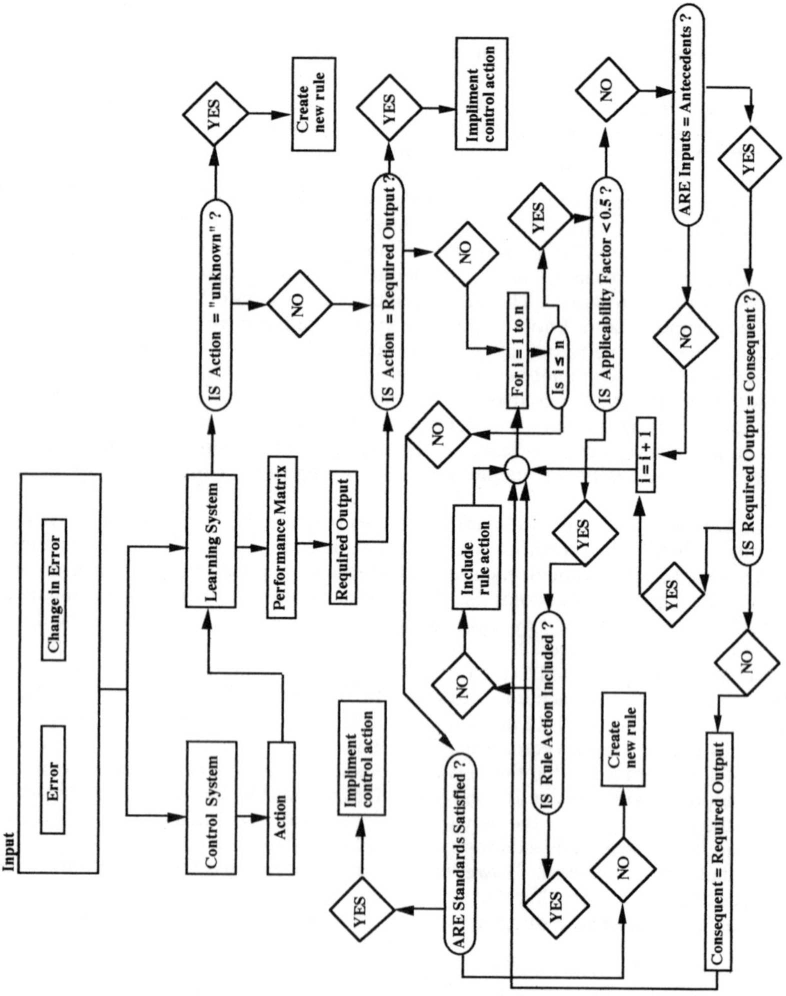

Figure 9. Block Diagram of a Performance Matrix Learning System

and location of control forces in order to suppress a given excitation. This, in fact, is one of the advantages of the neural pattern recognition unit introduced earlier. The network identifies the mode shapes closely influencing the overall displacement pattern of the system. Thus, identifying the location where control forces might be located in order to actively and efficiently control the performance of the system.

5. References

1. M.H.M. Hassan and B.M. Ayyub, *Proceedings of the Joint ISUMA'95 and NAFIPS'95*, (1995), pp. 559-563.
2. L. Faravelli and P. Venini, *Journal of Structural Control,* **Vol. 1, No. 1,** (1994), pp. 79-101.
3. F. Casciati, L. Faravelli and T. Yao, *Proceedings of the Second European Conference on Smart Structures and Materials*, (1994), pp. 206-209.
4. M.H.M. Hassan and B.M. Ayyub, *Civil Engineering Systems,* **Vol. 10**, (1993), pp. 37-53.
5. E. Tachibana, Y. Inoue and B.G. Creamer, *Microcomputers in Civil Engineering,* **Vol. 7,** (1992), pp. 179-189.
6. B.M. Ayyub and M.H.M. Hassan, *Civil Engineering Systems,* **vol. 9**, (1992), pp. 123-146.
7. B.M. Ayyub and M.H.M. Hassan, *Civil Engineering Systems,* **vol. 9**, (1992), pp. 179-204.
8. B.M. Ayyub and M.H.M. Hassan, *Civil Engineering Systems,* **vol. 9**, (1992), pp. 275-297.
9. K. Matsuoka and H.H. Tsai, *Proceedings of the 1992 Pressure Vessels and Piping Conference, ASME,* **Vol. 2,** (1992), pp. 67-73.
10. M.H.M. Hassan and B.M. Ayyub, *Annual Winter Meeting , ASME,* (1992).
11. B. M. Ayyub and K-L. Lai, *Research report for the US. Navy, University of Maryland, College Park*, (1991).
12. S-T Queck and A.H-S. Ang, *Journal of Structural Engineering, ASCE,* **Vol. 116, No. 10,** (1991), pp. 2656-2670.
13. M. Caudill and C. Butler, *Naturally intelligent systems*, (MIT Press, Cambridge, Massachusetts), (1990).
14. J.T.P. Yao, *Proceedings of ICOSSAR'89, the 5th International Conference on Structural Safety and Reliability,* **Vol. 1,** (1989), pp. 479-483.
15. A. D. Hall, *Metasystems methodology, a new synthesis and unification,* (Pergamon Press, New York, NY.), (1989).
16. R.P. Lippman, *IEEE ASSP Magazine*, (1989).
17. P.D. Wasserman, *Neural computing, Theory and practice*, (Van Nostrand and Reinhold, New York, NY.), (1989).
18. R. Rashedi and F. Moses, *Journal of Structural Engineering, ASCE,* **Vol. 114, No. 2,** (1988), pp. 292-313.
19. G. J. Klir and T. A. Folger, *Fuzzy sets, uncertainty, and information.* (Prentice Hall, Englewood cliffs, New Jersey), (1988).
20. T.T. Soong, *Technical Report, NCEER-87-0023*, (1987).
21. H. J. Zimmermann, *Fuzzy set theory - and its applications*, (Kluwer-Nijhoff publishing, Boston, MA.), (1985).
22. G. J. Klir, *Architecture of systems problem solving*, (Plenum Press, New York, NY.), (1985).
23. M. Sugeno, *Information Sciences*, **Vol. 36**, (1985), pp. 59-83.
24. G.J. White and B.M. Ayyub, *Naval Engineers Journal*, (1985), pp. 86-96.

25. B.M. Ayyub and A. Haldar, *Journal of construction Engineering and Management, ASCE,* **Vol. 110, No. 2**, (1984), pp. 189-204.
26. B. Wilson, *Systems: Concepts, Methodologies, and Applications*, (John Wiley & Sons, Inc., New York, NY.), (1984).
27. A. H-S. Ang and W. H. Tang, *Probability concepts in engineering planning and design - volume II,* (John Wiley & Sons, Inc., New York, NY.), (1984).
28. M. Mizumoto and H.J. Zimmermann, *Fuzzy Sets and Systems*, **No. 8**, (1982), pp. 253-283.
29. M. Mizumoto, *Cybernetics and Systems*, **No. 12**, (1981), pp. 247-306.
30. M. Abdel-Rohman and H.H. Leipholz, *Journal of the Structural Division, ASCE,* **Vol. 106, No. ST3,** (1980), pp. 663-677.
31. M. Abdel-Rohman and H.H. Leipholz, *Journal of the Engineering Mechanics Division, ASCE,* **Vol. 105, No. EM6,** (1979), pp. 1007-1023.
32. T.J. Procyk and E.H. Mamdani, *Automatica*, **No. 15**, (1979), pp. 15-30.
33. P.J. King and E.H. Mamdani, *Automatica*, **No. 13**, (1977), pp. 235-242.
34. E.H. Mamdani, *IEEE Transactions on Computers,* **Vol. C-26, No. 12**, (1977), pp. 1182-1191.
35. R.R. Craig, Jr. *Shock and Vibration Digest*, Naval Research Lab, Washington DC., **No. 9**, (1977), pp. 3-10.
36. E.H. Mamdani, *Int. J. Man-Machine Studies*, **No. 8** (1976), pp. 669-678.
37. R.W. Clough and J. Penzien, *Dynamics of structures*. (McGraw Hill, New York, NY.), (1975).
38. R.M. Hintz, *AIAA Journal*, **No. 13**, (1975), pp. 1003-1016.
39. E.H. Mamdani, *Proceedings IEE,* **Vol. 121, No. 12**, (1974), pp. 1585-1588.
40. L. A. Zadeh, *IEEE Transactions on systems, man, and cybernetics,* **Vol. SMC-3, No. 1**, (1973), pp. 28-44.
41. L.A. Zadeh, *J. of Dynamic Systems, Measurement and Controls*, (1972), pp. 3-4.
42. J.T.P. Yao, *Journal of the Structural Division, ASCE,* **Vol. 98, ST7,** (1972), pp. 1567-1574.
43. J. Stevenson and F. Moses, *Journal of the Structural Division, ASCE,* **Vol. 96, No. ST11,** (1970), pp. 2409-2427.
44. G. J. Klir, *An approach to general systems theory*. (Van Nostrand Reinhold Company, New York, NY.), (1969).
45. H. Chestnut, *Systems engineering methods*. (John Wiley & Sons, Inc., New York, NY.), (1967).
46. C.A. Cornell, *Journal of the Structural Division, ASCE,* **Vol. 93, No. ST1,** (1967), pp. 171-200.
47. F. Moses and D.E. Kinser, *Journal of the Structural Division,* **Vol. 86, No. ST12,** (1967), pp. 147-164.
48. H. Chestnut, *Systems engineering tools*. (John Wiley & Sons, Inc., New York, NY.), (1965).
49. L. A. Zadeh, *Information and control,* **No. 8**, (1965), pp. 338-353.
50. A. D. Hall, *A method for systems engineering*. (Van Nostrand Company, Inc., Princeton, New Jersey.), (1962).

Uncertainty Modeling in Vibration and Fuzzy Analysis of Structural Systems, pp. 233–251
edited by B. M. Ayyub, A. Guran and A. Haldar
Series on Stability, Vibration and Control of Systems Series B: Vol. 10

APPLICATION OF FUZZY LOGIC TO STRUCTURAL VIBRATION CONTROL

HITOSHI FURUTA
Dept. of Informatics, Kansai University, Takatsuki City, Osaka, Japan
HIROO OKANAN
Dept. of Construction Engineering, Osaka Prefectural Technical College
MASAYOSHI KANEYOSHI
Bridge Engineering Department, Hitachi Zosen Corp., Sakai City, Japan
and
HIROSHI TANAKA
Bridge Engineering Department, Hitachi Zosen Corp., Sakai City, Japan

ABSTRACT

In this chapter, an attempt is made to apply the fuzzy logic to structural vibration control of earthquake-excited and wind-excited oscillations. The numerical examples of single-degree-of-freedom and two-degree-of-freedom models show that the fuzzy control technique is useful for reducing the amplitude and acceleration of structural oscillation due to earthquakes. Experiments on structural vibration control are conducted using a system of prestressing tendons with spring connected to a stepping motor system with fuzzy control rules. The model is subject to base motions produced by a small shaking table, and a fluctuating flow generated by a fan. These experimental results confirm the efficiency of fuzzy control for structural vibration problems.

Thus, it has been proven that fuzzy active control can provide practical and robust control of structural vibration with simple algorithms. However, in order to make the fuzzy active controller more practical, it is necessary to tune the parameters and shapes of membership functions used in the fuzzy inference rules. In this chapter, an attempt is also made to apply the technique of genetic algorithms (GA) for the self-tuning of the inference rules and their parameters. To demonstrate the applicability of GA in the self-tuning, an experiment is conducted on a single degree of freedom vibration model with a simple pendulum actuator.

1. Introduction

Many structures, such as tall buildings, towers, and flexible bridges have already been designed using passive or active vibration suppression as part of the total design. Recently, active control has played an important role for the control of vibrations due to earthquake or wind.

It seems that many researchers are trying to reduce vibration response as much as possible even though the system of active control becomes complicated. On the contrary, the present study has been carried out aiming to develop a simple and practical system with sufficient efficiency and robustness, applying fuzzy control. Since the algorithm of fuzzy active control (FAC) is very simple, the calculation time is short. The simplicity of the structure of the system can provide high reliability and robustness. It is very easy to change and modify control parameters to adjust any vibration condition.

In order to make the fuzzy active controller more practical, it is necessary to tune the parameters and shapes of membership functions used in the fuzzy inference rules. However, in the field of structural vibration control, there does not exist any operator that establishes general control rules. Thus, decisions for developing control rules are made by trial and error based on data accumulated up to the present and by obtaining a firm understanding of the characteristics of the desired system. The result of tuning by trial and error must be optimal, however, it is difficult to guarantee this because in the fuzzy control, the tuning quality of the membership function greatly affects the control performance.

In this chapter, attempts are further made to tune membership functions automatically using Genetic Algorithms (GA) which has gained a lot of attention recently as an effective method for solving combinatorial optimization problems. Because GA features a high possibility to reach the optimal solution with multiple-points search, the position and shape of membership functions can be tuned simultaneously. Because there is no restriction on the fitness function, it affords freedom when selecting the object to be controlled.

In order to show the effectiveness of the automatic tuning method developed here, a vibration experiment on a single-degree-of freedom structure was conducted by using a simple pendulum-type actuator.

2. Application of Fuzzy Logic to Vibration Control

The numerical models employed here are SDOF (Single-Degree-of-Freedom) and 2DOF (2-Degree-of-Freedom) models; one of the 2DOF models has one control force, and the other one has two control forces as shown in Figure 1. The Min-Max-Height Method is applied (Figure 2) to fuzzy reasoning, and fuzzy production rules (If-Then rule) are utilized to express knowledge (Tables 1 and 2).

The triangular-type membership function (Figure 3) is used for the relative velocity V and the basement acceleration A in antecedent. The upper bounds (i.e., VV•VMAX and AA•AMAX) and lower bounds (i.e., -VV•VMAX and -AA•AMAX) are determined by multiplying VMAX and AMAX by the coefficients of VV and AA, respectively. VV and AA are constants to define the upper and lower bounds of velocity and acceleration, respectively. VMAX is the maximum absolute value of the relative response velocity; AMAX is the maximum absolute value of the acceleration of the basement. According to rule No. 1 shown in Table 1, the coefficient VV is tuned, changing its value from 0.1 to 1.

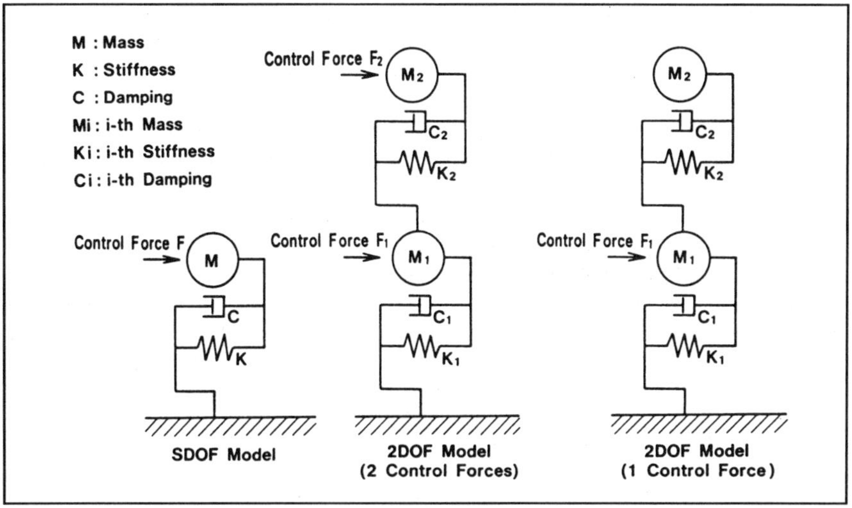

Figure 1 Vibration Models

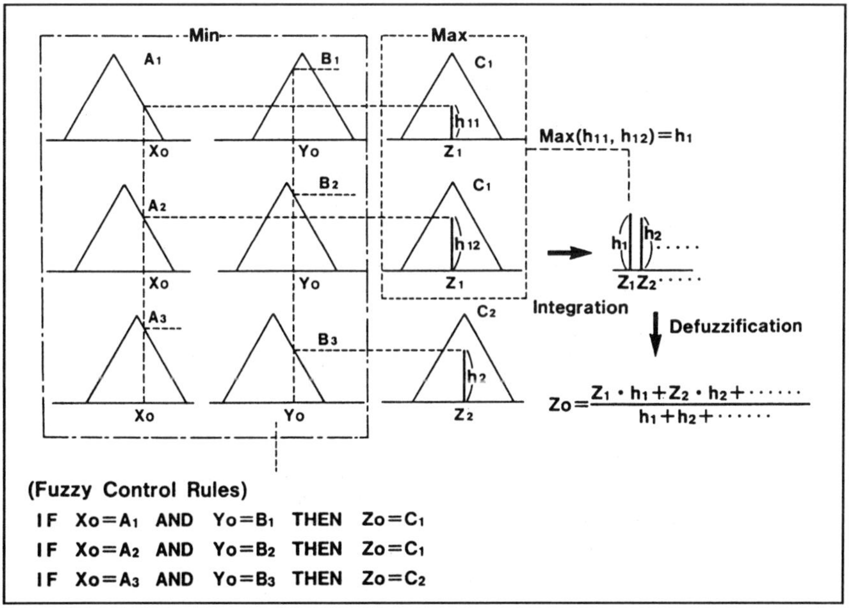

Figure 2 Min-Max-Height Method

It is found that the result is better if the value VV is smaller. In other words, it is better to take the wider part of Big in the membership function, which is divided into three parts: Big, Medium, and Small. Then in the particular case of the fuzzy control, rule No. 3 is made, which consists of only the Big part.

The Min-Max-Height Method is applied to determine the magnitude of control forces (Figure 2). It is not only simple but also very practical compared with other methods, e.g., Mini-Max-Gravity Center Method [1].

The shape of control forces is trapezoidal as shown in Figure 4. The control forces are calculated every 0.02 seconds. As no difference was observed when the time interval k in Figure 4 is changed from 0.002 to 0.02, k was taken to be 0.01 sec.

To determine what response factor is sensitive to active control, control rules were made in which one component of the response factor was specified. The rule of relative

Table 1 Rule No. 1(Only Relative Velocity Considered)

V: NB	NM	NS	ZR	PS	PM	PB
PB	PM	PS	ZR	NS	NM	NB

If relative velocity is negative and is not big, give positive and big control force.

IF V is NB THEN P is PB

• • • • • • •

Table 2 Rule No. 2 (Both Relative Velocity and Basement Acceleration Considered)

		V: NB	NM	NS	ZR	PS	PM	PB
A	NB	PB		NB		NB		NB
	NM	PB		NM		NM		NB
	NS	PB		NS		NS		NB
	ZR	PB	PM	PS	ZR	NS	NM	NB
	PS	PB		PS		PS		NB
	PM	PB		PM		PM		NB
	PB	PB		PB		PB		NB

If relative velocity is negative big and basement accelaration is also negative big, give positive big control force.

IF V is NB and A is NB THEN P is PB

• • • • • • •

displacement affected the most when the time lag between the control force and the relative displacement was three-fourths of its cycle; the rule of relative velocity affected the most when there was no time lag. The rule of relative acceleration affected the most when the time lag between the control force and relative acceleration was one-fourth. Therefore, the rule of relative velocity is convenient for the simulation because there is no time lag.

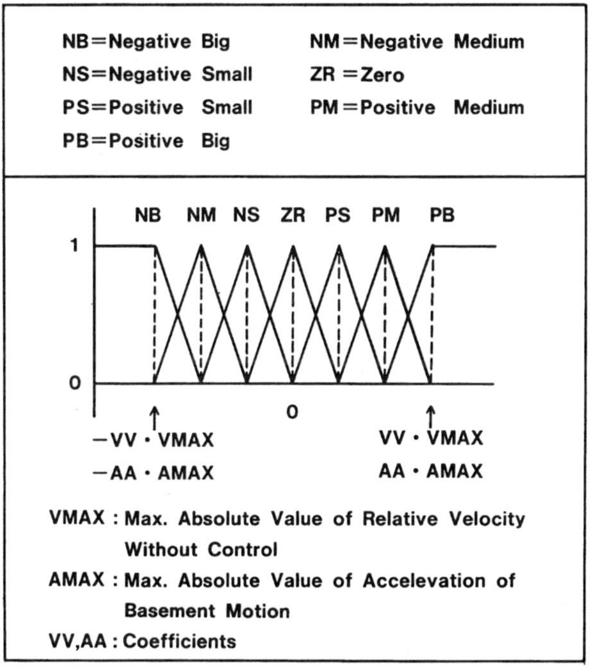

Figure 3 Membership Function (Antecedent)

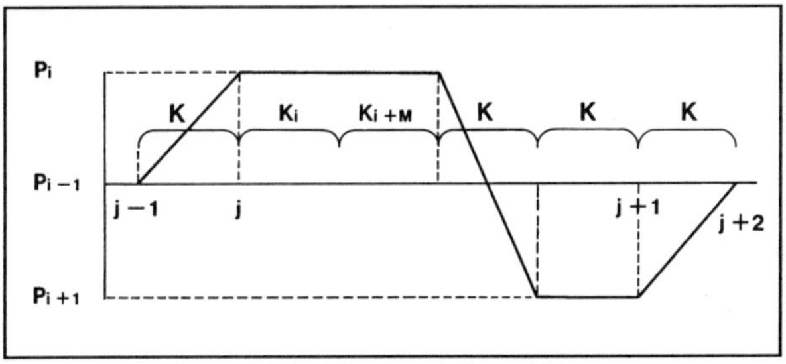

Figure 4 Classification of Control Forces

3. Numerical Simulation

Numerical simulations were studied under the following conditions; vibration system of the model is SDOF; earthquake is El Centro NS component (maximum accelcration is 341.6gal); maximum control force is 0.88% of model mass; period T is 2 sec.; logarithmic decrement δ is 1%. The result of numerical simulation is shown in Table 3. The result of the optimum control [2] was also compared by the ratio of the response under control and non-control in Table 3. Good result was given by the FAC concerning the control of relative displacement which is very harmful to structures. Rule No. 3 is considered to be a special case of fuzzy control (i.e., the membership function consists of only Big) and is very simple. However, it has almost the same efficiency as rule No. 1 and 2 as shown in Table 3.

From these comparative studies, rule No. 3 is selected as the best method. Therefore it was applied to the following experimental study.

Table 3 Comparison between FAC and Optimum Control
(Result of El Centro NS Component)

Max Control Force: 0.88 % of Model Mass T = 2.0 sec, K = 0.01 sec			
	maxX/maxX0	maxV/maxV0	maxA/maxA0
OPT.C.	0.46	—	0.53
RULE 1	0.63	0.62	0.98
RULE 2	0.63	0.62	0.62
RULE 3	0.61	0.61	1.01

N.B.)
X0 : Relative displacement without control
X : Relative displacement with control
V0 : Relative velocity without control
V : Relative velocity with control
A0 : Relative acceleration without control
A : Relative acceleration with control

4. Experiment

4.1 Models

Experimental study was carried out using the model controlled by the small stepping motor (Figure 5). The details of the experiment are as follows:

Three models were used applying rule No. 3:

Model 1: $m = 0.0269 kgf \cdot m^{-1} \cdot sec^2$, $f = 7.75Hz$, $\zeta = 0.017$

Model 2: $m = 0.0622 kgf \cdot m^{-1} \cdot sec^2$, $f = 5.09Hz$, $\zeta = 0.019$

Model 3: $m = 0.1010 kgf \cdot m^{-1} \cdot sec^2$, $f = 3.14Hz$, $\zeta = 0.020$

where m is the mass, f is the natural frequency, and ζ is the damping ratio.

The model frame is one story and is of a shear deformation type. The control force was introduced by the stepping motor. Its shaft has a pulley which conveys force in a step manner to the model through a coil spring (spring constant k = 21.7gf/cm or 83.3 gf/cm). Case 1 : one spring is installed diagonally with an angle of 25.7 degrees; Case 2: two springs are installed in a V-figure form with an angle of 35.2 degrees. The magnitude of control force of Model 1 in Case 1, for example, was 0.368% of the model mass with k = 21.7gf/cm, and was 1.41% of the model mass with k = 83.3gf/cm, respectively.

Furthermore, the two-story shear deformation type model was used for the experiments with the control force changing to be Big, Medium and Small. The mass of the first story M1 is $0.261 kgf \cdot m^{-1} \cdot sec^2$ and mass of the second story M2 is $0.113 kgf \cdot m^{-1} \cdot sec^2$. The natural frequencies and mode shapes of the first mode and second mode are 2.83Hz, 5.36Hz, $X1^1 : X2^1 = 1:1.75$ and $X1^2 : X2^2 = 1:-2.07$, respectively. The 32 bit personal computer (NEC PC9801-ES2) was utilized for the control. The models are subject to base motions produced by a small shaking table and a fluctuating flow generated by a fan. To produce earthquake excitation, an El Centro time history was used as a random excitation.

The magnitudes of the control forces were determined by the relative displacement, which was easy to measure. Following rule No. 3, the positive control forces were applied quarter cycle delay to the time when the relative displacement changed from negative to positive and vice versa. Using the delay time, analog-digital conversion, the control calculation and motor action were carried out and the balance of the time was consumed by the "FOR ~ NEXT (non-execution) loop" in the BASIC program.

4.2 Experimental Results

The experimental result of Case 1 by Model 1 (k = 83.3gf/cm) is shown in Figure 6. White circles indicate the control efficiency by the amplitude ratio (rms) between before and after control to the model of sinusoidal base motion corresponding to the natural frequency of the model. The black circles indicate the efficiency to the turbulent flow. The change of waiting time does not affect in a wide range (0 ~ 300). In other words, quarter cycle delay is not always necessary, particularly in turbulent flow, because structures vibrate in many natural frequencies. There is more room for determination of optimum delay (i.e., waiting time).

Figure 7 shows the efficiency for the fluctuating flow. The symbol of Z in Figure 7 indicates the number of "FOR ~ NEXT loop." The power spectral density (PSD) was measured for the lift force acting on the model (Figure 8). The magnitude of the control force is 2.82% of the model mass and is about 40% of the the wind load.

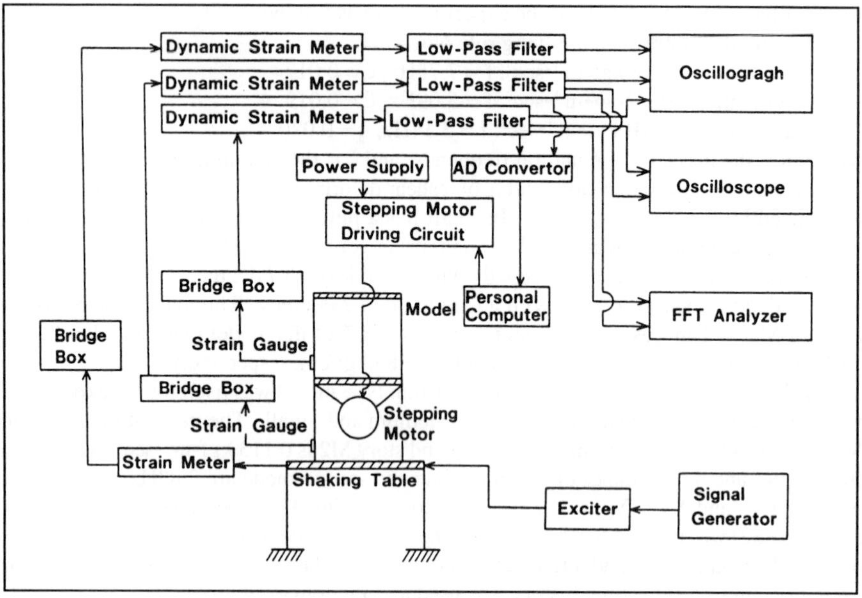

Figure 5 System Block-Diagram

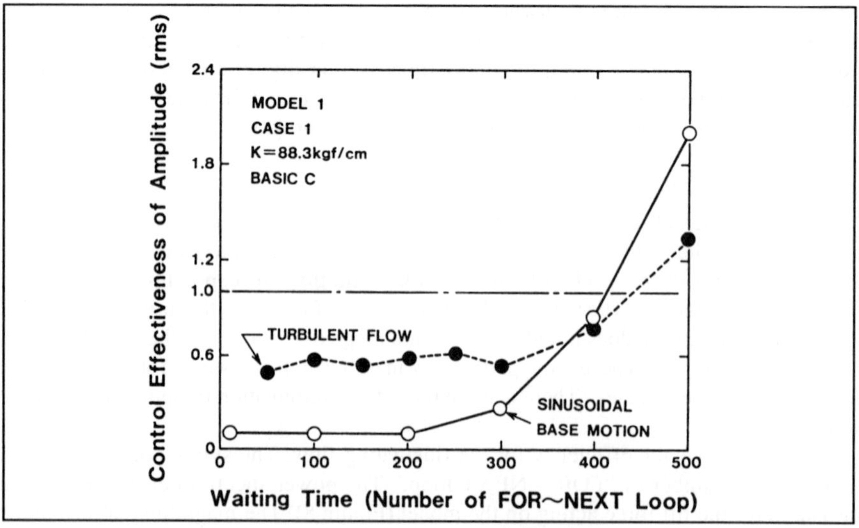

Figure 6 Control Effectiveness vs. Waiting Time

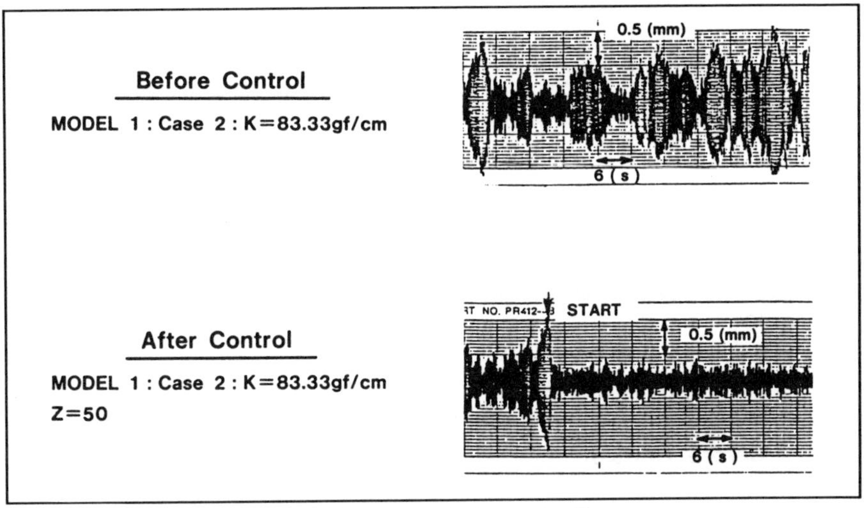

Figure 7 Efficiency for the Fluctuating Flow

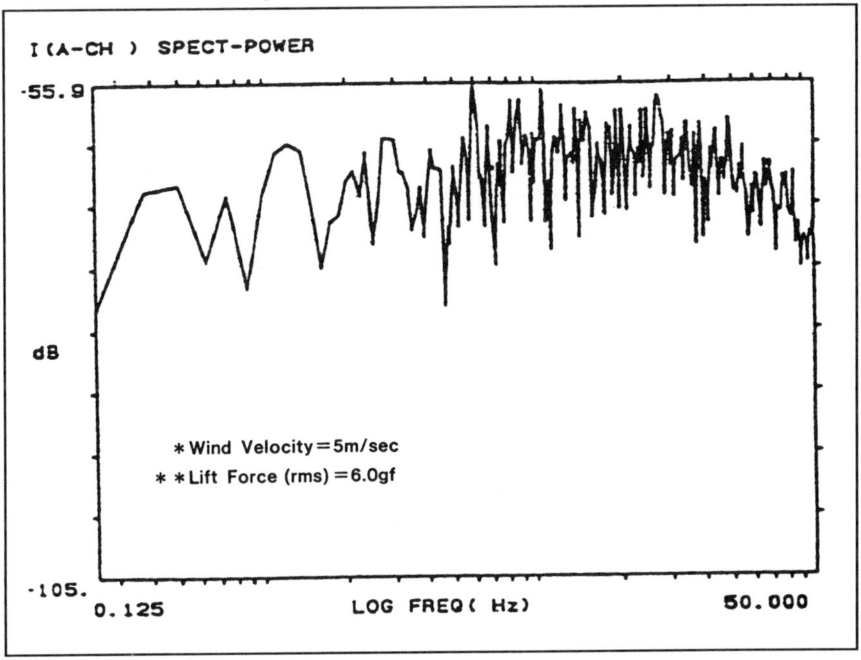

Figure 8 PSD of Fluctuated Lift Force

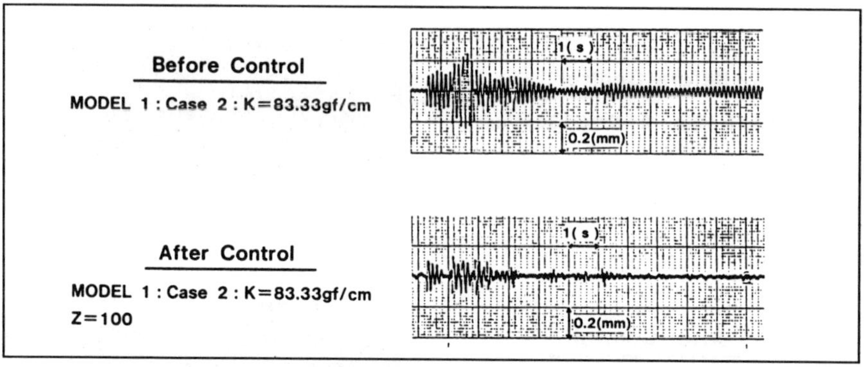

Figure 9 Efficiency for El Centro Wave

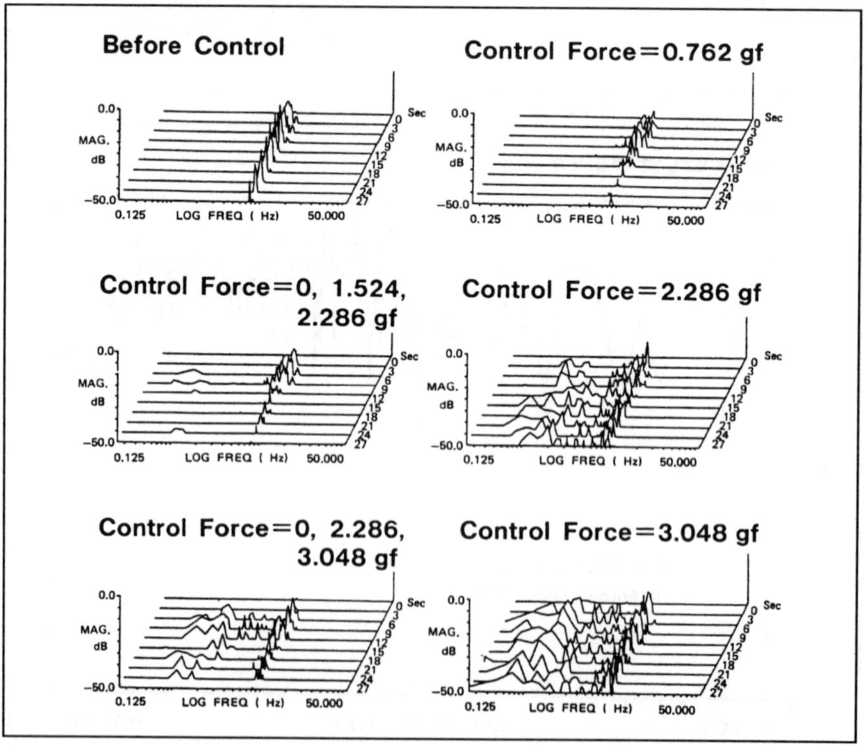

Figure 10 Comparison of Various Controls for SDOF (Linear PSD of Relative Displacement).

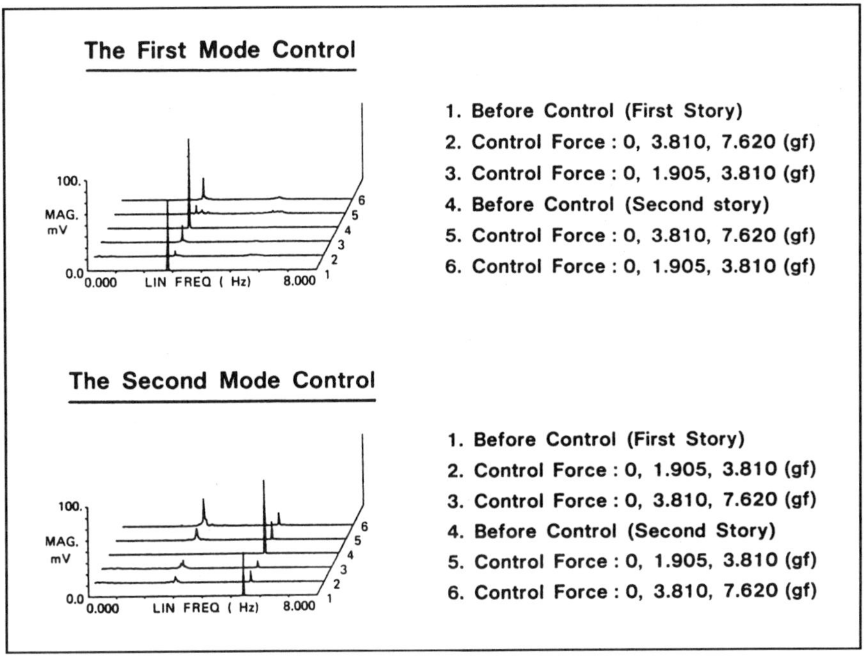

Figure 11 Comparison of Various Controls for 2DOF.

Figure 9 shows the efficiency of the vibration due to El Centro wave. The magnitude of the control force is 2.82% of the model mass. The amplitude ratio between before and after control is 0.42, and this may be thought effective in spite of the simple method.

Figure 10 shows the comparison between the control of constant force and that of various force (i.e., Big, Medium and Small) to El Centro wave. The control by constant force is more effective with bigger force; however, it causes a small vibration of low cycles. On the other hand, this tendency is small due to the control by various forces. A small control force may be effective when the amplitude of vibration is small. Numerical simulations sometimes bypass this fact; therefore, experimental backing is very important.

Figure 11 shows the linear PSD of the relative displacement of 2DOF systems. The vibration frequency of the shaking table was controlled to correspond with each natural frequency of the model. The magnitude of the control force was given in the range of Small, Big and Zero. The larger control force is more effective. However, the larger force may evoke the vibration of the first mode when the vibration of the second mode is suppressed. In this experimental study, only one control force was applied on the first story. Therefore, the control efficiency depends on the precise information of amplitude ratio of modes.

5. Self Tuning of Fuzzy Active Control

In order to make the fuzzy control more effective, it is necessary to tune the parameters and shapes of membership functions used in the fuzzy inference rules. Here, genetic algorithms (GA) is applied for the self-tuning of the inference rules.

5.1 Outline of Genetic Algorithms

The genetic algorithms (GA) [3] is a means of solving optimization problems which imitates evolutionary process of living beings. The process is executed according to steps 1 through 7 of the flow chart shown in Figure 12.

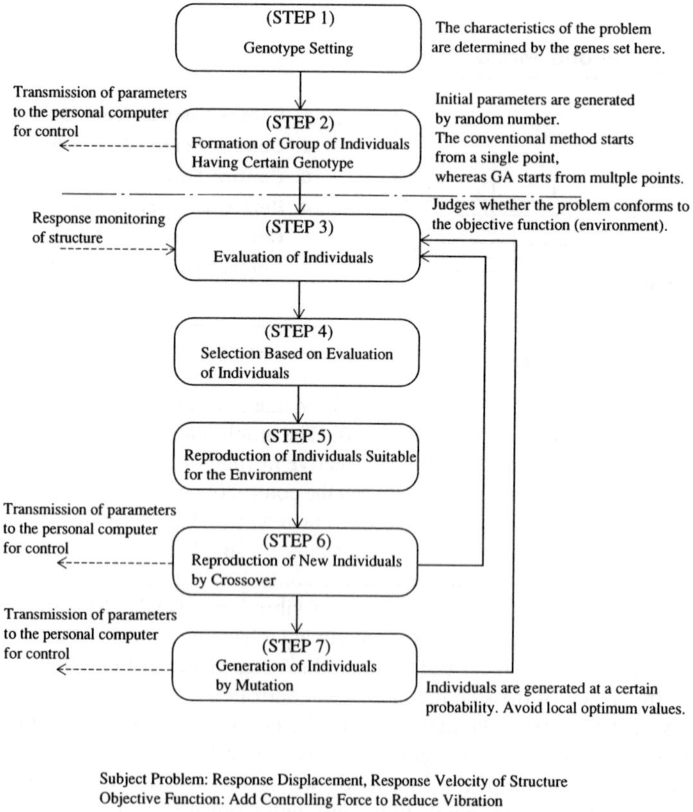

Figure 12 Implementing Procedure for GA

5.2 Modelling of Fuzzy Control Genotype

Fuzzy Control The system of fuzzy control used here is carried out using seven triangular membership functions shown in Figure 13. For the vibration problem under construction, the following rules are employed: If the response velocity is zero, and

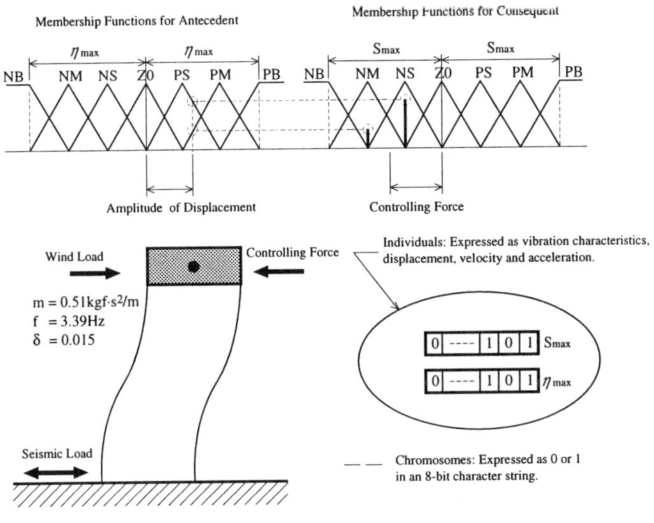

Figure 13 Genotype Model for Fuzzy Control

displacement is positive and small (PS), and negative-small (NS) controlling force is applied. If the response velocity is zero, and displacement is positive and medium (PM), then negative-medium (NM) controlling force is applied.

The controlling force applied to the model is decided by the height method determined from the position and grade of the consequent membership functions.

Coding The success of GA depends on the coding which assembles the chromosomes that decide the traits of actual individuals to symbol strings within the computer. If this is done well, the problem is considered to have been solved.

Coding is carried out in step 1 described in section 5.1. Since this research aims to establish an active control of structural vibration by fuzzy control, the problem results in an optimization problem in which number, shape, position, and normalizing constant of membership functions are optimized. Here, triangular membership functions used occasionally are immobilized at a position such as shown in Figure 13, and the normalizing constants η_{max} and S_{max} of the antecedent and consequent parts of the control rules selected by trial and error are used as the search parameters. Coding is carried out with η_{max} and S_{max}

in the form of chromosomes. These two chromosomes are expressed as 8-bit character strings in which η_{max} and S_{max} are represented by "0" and "1", respectively.

Because the antecedent parts of the control rules take displacement as its subject, and the consequent parts take controlling force, the character strings and normalizing constants are related by proper conversion functions. Figure 14 shows an example of a character string modeled from a chromosome together with its crossover structure. If the normalizing

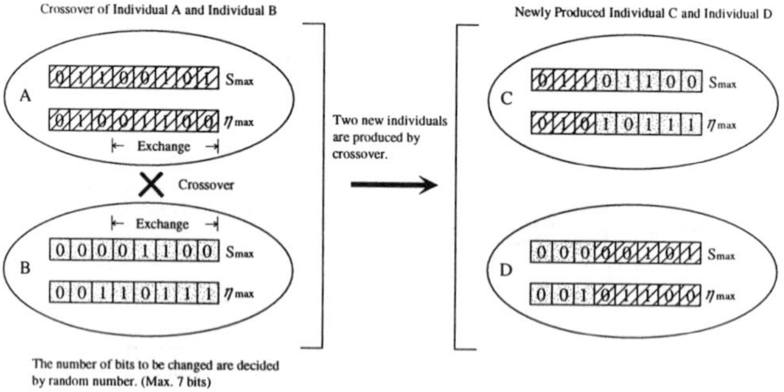

Figure 14 Model for Switching Bits by Coding and Crossover of Chromosomes

constant of the antecedent and consequent parts shown as a pair of chromosomes is changed, the control characteristics are altered with the state of vibration also being changed. If the vibration is considered as the behavior of an individual, the traits of the individual are determined by the pair of chromosomes, and are inherited by the next generation. The state of vibration can be quantitatively grasped by the rms value of response, maximum amplitude within a certain time period, and attenuation factor immediately subsequent to the start of control.

Figure 14 shows the structure of switched chromosomes when new individuals are produced by crossover of the two individuals whose traits (nature of vibration) are decided by the pair of chromosomes. The number of bits to be switched and the combinations of crossover individuals are determined by random number on a computer, with the crossover ratio among groups of individuals set as data to a certain value before starting calculation.

6. Automatic Tuning by Experiment

The system configuration for conducting the experiment is shown in Figure 15. The model is a single-layer, shear-type model. Its mass, its natural frequency f, and logarithmic decrement ratio of damping are 0.51kgf•s/m, 3.39Hz and 0.015, respectively.

The control actuator is a simple system which consists of a stepping motor (made by

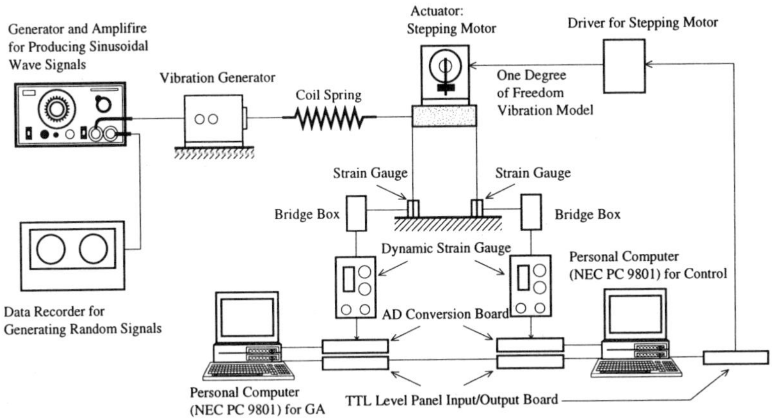

Figure 15 Block Diagram of Automatic Tuning System

Nihon Servo K.K., KY56KM0-551) with an eccentric mass (2.3% of model weight) attached to its shaft, and a stepping motor driver (made by Nihon Servo K.K., SBD-443OPS). Controlling force is induced by inertia force generated by the eccentric mass.

In general, wind load and earthquake load can be driven by a wind tunnel and a shaking table to vibrate the model. Nevertheless, a simple method is employed here, which consists of a vibration generator and a spring as shown in Figure 15. A sinusoidal wave signal and the signal used to record fluctuating wind velocity caused by grid turbulence on the data recorder were used as signals input to the vibration generator.

The control program and parameter search program were executed simultaneously using two personal computers (32-bit Personal Computer NEC PC-9801). This is because it is necessary to run the system continuously without interrupting control when the program shifts to parameter search in the case of an actual system.

Control is performed by a system based on fuzzy inference mentioned in the previous section. Model displacement is input to the personal computer via an AD conversion board and the program is executed. In the beginning of control, the normalizing constants in the antecedent and consequent parts must first be set in the program for control. This step corresponds to setting of initial parameters in step 2 explained in section 2. The parameters are generated by random number on the personal computer for genetic algorithms as shown in Figure 15 and are set by being transmitted to the personal computer for control via a TTL level parallel input/output board. The personal computer then performs control in accordance with the transmitted parameters. At this time, the personal computer for genetic algorithms measures response (displacement), and determines the rms value which is the subject of evaluation, and files it as the evaluation value for parameters. This corresponds to generation of individuals having pairs of chromosomes.

When thc initial parameters have been set, the first generation is evaluated, selected and reproduced. Then, crossover is performed to produce the next generation. When crossover is complete, a new generation is produced by transmitting the newly created parameters in sequence to the computer force control. By further repeating evaluation, selection and reproduction of the new generation, parameters which are most effective for control are sought and discovered. During the search process, mutation is manipulated for producing different chromosomes at a certain probability. As stated previously, this is the step where the possibility of turning a more highly evaluated state of vibration is investigated.

The processing described above is executed repeatedly by a personal computer, with the transition of mean evaluation value of each generation being displayed on its computer display. The system stops at the point where the mean evaluation value of all generations no longer changes for each generation. In this paper, the convergence of mean value of all evaluation values is evaluated through change on the display monitor. The parameters set as chromosomes at the time of completion are the normalizing constant of the membership function, that is, the desired result of searching.

7. Experimental Results

The results of the load using a random signal caused by grid turbulence are shown in Figures 16 and 17. Figure 16 is the response displacement wave when GA was executed as twelve individuals. The top level represents the results during non-control. The next level

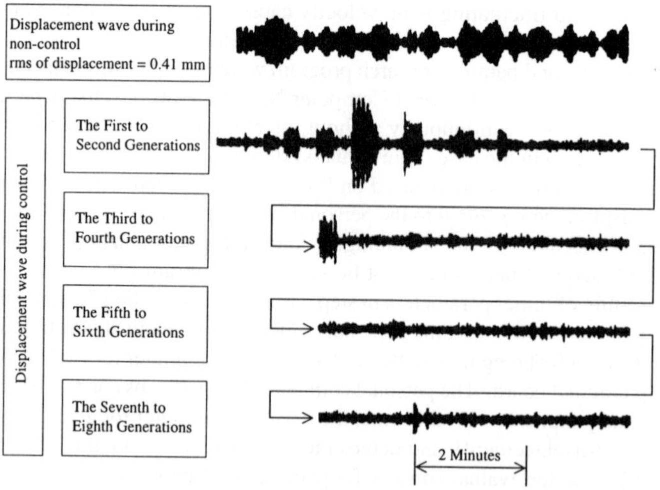

Figure 16 Transition of Vibration State for Each Generation

and those subsequent show the convergence process of response displacement after executing GA. With the results of this experiment, control is carried out for a certain amount of time after setting the control parameters, and the rms value of response displacement which is thought to be somewhat removed from the transient area is considered as the evaluation value.

The first generation is the initial setting step, and parameters in the form of pairs of chromosomes are the result of producing the number of individuals by random number. For this reason, it is seen that ranges from small to quite large excitation for response displacement amplitude occur at the first generation.

With the second generation, as a result of selection, reproduction, and crossover in accordance with evaluation results, the individual with low evaluation is considered to have been eliminated. With the progress of generations, it is known that response displacement becomes smaller and control becomes more effective. A somewhat larger amplitude was seen at the seventh generation, however this thought to have appeared as a result of mutation. Because the objective is to find the optimal normalizing constant of membership functions by GA, it is not a problem even if results of poor evaluations appear.

Whenever GA is introduced to the system, a learning function for optimization works according to input. At this time, if parameters that generate more intensive vibration than the present by mutation are set and control is started, control is stopped immediately by judgment of the transient state subsequent to control, and one may move on to the next step. Figure 17 shows the results of evaluation by displacement and rms value of velocity. The transition of mean evaluation value of each group is shown and it should be noted that the evaluation values converge on a certain value as the generations proceed. With evaluation values during non-control, because it has been made dimensionless, if control is not executed

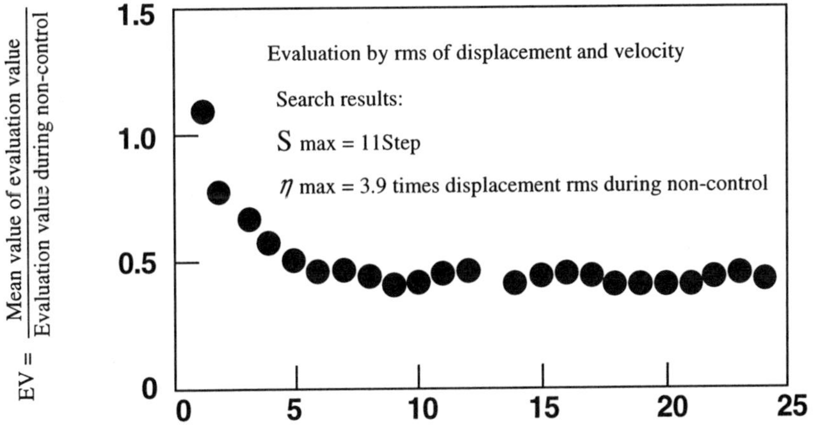

Figure 17 Convergence Process of Evaluation Values by GA

with values larger than one, it indicates that the state of vibration has become worse.

The first generation shows a value larger than one, however as explained before, this is the stage where initial parameters are set. Because the actuator of the control system is simple, the accuracy of controlling force is not effected. Thus, controlling force from the results of the discovered parameters shown in Figure 17 is not large and is a suitable value, and it is therefore assumed that optimal control will be executed.

8. Conclusions

Active control of structural vibration has recently received a great deal of attention. The basic philosophy of active control is, in many cases, based on the "optimum control theory"[4] developed in the fields of mechanical, electrical and aeronautical engineerings. The majority of such conventional vibration control is carried out by solving differential equations. However, there sometimes arise problems concerning uncertainty toward random dynamic behavior of structure or actuator.

On the other hand, fuzzy control which has been developed by Mamdani [5], features an extremely simple control program, quick response, flexibility, and robustness. The fuzzy control has shown remarkable results in such fields as subway, crane and elevator operation which had required expert control by experienced operators or specialists [6].

In this chapter, the fuzzy control was applied for the structural vibration control of earthquake-excited and wind-excited oscillations. Through the numerical and experimental investigations, the following conclusions were derived :

1) For such simple structural models as SDOF and 2DOF, the control rules concerning velocity or displacement are enough. However those of acceleration are necessary to minimize the amplitude if comfortable living circumstances are to be kept even under severe earthquake excitations.
2) If the control forces are small compared with the excitation forces, there is no difference between the fuzzy active control and the constant force control. However one must be very careful in that the latter method may evoke the second mode when the first mode is suppressed.
3) Reasonable control efficiency is observed when the fuzzy rule for the SDOF is applied to the first and second stories of the 2DOF model, respectively, assuming that these stories are independent. The control efficiency will be improved by the precise information of amplitude ratio of modes when one control force is applied to the 2DOF.

Thus, it is confirmed that the fuzzy control is useful in the vibration control of structural systems subject to earthquake and wind excitations. However, the tuning of the fuzzy inference rules is desirable so as to make it more practical. In this chapter, attempts were also made to apply the technique of GA for the self-tuning of the fuzzy control rules. Through a simple experiment, it is proved that the GA can automatically tune the membership functions without complicated procedures.

When controlling force is not appropriate, it is ineffective for damping and may oppositely cause excitation. If genetic algorithms is applied to systems containing such uncertain factors, parameters of proper value which would probably have good controlling effect, can be searched for and discovered within practical processing time. In addition, if further detailed study of genetic algorithms is conducted, it will yield some very interesting results concerning the subject problems mentioned in this paper.

9. References

1. M. Mizumoto, *Comprehensive Fuzzy Set Theory - Fuzzy Logic and Control* (Corona Press, Tokyo, 1989) 32-45. (in Japanese)
2. H. Abe and K. Fukao, *Proceedings of AIJ*, (1990) 825-826. (in Japanese)
3. D.E. Goldberg, *Genetic Algorithms in Search Optimization and Machine Learning* (Addison Wesley, 1989).
4. R.K. Miller, S.F. Masri, T.J. Dehghangyar, and T.K. Caughey, *Journal of Engineering Mechanics Division, ASCE.* **114 (8)** (1988) 1543-1570.
5. E.H. Mamdani, *Proceedings IEEE.* **121 (12)** (1974) 1585-1588.
6. M. Sugeno, *Information Sciences*, **36** (1986) 59-83.

.Uncertainty Modeling in Vibration and Fuzzy Analysis of Structural Systems, pp. 253–318
edited by B. M. Ayyub, A. Guran and A. Haldar
Series on Stability, Vibration and Control of Systems Series B: Vol. 10

EXPERIMENTAL COMPARISON OF PROBABILISTIC METHODS AND FUZZY SET BASED METHODS FOR DESIGNING UNDER UNCERTAINTY

GEORGE MAGLARAS
Department of Aerospace and Ocean Engineering, Virginia Tech
Blacksburg, VA 24061, USA

EFSTRATIOS NIKOLAIDIS
Department of Aerospace and Ocean Engineering, Virginia Tech
Blacksburg, VA 24061, USA

RAFAEL T., HAFTKA
Department of Aerospace Engineering, Mechanics and Engineering Science,
University of Florida
Gainsville, FL 32611, USA

and

HARLEY H., CUDNEY
Department of Mechanical Engineering, Virginia Tech
Blacksburg, VA 24061, USA

ABSTRACT

Recently, probabilistic methods have been used extensively to model uncertainty in many design optimization problems. An alternative is to use fuzzy set based methods for modeling uncertainties. These methods use possibility theory for assessing the safety of a system. Fuzzy set based methods usually require much less information than probabilistic methods and they can utilize expert opinion. The first objective of the study described in this chapter is to understand how each approach maximizes safety. The second objective is to experimentally compare designs obtained using each method.

A cantilevered truss structure is used as a test case. The truss is equipped with passive viscoelastic tuned dampers for vibration control. The structure is optimized by selecting locations for tuning masses added to the truss. The design requirement is that the acceleration at given points on the truss for a specified excitation be less than some limit. The properties of the dampers are the primary sources of uncertainty. They are described by their probability density functions in the probabilistic analysis and by their possibility functions in fuzzy sets based analysis.

A pair of alternate optimal designs are obtained from the probabilistic and the fuzzy set based optimizations, respectively. The probabilistic optimization minimizes the system *probability* of failure. Fuzzy set based optimization minimizes the system *possibility* of failure. Problem parameters (*e.g.*, upper limits on the acceleration) are selected in a way that the probabilities of failure of the alternate designs differ significantly, so that the

difference can be measured with a relatively small number of experiments in the laboratory.
The main difference in the way each method maximizes safety is that probabilistic optimization tries to mostly reduce the probabilities of failure of the modes that are easier to control, whereas fuzzy set based optimization tries to equalize the possibilities of failure of all failure modes.
The resulting optimum probabilistic and fuzzy set based designs are compared in the laboratory. The results confirm that, for the selected problem, probabilistic methods can provide designs that are significantly more reliable than designs obtained using fuzzy set based methods.

1. Introduction

This section, first, explains the motivation that led us to this research project. Then, it reviews the literature on methods for design under uncertainty. Following that, it contrasts the two methods that will be the focus of this study (probabilistic methods and fuzzy set based methods). Next, it states the objectives of this work and the approach followed. Finally, it presents the outline of the rest of this chapter.

1.1. Motivation

A designer often has to design a system in the presence of uncertainties in geometry, loading and material properties, as well as uncertainties in the operating environment. Uncertainties are often classified as *imprecision, modeling* and *random:*
1) Imprecision is due to vagueness in characterizing performance with terms such as "good" or "unacceptable",
2) Modeling uncertainty is due to idealizations in modeling a system and simplifications in analyzing models to predict performance,
3) Random uncertainty reflects variations in the operating environment and the lack of control of a process by a designer.

There exist several methods for designing systems that must perform well in spite of uncertainties. These include safety factors, worst-case scenario, Taguchi methods, probabilistic methods and fuzzy set based methods.

This study focuses only on the treatment of random uncertainties. The most widely used methods for modeling random uncertainties are based on probability theory and on fuzzy set theory. Fuzzy set based methods use possibility distributions to model uncertainties and assess safety. Possibility distributions are estimated using both numerical data or expert opinion. In theory, probabilistic methods should be more effective for problems involving only random uncertainties, because they account for more information about these uncertainties than the other methods. However, to be applied, probabilistic methods may require more information than is available. On the other hand, fuzzy set based techniques require less data than probabilistic techniques.

It is important to develop guidelines that, depending on the amount of information on uncertainties and the accuracy of the predictive model used to solve a given problem, recommend what method to use for design under uncertainty. Such a comparison should use experiments to compare designs obtained using each method because: 1) different methods use different approximate models of uncertainties and also approximate models for predicting a system's performance, and 2) the ultimate test for a method is if it yields designs that behave well in the field. To our knowledge, very little has been done in this direction. This study on experimental comparison of probabilistic methods and fuzzy set based methods for designing under uncertainty aims to be a first step towards understanding how each method maximizes safety and developing guidelines on which method is better for a given problem.

1.2. Review of Methods for Designing Under Uncertainty

Traditionally, minimizing the risk of failure has been accomplished using *safety factors* or *safety margins.* Other methods for designing under uncertainty include:

- Methods that rely on the *worst case scenario* concept for improving safety and reducing sensitivity to errors [1, 2, 3, 4, 5, 6, 7, 8, 9]. In many cases these methods lead to overdesign.
- Taguchi methods [10, 11,12] find the values of some parameters for which a system performance is close to a target and the performance is insensitive to uncertainties. Taguchi methods do not account explicitly for the probability distributions of the uncertainties and also for constraints.

1.2.1. Probabilistic Methods

Probabilistic methods explicitly include uncertainties of the input parameters in the analysis. For given statistical data about geometry, loading and material properties, the statistics of the response quantities are evaluated using various analytical techniques. In problems involving optimization, the estimated system and/or component failure probability is used in the objective function or in the constraints of the optimization formulation.

The first direct reliability based formulation was presented by Charnes and Cooper [13]. They transformed the stochastic optimization problem into an equivalent deterministic formulation by linearly expanding the objective and constraints around the mean value of the random parameters. This approach was called *chance constrained programming (CCP)* technique. Since then, there has been a very large number of publications on reliability-based design in various engineering disciplines. A review of these publications can be found in Frangopol and Moses [14].

However, probabilistic methods have not enjoyed great popularity in industry. One of the reasons is the long experience with the traditional safety factor approach. Another reason is that probabilistic design methods have a number of limitations:

1. In real life applications, there is rarely sufficient data to accurately estimate the statistics of the random parameters. This is particularly true at the tails of the distributions, because they correspond to extremely rare observations, which do not provide substantial evidence to support any particular choice of a distribution. On the other hand, some researchers [15, , 16] maintain that it is more important to estimate accurately the standard deviations of uncertainties than to choose the right distribution. Ben-Haim and Elishakoff [1] and Fox and

Safie [17] showed that, in some cases, the predicted probability of failure is very sensitive to the choice of the probability distribution.
2. When approximate analytical techniques are used to evaluate the output scatter, the probability of failure can only be calculated by assuming a particular distribution for the output quantities. A common choice is the normal distribution. This choice is theoretically justified by the central-limit theorem, for linear problems in which the response is the sum of numerous random variables. However, for problems with pronounced non-linearities or with a small number of uncertain parameters dominating the output scatter, the distribution of the response quantities can be very different from normal. To avoid making an assumption on the distribution of the response quantities, some optimization methods are based on safety indices [18, 19]. Safety index is a normalized measure of the distance from the nominal design point to the most probable failure point (MPP).
3. Modeling errors can greatly affect the predicted failure probability. Ben-Haim and Elishakoff [1] present examples of simple structures, where the predicted failure rates largely depend on the theoretical assumptions used in the model, the effect of initial imperfections, etc. Ideally, modeling errors should be included in the probabilistic formulation as additional uncertainties. However, estimating the statistics of the modeling error is extremely difficult, because it requires data about analytical-experimental mismatch observed over a large number of systems of the same type. For example, Hasselman and Chrostowski [20] [21] accumulated 22 sets of analysis/test data for dynamic analysis of conventional space structures. However, these results are applicable only for the particular type of analysis and structure.

Ponslet, *et al.* [22] demonstrated experimentally the superiority of probabilistic optimization over deterministic optimization. However, the conclusions of this study may not apply to all real life problems because

- This study relied on almost complete knowledge of the statistical properties of the input random parameters.
- Modeling error was minimized by using experimental results to improve the analytical models, which may not always be the case in real life design.

Because of the above limitations, only an idealized, *nominal* probability of failure can be predicted, which can be significantly different from the actual one [23]. It is not clear if probabilistic optimization, which relies on this nominal value, can yield better designs than other design methods.

1.2.2. Review of Studies on Fuzzy Sets

Including explicitly the uncertainties of the input parameters in the analysis makes probabilistic methods a useful tool. Nevertheless, there are types of uncertainties that cannot be accounted by probability theory, because probability theory deals with events that are collections of outcomes of well-defined actions. These events must be subjected to repeatable testing and observations. Real-life problems are usually more complex than their corresponding mathematical models. Occasionally, we need to add a verbal explanation to the results obtained through models. The concept of *fuzzy sets* has been developed to deal with verbal information, that is usually meaningful but not clearly defined (imprecision).

The initial theory of fuzzy sets was formulated by Zadeh [24]. A fuzzy set is a set with boundaries that are not sharply defined. A function, called *membership function*, signifies the degree to which each element of a universe belongs to the fuzzy set. Later, Zadeh [25] related the *theory of possibility* to the theory of fuzzy sets, by defining the concept of a *possibility distribution* as "a fuzzy restriction that acts as an elastic constraint on the values that may be assigned to a variable, called *fuzzy variable*." In the same publication, the author introduced the *possibility/probability consistency principle*. This principle constitutes a weak connection between possibility and probability for a variable that can be associated both with a possibility function and a probability function.

Fuzzy arithmetic is based on the *extension principle* [26], which permits the derivation of the membership functions of functions of fuzzy variables. Unfortunately, the extension principle is not trivial to implement directly. Researchers have proposed approaches such as the *α-cut approach* [27], and the *vertex method* [28], which permit the approximation of a fuzzy set as a collection of intervals using interval analysis concepts.

Fuzzy numbers and their associated arithmetic and calculus have been the subject of many publications and several textbooks [29, 27, 30] and will not be presented here.

Since their establishment, fuzzy sets have been used in a variety of engineering and other applications. Works by Wood, *et al.* [31 32], Thurston and Carnahan [33] and Buckley [34] in multiattribute decision making, Kubic and Stein [35] in designing chemical engineering systems under random and modeling uncertainty, Allen, *et al.* [36] in solving hierarchical design problems and Fang and Chen [37] in geology are only a few examples.

In structural design, Brown [38] applied fuzzy sets to structural safety assessment. He observed that the actual probability of failure of most structures appeared to be much higher than the value obtained if only objective information (statistics, probabilities) were taken into account. He concluded that subjective information (expert knowledge) was as important as objective information and had to be taken into account for safety measurement. Ayyub and Lai [39] examined the effect of the vagueness in the perception of damage on structural reliability. Instead of using a crisp definition of failure, they used a fuzzy definition of failure. They concluded that by introducing vagueness in the definition of failure, the resulting average probability of failure was larger than the probability of failure based on the crisp failure definition.

Fuzzy sets were used by Dong, *et al.* [40] to model linguistic and qualitative factors, in evaluating the safety of existing buildings. In the same publication the authors introduced the notion of *failure possibility* as a measure of structural safety. There were numerous other applications of fuzzy set theory in the past decade in civil engineering. An extensive review of relative publications can be found in Chou and Yuan [41].

1.2.3. Studies Comparing Fuzzy Set Based Methods and Probabilistic Methods

One of the biggest issues seems to be how fuzzy set based methods and possibility theory compare to probability theory. There have been several debates concerning whether probability theory or fuzzy analysis is the appropriate basis for addressing uncertainty. Recently, Laviolette and Seaman [42] criticized the argument that fuzziness represents another type of uncertainty distinct from probability. In their conclusions, they emphasized that, although fuzzy set theory is applicable to some real life problems, probability theory provides a complete and uniquely optimal means for solving problems and managing uncertainty. Dubois and Prade [43], Wilson [44] and Klir [45] disagreed with the statements made by Laviolette and Seaman [42], whereas Lindlcy [46] fully agreed.

Other studies tried to establish some common theoretical ground for possibility theory, which is based on fuzzy set theory, and probability theories. Gaines [47] established a common theoretical basis for probability and fuzzy logic. Natvig [48] indicated that, at least in some applications, *possibility distribution* can be interpreted as a *family of probabilities*. Henkind and Harrison [49] concluded that Bayesian calculus is well suited for applications where probabilities are known or can be acquired with a reasonable effort. They also concluded that fuzzy set calculus is well suited for applications where the evidence itself is fuzzy in nature. Fuzzy set based methods are also advantageous in situations with little information. Bordley [50] concluded that there is a convergence between probability theory and possibility theory in *observer bias* situations. Dubois and Prade [51, 52] tried to clarify some "classical misunderstandings" between fuzzy sets and probability. They also presented examples that "... should convince us that instead of considering probability and fuzzy sets as conflicting rivals, it sounds more promising to build bridges and take advantage of the enlarged framework for modeling uncertainty and vagueness they conjointly bring us to."

Chou and Yuan [41] present such an approach, where fuzzy set and Bayesian theories are combined in evaluating the reliability of existing structures. Wood and Antonsson [31, 32] also combined the two theories in modeling imprecision and random uncertainty in preliminary engineering design. They used the notion of hybrid numbers (proposed by Kaufmann and Gupta [27]). A hybrid number is similar to a complex number except that, instead of having a real and an imaginary part, a hybrid number has a fuzzy component (representing imprecision) and a random component (representing random uncertainty).

Vadde, *et al.* [53], compared fuzzy set based and probabilistic optimization. They observed that the fuzzy set based design was more conservative.

Other analytical studies have compared methods for design under various types of uncertainty, but it is not always clear which method is better. Wood, *et al.* [31 32] compared probability calculus and fuzzy set calculus for handling imprecision. They concluded that fuzzy set calculus were more suitable than probability calculus in handling the imprecision aspect of uncertainty in design.

There is no consensus on how to treat modeling uncertainty. Most studies have used probability theory [54, 55], but a few have also used fuzzy set based methods [31, 32, 35, 56].

Concerning random uncertainty, which is the focus of this study, most researchers [31, 32, 49] agree that it is better represented by probability theory, when there is sufficient statistical information about the random variables. However, when there is little information on the statistics of the random parameters, it might be better to use possibility theory, which is based on fuzzy set theory. Chiang and Dong [57], Hasselman, *et al.* [58] and Dong, *et al.* [59] compared analytically fuzzy set based methods and probabilistic methods for deriving the uncertainty in the response of a system due to random uncertainties in loads or properties of the system. Chiang and Dong [57] and Dong, *et al.* [59] concluded that fuzzy set based methods should be used in cases where there is little information about uncertainties, because they are simpler and more efficient than probabilistic methods. However, this conclusion was based on few simple examples in which the predictions of fuzzy set based methods and probabilistic methods were quite similar.

The following are three important issues in comparing probabilistic methods and fuzzy set based methods, which, to our knowledge, have not been addressed:
1) How can we determine if fuzzy set based methods are better to model random uncertainty when little information is available,
2) How much information on random parameters is *little enough* to justify switching from probability theory to possibility theory for representing random uncertainty,
3) How can we compare experimentally *designs* obtained using these two approaches.
This study will primarily focus on the third issue, that is experimentally compare designs obtained from fuzzy set based methods and probabilistic methods.

1.3. Contrast Between Probabilistic Methods and Fuzzy Set Based Methods

In a problem where only random uncertainties are involved, if we have accurate models of these uncertainties, accurate models of the system involved, and a crisp definition of failure, we should expect that probabilistic design will yield better designs than fuzzy set based design, because it accounts for more information about uncertainties than all other methods for design under uncertainty. However, in many real life problems, probabilistic design may not be the best, because it requires too much information and in many cases it is sensitive to lack of information. This includes detailed information on the joint probability distributions of the random variables, as well as information on modeling and human errors. Because we rarely have all this information, we often need to guess or estimate it. For example, in most real life applications, very little is known about the correlation between different variables, and it is common practice to assume that variables are uncorrelated. Design optimization can often exploit model weaknesses. Therefore, we may question the utility of probabilistic optimization procedures, that rely on inaccurate models of uncertainties.

Fuzzy set based methods also maximize reliability. They model uncertainties using the *possibility function*, which measures the degree to which it is possible for a variable to take a certain value. For example, assume that the nominal value of a length of a panel is 1 meter and experts say that the length can vary between 0.9 and 1.1 meters. The possibility function of the length may have a value of one between 0.99m and 1.01m and then taper linearly to zero at 0.9m and 1.1m. Possibility functions are numerically equal to membership functions. Suppose for example that an expert said that the length is small. The value of the possibility function for a particular value of the length would be equal to the membership of a panel whose length is equal to the particular value to the fuzzy set "small". Possibility functions are do not appear to be so different from probability distributions. However, possibility functions are often constructed without precise information. Also, rules for calculating the possibility function of the system response from the possibility functions of the system parameters are different from the rules used to

calculate probability distributions. These rules, which are based on fuzzy set calculus, tend to cater more to worst case scenarios because they reflect:
1) The fuzzier nature of the possibility functions.
2) The fact that the possibility of an event is always larger than the probability of this event (consistency principle [25]).

Possibility and probability distributions are alternate representations of random uncertainty. Both possibility and probability distributions can be estimated using numerical data or subjective information [29, 52]. Moreover, one can transform a probability distribution to a possibility distribution and vice versa provided that the transformation respects a principle of uncertainty and information invariance [29].

The fuzzy set approach for structures has been found to provide more conservative (heavier) designs than probabilistic approach [53]. Because fuzzy set calculus assumes that models of uncertainty are approximate and produces more conservative results than probabilistic calculus, it might be better than probabilistic design for problems in which little information is available.

The definition of failure is usually crisp in the probabilistic design. A fuzzy set approach, however, can select a fuzzy definition of failure [39]. That is, failure can have a membership function between zero and one expressing a transition between a totally acceptable and a totally unacceptable design.

From the above discussion we conclude that the relative merits of probabilistic design and fuzzy set based design may depend on:
1) The amount of information available on the uncertainty.
2) How crisply failure is defined.
3) The accuracy of the deterministic models used to predict the system performance.

The work presented here is part of a study whose long term goal is to understand how these three factors affect each method and determine which method is better for a given problem.

1.4. Objectives

This study intends to take a first step in the direction of comparing probabilistic and fuzzy set based methods by:
- Understanding how each method maximizes safety.
- Finding problems for which fuzzy set based methods and probabilistic methods yield alternate designs that have significantly different probabilities of failure, which can be measured with a small number of experiments in the laboratory.

1.5. Approach

In this study, we consider a problem where, <u>using the same resources</u>, we design two alternate systems whose properties are uncertain, to maximize safety. The properties of one design, called *probabilistic design*, are modeled as random variables. The properties of the alternate design, called *fuzzy set based design*, are modeled as fuzzy variables. Using optimization we find an optimum probabilistic and an alternate optimum fuzzy set based design. A large number of realizations is built and tested in the laboratory to compare the relative frequency of failure resulting from each alternate design approach.

Experimental study of the effects of uncertainty on safety requires building and testing many nominally identical designs and measuring how many fail. This can be difficult, if we have to build, compare and possibly destroy a large number of sample structures. Therefore, we need a low cost approach for the experimental comparison of methods for designing under uncertainty.

The key idea is to consider a structure with failure modes that do not imply the destruction of the structure or its members. The configuration of the structure must allow us to construct and test a large number of identical random samples of the same design at low cost.

We selected a dynamic system and defined failure as excessive vibration. We consider the optimum design of a small cantilevered truss structure equipped with tuned dampers for reducing the vibration amplitude. The truss is made of aluminum members and nodes. The dampers are made by hand from widely available plastic and viscoelastic materials. The performance of those dampers is controlled by tuning their natural frequency to a natural frequency of the truss. The variability of material properties and the manual fabrication induce uncertainties in the properties of the dampers.

In designing the truss, only the uncertainties in the properties of the dampers are considered, because they are much larger than the other uncertainties. We manufacture a large number of dampers, measure their properties and estimate the statistics of the properties. We create many realizations of the damped truss by attaching different nominally identical dampers to the truss. The dampers are not perfectly tuned to the natural frequencies of the structure so the vibration amplitudes are high. The optimizer tries to correct the mistuning by adding tuning masses to the truss. If the dampers are undertuned (*i.e.*, their natural frequencies are lower than those of the structure), the added masses reduce the natural frequencies of the truss, thereby increasing the effectiveness of the dampers.

The optimization problem consists of finding the best locations of a maximum of 10 tuning masses on the truss. One damper per mode is used to control vibration modes 1 and 3 of the truss. Failure is defined as the event of the *peak dynamic acceleration* of the first or third mode (or both) exceeding given maximum allowable limits. We create two alternative formulations of the optimization problem. The probabilistic formulation minimizes the probability of failure while the fuzzy set based formulation minimizes the possibility of failure. The same resources are available for both formulations (10 tuning masses).

In this study we account for random uncertainties only. To minimize *modeling uncertainties* we use experimental results to improve the analytical models. Specifically, we use a correction equation to make the analytically predicted peak accelerations closely match experimental measurements (see Paragraph 4.2.2). Because the experimental truss is available in the laboratory, we can update that analytical-experimental correction at any time. However, in real life a designer has seldom the luxury of experimental measurements during the design phase. Instead, the designer has to rely on previous results obtained by others. To simulate this situation we use in our original analysis analytical-experimental correction equations that were estimated in a previous study [22].

Because the design variables in this optimization problem (locations of added masses) are discrete, a genetic algorithm is used for the optimization. Detailed description of the algorithm can be found in Ponslet [60]. All probabilistic analyses are performed using Monte Carlo simulation. In the fuzzy set analysis, the membership function of the response is evaluated from the membership functions of the input parameters using the vertex method [28].

As mentioned earlier, fuzzy set based methods use the failure possibility as a metric of how unsafe a design is. Note that there are several definitions of *failure possibility*. For example, Dong, *et al.* [40] present three different definitions of failure possibility, depending on three different criteria of ranking two intervals that contain the load effect and the strength of a system. In this study, we define the failure possibility as being equal to the *possibility measure* of the interval containing all the values that correspond to structural failure. Zadeh [25] defines the possibility measure $\pi(A)$ of a nonfuzzy subset A of a universe U as

$$\pi(A) \equiv \sup_{u \in A} \pi_x(u) \tag{1}$$

where $\pi_x(u)$ is the possibility distribution function associated with a variable X, which takes values in U. This number, then, is interpreted as the possibility *that a value of X belongs to A*. Symbol "sup" denotes the supremum, which is the smallest upper bound.

The optimal designs are compared analytically on the basis of their probabilities and possibilities of failure. To validate the results experimentally, 29 realizations of each design are tested in the laboratory (We constructed only 29 realizations, because of shortage in viscoelastic foam). The experimental comparison is based on the number of realizations of each design that fail. To compare the alternate probabilistic and fuzzy set based designs using a relatively small number of experiments, some problem parameters (such as the failure limits and the scatter of the natural frequencies of the dampers) have to be adjusted in a way that the difference in the probabilities of failure between the probabilistic and fuzzy set optimum designs is large. This procedure is along the lines of *anti-optimization* or *contrast maximization* [61, 62, 63].

Two points must be made regarding the fairness of the comparison:

a) Because the definition of failure is crisp, it is appropriate to compare the probabilistic and fuzzy set based design using the relative frequency of failure.

b)By maximizing the difference between the probabilities of failure of the probabilistic and fuzzy set based design, we come up with a problem that is favorable to probabilistic design. This helps understand some of the differences between the two design methods because it makes these differences stand out but obviously makes the comparison biased toward the probabilistic design. A complete comparison should also consider problems that are favorable to fuzzy set based design.

1.6. Outline

Section 2 briefly describes the hardware used in this study and the finite element models used to model it. In Section 3, we present alternate probabilistic and fuzzy set approaches for designing a damped truss. The corresponding optimization problems are formulated and the probabilistic and fuzzy set analysis methods are described. Probabilistic analysis refers to the calculation of the probability of failure of a system when uncertainties in the parameters are described using their probability distributions. On the other hand, fuzzy set analysis refers to the calculation of the possibility of failure of a system when uncertainties in the parameters are described using their membership functions. Finally, we select a design problem for which probabilistic and fuzzy set approach yield alternate designs that have considerably different failure probabilities. For this purpose we determine the statistics (mean values and standard deviations) of dampers.

Section 4, compares analytically probabilistic and fuzzy set optimization results. First, dampers whose natural frequencies, mean values and standard deviations are approximately equal to those found in Section 3 are constructed. Once the dampers have been constructed, they are measured again and the probability distributions of their properties are used to find a probabilistic and a fuzzy set optimum design. Then, the probabilities and possibilities of failure of the two designs are calculated and compared.

Section 5, first, explains how the 29 samples of dampers for the experimental validation were obtained. These 29 realizations are then tested in the laboratory and the numbers of failures of the two designs are compared. The measured relative frequencies of failure are also compared to the corresponding analytical predictions. Then some sources of error that can be responsible for the discrepancy between analytical and experimental failure probabilities are examined.

2. Description of the Selected Problem

This section describes the truss structure, the tuned dampers and the tuning masses used in this study and the corresponding finite element models. Next, it describes the instrumentation used in our experiments. Finally, it gives the definition of failure for this problem. The description is short and limited to what is necessary for the reader to follow the rest of the chapter. Detailed descriptions can be found in Ponslet [60].

2.1. System Description

2.1.1. Truss Structure

The structure used in this study is shown in Figure 1. It is a short, beam-like truss assembled from 30 tubular aluminum members with steel end fittings, connected through 12 spherical aluminum nodes. The truss is about 1 meter long and weighs about 4.4 kg. The two middle bays of the truss are pyramids with 0.254 m square bases. Two half bays attached to the ends of the middle section complete the structure. The 26 non-diagonal members are 0.254 m long from node center to node center, while the 4 diagonals are $\sqrt{2}$

times longer. Three nodes are attached to a base made of thick steel and aluminum plates mounted on the laboratory wall.

Figure 2 shows a plot of the magnitude of a measured frequency response function (FRF) from excitation force to response acceleration for that truss.

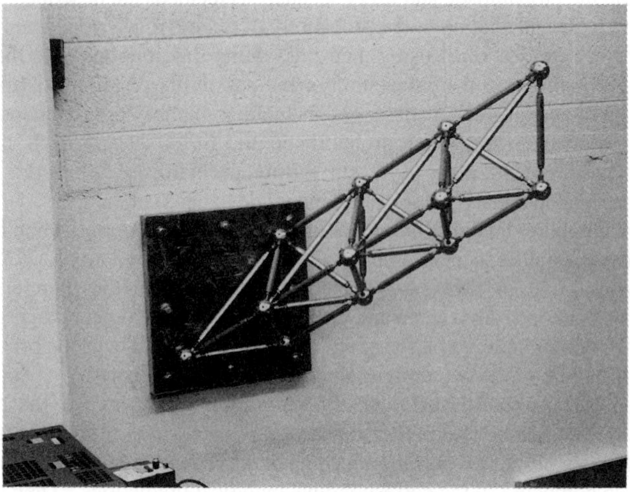

Figure 1 Laboratory truss.

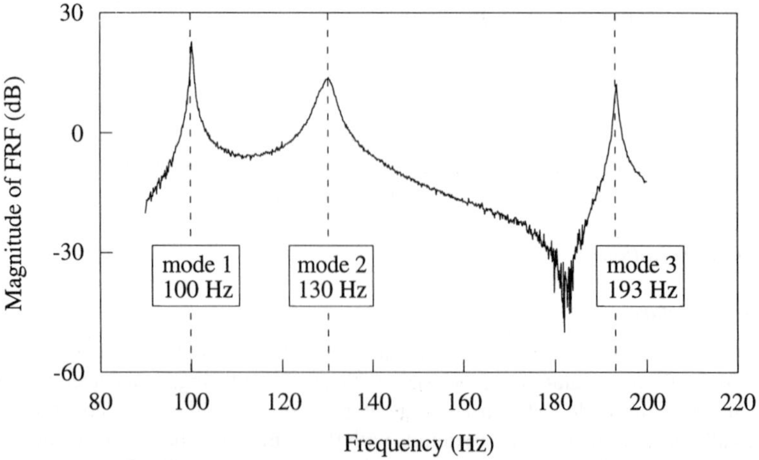

Figure 2 Measured frequency response function of the laboratory truss (magnitude of acceleration per unit of force).

The locations of the excitation and response measurements are shown in Figure 3. The first three modes are well separated and clearly identified. Their natural frequencies are 100, 130 and 193 Hz. Local bending modes in the members occur at frequencies 280 Hz and higher. The truss model neglects bending (see Paragraph 2.2.2). Therefore, it is useful only for the first three modes. For this reason, only the first 3 modes of the structure will be considered.

The first and third modes of the truss are very lightly damped; their measured damping ratios are only 0.13% and 0.08%, respectively. The damping ratio of the second mode is significantly higher (1.05%). This is believed to be due to coupling with the dynamics of the wall.

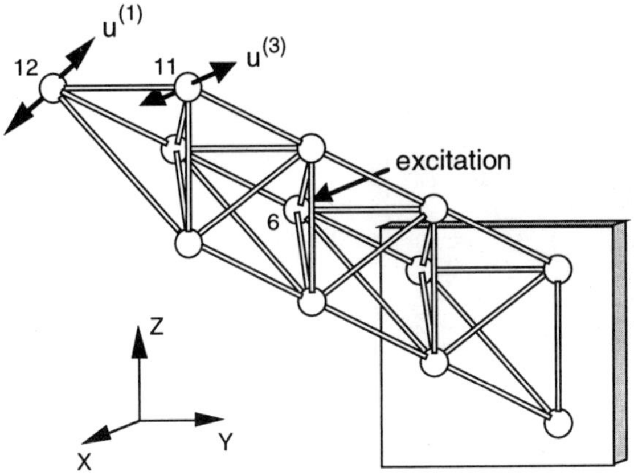

Figure 3 Locations of excitation and response measurements on the laboratory truss.

2.1.2. Tuned Dampers

We designed and manufactured viscoelastic tuned dampers (Figure 4) to reduce the dynamic response of the first and third modes of the truss. The dampers consist of symmetric cantilevered beams that carry adjustable tip inertias and are attached at their middle point to a node of the structure. The first bending mode of the damper beams is tuned to a natural frequency of the truss to maximize the energy absorbed by the tuned dampers.

Figure 5 is a schematic sketch of a tuned damper. The beam is a sandwich of 2 thin plastic sheets and an inner core of viscoelastic acrylic foam. This viscoelastic foam is available in the form of a 1.14 mm thick, double sided self-adhesive tape. A plastic tip block is glued to one of the plastic layers at each end of the beam. A steel tuning screw is threaded into each of these blocks. This allows the fundamental frequency of the damper to be adjusted by moving the center of gravity of the tip masses. A 6.35 mm hole is drilled in the middle of the sandwich plate. The damper is attached to a node of the structure with a 6.35 mm nylon screw passing through that hole.

Figure 4 Tuned damper attached to the truss.

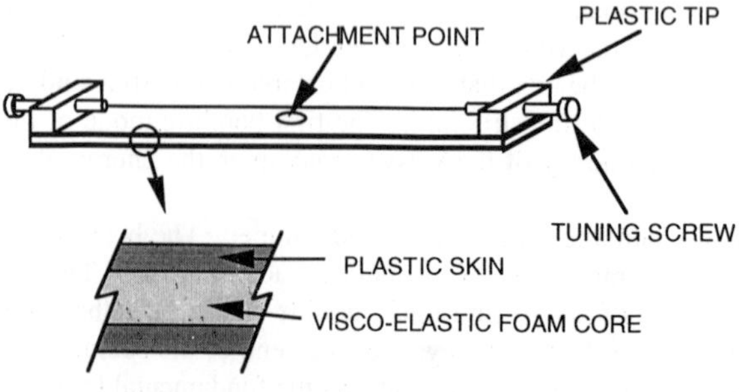

Figure 5 Design of the tuned damper.

Two slightly different versions of this damper design (we will refer to them as *type-1* and *type-3*) are used to target the first and the third natural frequencies of the structure. Type-1 dampers use 0.55 mm thick plastic sheets and are 108 mm long and 25.4 mm wide. Type-3 dampers use 0.82 mm thick plastic sheets and are 92.3 mm long. Their width is the same as type-1 dampers. Type-1 dampers can be adjusted from about 98 Hz to about 116 Hz; the range for type-3 dampers is about 170 Hz to 204 Hz.

The mass of the dampers is very small compared to the mass of the structure. Both types weigh about 10 g, which represents about 0.2% of the total mass of the truss. Despite their size, these dampers provide very significant damping to the truss. Figure 6 shows the frequency response function of a tip node acceleration, before and after adding one type-3 damper to the structure. The damper was tuned to the frequency of mode 3 and located at the node and the direction that correspond to the largest acceleration amplitude of vibration in the third mode shape. The reduction in acceleration amplitude achieved in mode 3 is more than 25 dB.

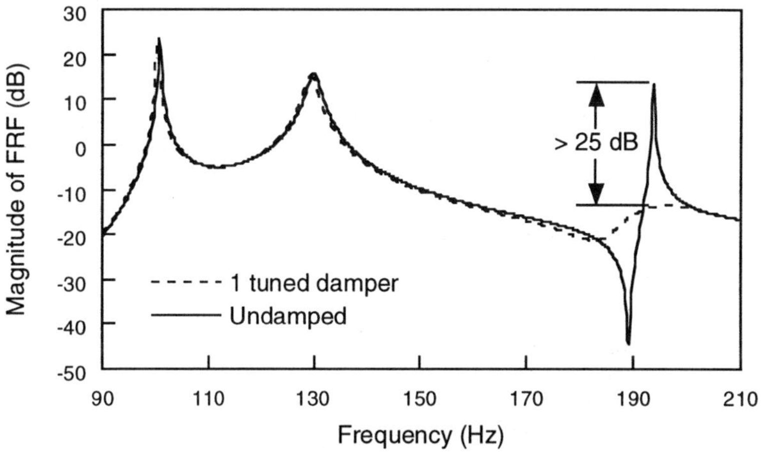

Figure 6 Effect of a type-3 tuned damper on the response of the laboratory truss.

The natural frequency of the damper most directly determines its effectiveness. The tuned damper does not significantly affect modes of the truss that are far from the tuned frequency of the damper. Therefore, at least one damper per target mode is needed.

2.1.3. Tuning Masses

We used tuning masses that can be attached to the nodes of the structure to modify its natural frequencies. They consist of standard steel screws and nuts. These screws are compatible with the standard holes in the nodes of the structure so that they can be easily attached to the truss. Ten screws were selected and weighed. Their average mass is equal to 16.61 g with a sample coefficient of variation of about 0.4%. Even when all 10 tuning masses were added to the truss, the added mass represented less than 3.8% of the mass of the original truss.

2.2. *Finite Element Models*

2.2.1. *Truss Structure*

The finite element model of the truss is shown in Figure 7. We define a global axis system the following way: positive z-axis is pointing vertically up, positive x-axis is orthogonal to and pointing away from the wall and positive y-axis completes the right-handed coordinate system.

Each member of the structure is modeled as a 6-degree of freedom, 3D rod finite element defined by its mass and complex stiffness $k(1+i\eta)$, where k is the stiffness and η is the loss factor of the member used to model inherent damping. The bending stiffness of the rod elements is zero. Consistent mass matrices are used to represent the inertial properties of the members. Each node is modeled as an infinitely stiff concentrated mass equal to the measured mass of the physical node.

The values used in the model for the masses of the members and nodes and for the stiffnesses of the members are the mean values of series of measurements performed on a large number of members and nodes. Details on these measurements can be found in Ponslet [60]. The flexibility of the base (wall) is simulated in the finite element model of the structure using 9 springs with complex stiffnesses (3 per attached node, one normal to the base and two in the plane of the base), as shown in Figure 7. With these support springs, the finite element model of the truss contains 36 degrees of freedom.

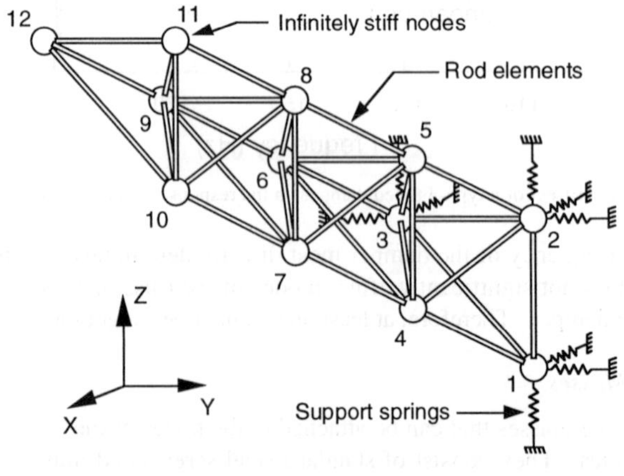

Figure 7 Finite element model of the laboratory truss.

2.2.2. Tuned Dampers

The following simplifying assumptions are made about the dampers: First, the attachment point is assumed to move in pure translation, neglecting any rotation of the supporting node (note that the truss model does not include these rotations). Second, the two halves of the damper are assumed identical, so that their deflections are identical in magnitude and phase, and the contributions from the two halves can be superimposed. Finally, the damper is assumed to deform only in its first bending mode (higher modes are neglected as well as torsion along the axis of the beam).

With these simplifications, a two degree of freedom (D.O.F.) model was devised for the tuned dampers. This simplified model is shown in Figure 8. It consists of two masses (a base mass m_o attached to the node and a tip mass m) that are connected with a spring of complex stiffness $k(1+i\eta)$. Four parameters are needed to completely define the damper model. The total mass m_T, the tip mass, m, the natural frequency, f_n, and the loss factor, η, representing the damping effect of the viscoelastic layer will be used in this study. The total mass is measured on a scale. The remaining three parameters are identified by a three-parameter least-squares fit on the imaginary part of the measured transfer function from base acceleration to base force. Details on the experimental setup used for measuring the parameters of the tuned dampers are given in Ponslet [60].

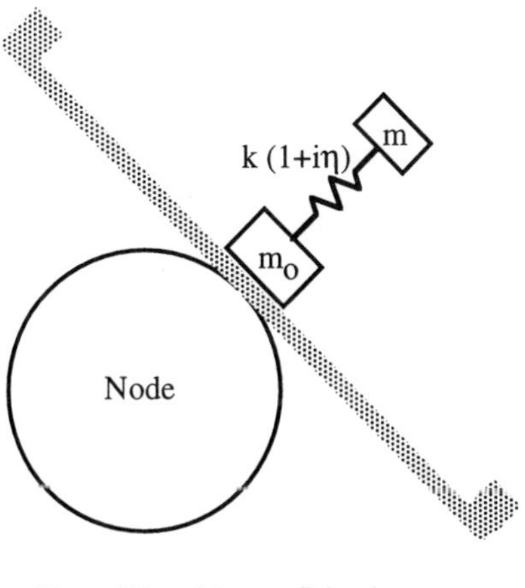

Figure 8 Tuned damper finite element.

To include the tuned damper in the finite element model of the truss, we created a special 4 degrees of freedom (D.O.F.) element. This element models the dynamics of the damper in the direction orthogonal to the sandwich beam, as well as the added mass effect in the other directions. Each damper adds one degree of freedom to the model of the structure. The remaining 3 D.O.F. correspond to the 3 components of displacement of the node to which the damper is attached and are shared with the existing model of the truss. With two dampers on the truss, the number of degrees of freedom increases from 36 to 38.

2.3. *Instrumentation*

To measure the dynamic response of the truss structure an excitation is provided by an electromagnetic shaker (Ling Dynamics, model 102) attached to the steel base plate and connected to node 6 of the truss through a stinger orthogonal to the wall and parallel to the x-axis (see Figure 7). A piezoelectric load cell (PCB model 208B) measures the excitation force.

The shaker is driven by the signal generator of a Tektronix model 2630 FFT analyzer through a power amplifier (KEPCO, model BOP 50-2M). The response accelerations are measured at nodes 11 and 12 by sub miniature piezoelectric accelerometers (PCB model 353B17). A sine dwell technique is used to measure the transfer functions in the frequency windows around modes 1 and 3. The locations of the excitation and the response measurement were selected so that the first 3 modes could be identified simultaneously. The data acquisition and FFT analysis were performed on the Tektronix model 2630, PC controlled analyzer.

To estimate the tip mass, the natural frequency, and the loss factor of a damper, the damper is attached to the moving coil of an electromagnetic shaker (MB Electronics, model EA1500) and the transfer function from base acceleration to base force is measured using a sine dwell technique. A 3 parameter least-squares fit is then performed using MATLAB [64] on the imaginary part of the measured transfer function and provides estimates for the mass, frequency and loss factor.

2.4. *Definition of Failure*

In this study, failure is associated with excessive vibration. In structural analysis, the damping ratios associated with the vibration modes are often used as measures of the damping in the structure. However, the use of tuned dampers produces pairs of very closely spaced modes. Because the mass ratio of the tuned dampers to truss is so small (3.8%), there is only one identifiable resonance peak in the response of the truss. The damping ratios associated with two closely spaced modes are very sensitive to small changes in the properties of the system. Under these conditions, the damping ratios of the individual modes are not a good measure of the damped response of the structure.

For a given location of the excitation, a less sensitive measure is the largest magnitude (over a frequency range) of the accelerance transfer function between the excitation force and the response acceleration at specified points on the structure. This measure is referred to as *peak accelerance*.

3. Problem Formulation and Solution Techniques

In this section we design a damped truss subject to upper limits on the response (acceleration) to a given dynamic excitation. Two alternative design optimization procedures are presented; a probabilistic and a fuzzy set approach. The probabilistic optimization minimizes the *probability* of exceeding given acceleration limits, while fuzzy set based optimization minimizes the corresponding *possibility*. Note that the same resources (added mass and dampers) are used in both formulations.

First, we present a general procedure for designing a damped truss structure. Then, we formulate the probabilistic and fuzzy set optimizations for finding the safest design for given resources. Next, we describe the methods used for assessing safety in the probabilistic and fuzzy set optimizations. Following that, we explain how we selected a problem that provided a large difference in the probabilities of failure between probabilistic and fuzzy set optimum designs. Finally, we present an approximate solution technique for estimating the frequency response function used to reduce computational cost during the optimizations.

3.1. Probabilistic and Fuzzy Set Based Approaches

The design requirements are expressed as upper limits on the acceleration at given points on the structure and within prescribed frequency ranges, for a given excitation. These requirements are formulated as a series of upper limits $A_{lim}^{(m)}$ on the acceleration at prescribed locations within a series of n_m frequency "windows" as illustrated in Figure 9. The system "fails" if acceleration exceeds these limits. The locations, frequency content and amplitudes of the excitation forces are assumed to be specified.

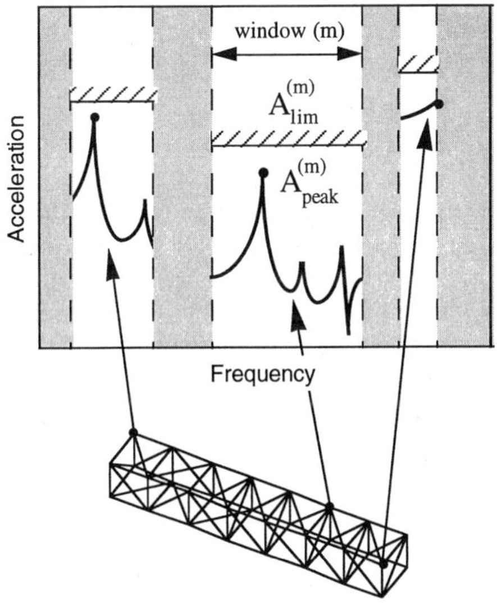

Figure 9 Performance requirements for a damped truss.

Note that the locations on the structure where acceleration limits are imposed need not be the same in each frequency window. This is the case, for example, when sensitive devices, each of which could be sensitive to a particular range of frequencies, are attached at given locations on a structure.

The design problem consists of bringing the peak acceleration $A_{peak}^{(m)}$ in each window m below the prescribed limit $A_{lim}^{(m)}$ by adjusting the design variables between bounds x_l and x_u, while satisfying r resource limits $g_i \leq 0, \quad i = 1,\ldots,r$ (for example, limit on the total weight of the structure, limit on gains in active systems, etc.).

Design variables **x** can include parameters of the tuned dampers and the truss itself, as well as the locations of the tuned dampers on the structure. In the actual structures, some or all design variables as well as other parameters **p** of the model are uncertain. In the probabilistic approach they are considered random variables. In the fuzzy set approach they are represented as fuzzy variables.

We can replace random variables, which would be design variables in deterministic optimization problem formulation, with parameters that define their probability distributions (*e.g.*, means and standard deviations). Then, the probability of failure can be computed and minimized using a model of the truss. If we use the means of the design variables, probabilistic optimization minimizes the probability P_f that any of the acceleration amplitudes will exceed the corresponding limits, and can be formulated as,

$$\begin{array}{ll} \textbf{Find} & \bar{\mathbf{x}} \quad \text{to} \\ \textbf{Minimize} & P_f = P\left(\bigcup_{m=1}^{n_m} \left[A_{peak}^{(m)}(\mathbf{x},\mathbf{p}) \geq A_{lim}^{(m)} \right] \right) \\ \textbf{such that} & \mathbf{x}_l \leq \bar{\mathbf{x}} \leq \mathbf{x}_u \\ \textbf{and} & g_i(\bar{\mathbf{x}},\bar{\mathbf{p}}) \leq 0, \quad i = 1,\ldots,r \end{array} \tag{2}$$

where $\bar{\mathbf{x}}$ and $\bar{\mathbf{p}}$ denote the mean values of the design and the other uncertain variables respectively, ***U[•]*** denotes the union of events, and *P[•]* denotes the probability of an event.

An alternate formulation can be produced if each uncertain variable is represented as a fuzzy variable. In this case, we can obtain a possibility function of the structural response. Then, optimization can be used to minimize the possibility that the peak acceleration will exceed the limits $A_{lim}^{(m)}$ in the frequency windows of interest. A fuzzy set optimization problem can be formulated as follows:

$$\begin{array}{ll} \textbf{Find} & \bar{\mathbf{x}} \quad \text{to} \\ \textbf{Minimize} & \Pi_f = \Pi\left(\bigcup_{m=1}^{n_m} \left[A_{peak}^{(m)}(\mathbf{x},\mathbf{p}) \geq A_{lim}^{(m)} \right] \right) \\ \textbf{such that} & \mathbf{x}_l \leq \bar{\mathbf{x}} \leq \mathbf{x}_u \\ \textbf{and} & g_i(\bar{\mathbf{x}},\bar{\mathbf{p}}) \leq 0, \quad i = 1,\ldots,r \end{array} \tag{3}$$

where Π stands for possibility. To compare designs obtained by each approach fairly, the same resources are used in both formulations. For this reason, the constraints in both formulations are identical.

3.2. Problem Description

In Section 2, we mentioned that the damping of the second mode of the truss is much larger than that of modes 1 and 3, probably because of coupling with the dynamics of the support plates and wall. These dynamics are not included in our model. Hence, we will ignore mode 2 and only add damping to modes 1 and 3. For simplicity, only one tuned damper will be used for each mode, adding a total of two dampers to the truss.

3.2.1. Uncertainties

We neglected uncertainties in the truss elements because we found experimentally that these uncertainties are negligible compared to uncertainties in the properties of the dampers. The probability distributions of the properties of the dampers were estimated from measurements.

3.2.2. Design Requirements

The locations and directions of the excitation and the response measurements used in this study are shown in Figure 3. The locations of the response measurements ($u^{(1)}$ and $u^{(3)}$) were chosen at the nodes and directions that correspond to the largest acceleration amplitudes in the first and third modes, respectively, in the undamped truss.

The accelerance transfer functions $H^{(1)}$ and $H^{(3)}$ corresponding to the two measurement points are defined as:

$$H^{(1)}(i\omega) = \frac{\ddot{u}^{(1)}(i\omega)}{f(i\omega)} \tag{4}$$

$$H^{(3)}(i\omega) = \frac{\ddot{u}^{(3)}(i\omega)}{f(i\omega)} \tag{5}$$

where $u^{(1)}$ and $u^{(3)}$ are displacements, f is the excitation force, ω is the frequency, $i = \sqrt{-1}$, and $(\cdot)$ denotes differentiation with respect to time. The upper limits on $H^{(1)}$ and $H^{(3)}$ in two frequency windows, covering the two modes of interest (mode 1 and mode 3) are $H^{(1)}_{lim}$ and $H^{(3)}_{lim}$, respectively. This is illustrated in Figure 10. The limits $H^{(1)}_{lim}$ and $H^{(3)}_{lim}$ are selected in a way that the difference in probabilities of failure between probabilistic and fuzzy set approaches is large. This is explained in detail in Subsection 3.5.

3.2.3. Design Scenario

We consider the following scenario: The truss has been designed with one damper for each mode (modes 1 and 3 only) to limit dynamic response accelerations. The locations of the dampers have been determined experimentally to be the most effective for each mode of the undamped truss. Dampers whose properties are random, are used to control vibration. Tests have revealed a significant mistuning of the dampers that will result in poor overall performance of the damped structure. We assume that the dampers cannot be modified to improve their tuning. However, *tuning masses* (described in Paragraph 2.1.3) can be easily added to the nodes of the truss to modify its natural frequencies, thereby improving tuning. The tuning masses have a fixed mass (16.6 g), and, to limit the added weight, a maximum of 10 masses can be used. For both probabilistic and fuzzy set based design, the problem consists of optimally redesigning the system by adding tuning masses to maximize performance.

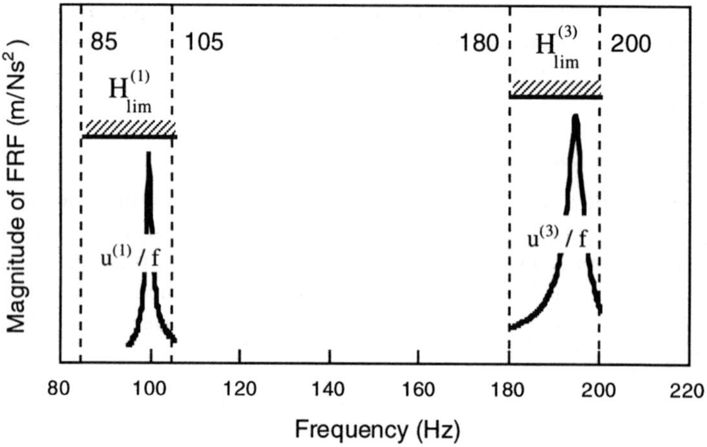

Figure 10: Frequency windows and amplitude limits.

3.2.4. Formulation of the Probabilistic and Fuzzy Set Based Optimizations

As explained in Paragraph 3.2.2, the design requirements consist of upper limits on the peak acceleration of modes 1 and 3. The probabilistic and fuzzy set optimizations use the same 10 design variables: the locations of a maximum of 10 tuning masses on the structure. To reduce the CPU time needed for the optimizations, we restricted the candidate locations for the masses to nodes 7 to 12 (Figure 7). Note that these are the most effective locations for the masses for altering the natural frequencies of modes 1 and 3.

The general probabilistic formulation of Eq. (2) can then be applied to the particular problem as follows:

Find	Tuning Masses Locations to
Minimize	$P_f = P(H^{(1)}_{peak} \geq H^{(1)}_{lim} \text{ OR } H^{(3)}_{peak} \geq H^{(3)}_{lim})$
such that	number of tuning masses ≤ 10 (6)

The corresponding fuzzy set formulation minimizes the *failure possibility* that the acceleration will exceed any of the two acceleration limits $H_{lim}^{(1)}$ and $H_{lim}^{(3)}$:

$$\begin{array}{ll} \textbf{Find} & \text{Tuning Masses Locations to} \\ \textbf{Minimize} & \Pi_f = \Pi(H_{peak}^{(1)} \geq H_{lim}^{(1)} \ \text{OR} \ H_{peak}^{(3)} \geq H_{lim}^{(3)}) \\ \textbf{such that} & \text{number of tuning masses} \leq 10 \end{array} \tag{7}$$

According to the definition of the possibility of a union of events (Zadeh [25]):

$$\Pi(H_{peak}^{(1)} \geq H_{lim}^{(1)} \text{OR } H_{peak}^{(3)} \geq H_{lim}^{(3)}) = \max[\Pi(H_{peak}^{(1)} \geq H_{lim}^{(1)}), \Pi(H_{peak}^{(3)} \geq H_{lim}^{(3)})] \tag{8}$$

Note that alternate definitions of the possibility of the union of events have been proposed. Some include parameters that are calibrated for the particular problem considered. Definition (8) was selected for this study because is a simple, standard definition [25, 65, 66].

3.2.5 Optimization Using a Genetic Algorithm

The locations of the masses are restricted to the nodes of the truss so that all 10 design variables are discrete. For this reason, we use a genetic algorithm for the optimizations. The same algorithm is used for the probabilistic and fuzzy set cases; only the objective functions differ.

The genetic algorithm uses the three classical genetic operators (selection, crossover, and mutation) and an elitist strategy, where the best individual of a population is always cloned into the next generation. The selection uses the ranking technique, where the probability of selecting an individual is proportional to one plus the population size minus its rank in the population. The genetic optimization algorithm details can be found in Ponslet [60].

3.3 Probabilistic Analysis

For the probabilistic optimization, we need to repeatedly evaluate the system probability of failure, that is, the probability that the peak acceleration at given points on the structure exceeds a prescribed limit. The vibration amplitude (or acceleration) of a structure near resonance is a highly non-linear function of the structural parameters. For that reason, a mean value based, first order covariance propagation method cannot be used [58]. In addition, because of the relationship between the parameters of a tuned damper and the magnitude of the peak response of a truss equipped with that tuned damper, failure can occur at both tails of the damper parameter distribution. This means that we have multiple "most probable failure points" (MPFP). This is illustrated in Figure 11, which shows a hypothetical distribution of the natural frequency of a tuned damper. The other curve in the figure shows the peak acceleration of a structural mode as a function of the damper natural frequency. The total failure probability is the sum of the probability of failure of the two

MPFP. However, a second moment method may grossly underestimate the probability of failure, because it finds only one MPFP. Moreover, there is no guarantee that a second moment method will find the MPFP that has the largest probability of failure. Therefore, we used Monte Carlo simulation to evaluate the probabilities of failure.

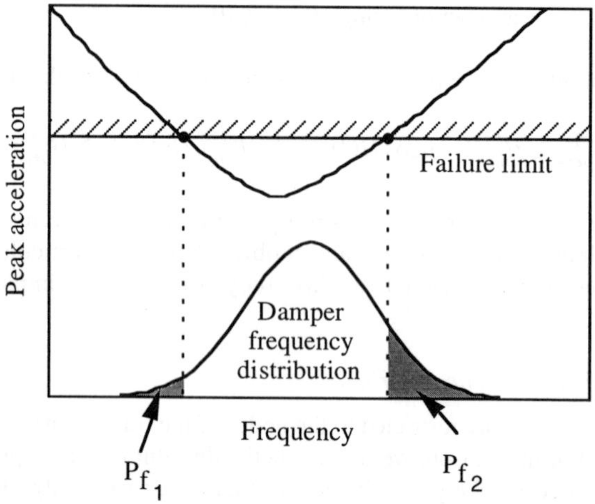

Figure 11 Existence of multiple failure regions for a truss equipped with tuned dampers with random parameters. Note: $P_f = P_{f1} + P_{f2}$.

In general, Monte Carlo simulation is computationally expensive and practical only when the cost for each analysis is small. To reduce computational cost for each analysis, we used an approximate solution technique for predicting the response (see Paragraph 3.6.2). For more complicated problems, other methods such as *integrated analysis and design* [67] and methods using *response surface polynomials* to estimate the response of the structure for many values of the random and design variables [68] can be used.

To determine the required sample size, N, for the Monte Carlo simulations, we used the formula:

$$N = \frac{1-P}{P\,COV_P^2} \tag{9}$$

where P is the anticipated probability of failure and COV_P the desired coefficient of variation (defined as the standard deviation divided by the mean) of the probability of failure. In this study, we work with probabilities of failure of the order of 0.1 and we accept a coefficient of variation of 0.1. Substituting these values in the above formula we find that the minimum value of N must be 900. We chose a sample size of 1000. The standard deviation in the evaluated probabilities of failure associated with that sample size is then [23]:

$$\sigma_P = \sqrt{\frac{P(1-P)}{1000}} \tag{10}$$

P is the estimated failure probability, which is equal to the relative frequency of failure obtained from Monte-Carlo simulation.

Because the total mass of the dampers has very little scatter, it was assumed deterministic in the simulations. The 3 remaining parameters -- natural frequency (f_n), tip mass (m) and loss factor (η) -- of each damper were found to be approximately normally distributed random variables (see Subsection 4.1). Their mean values, standard deviations and correlation coefficients were estimated from tests.

3.4. Fuzzy Set Based Analysis

The objective of the fuzzy set based analysis is to evaluate the system possibility of failure, using the same uncertain parameters as for the probabilistic analysis (natural frequency, the tip mass and loss factor for each type of dampers). Each uncertain parameter is represented as a fuzzy variable. Determining the possibility functions of the fuzzy variables is critical. As mentioned earlier this is equivalent to determining membership functions. A review of methods for practical estimation of membership functions can be found in Dubois and Prade [29]. However, there is no general method for selecting the appropriate type of membership function for given information. Nevertheless, in many practical applications, [28, 40, 69] there is a strong trend to use triangular membership functions to represent fuzzy sets. This study used triangular membership functions because they involve few parameters and are easy to manipulate.

Triangular membreship functions have a linear transition from a zero level to a maximum value of membership. In this study, we needed to determine how to use the available statistical data to generate the membership functions. The apex of the triangle representing a membership function corresponds to the statistical mean of the uncertain parameter. The points where the membership function becomes zero correspond to $\pm 3\sigma$ (σ is the standard deviation of the uncertain variable). This is illustrated in Figure 12. This membership function is consistent with the following scenario. We ask an expert who has manufactured and measured a large number of tuned dampers to give three values for, say, the mass; the most common one and upper and lower limits. These values define a range which contains all possible values of the mass. Because the uncertain parameters in this study are approximately normal (see Subsection 4.1), it is reasonable to assume that the expert would give a range that is centered about the mean and extends plus and minus a number of standard deviations from the mean. This number can be from two to four. For a normally distributed variable, 99.74% of its values are contained in the interval between m-3σ and m+3σ, where m is the mean and σ the standard deviation. Note that in this study we will be dealing with probabilities of failure in the range of 20-30%. These levels of failure probability are not altered by extremely rare events. However, in a study involving small probabilities of failure, rare events are important, and wider interval for the membership function might be needed.

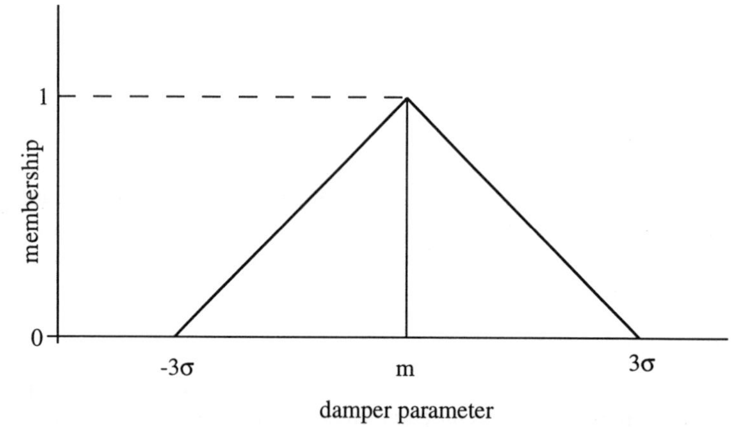

Figure 12 Fuzzy set representation of an uncertain damper parameter (m, σ are the statistical mean and standard deviation respectively).

When the same variable is modeled using both probability and possibility distributions, the probability - possibility *consistency principle* must be satisfied [25, 29]. This principle states that the possibility of an event is always greater than or equal to its probability. Figure 13 shows the probability density function of a standard normal variable (*i.e.*, with zero mean and unit standard deviation) and the membership function created the way we described in the last paragraph.

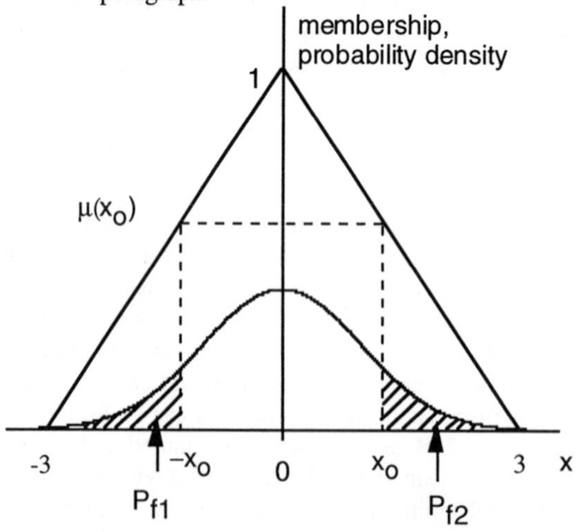

Figure 13 Validation of consistency principle:

$$P(x \le x_0 \cup x \ge x_0) \le \Pi(x \le x_0 \cup x \ge x_0) = \mu(x_0) \Leftrightarrow P_{f_1} + P_{f_2} \le \mu(x_0)$$

For any $x_o \geq 0$, any combination of events of the form $\{x \geq x_o$ or $x \leq x_1\}$, where x_1 is less or equal to $-x_o$, has by definition [25] a possibility equal to the maximum value of the membership function for all $\{x \geq x_o\}$ and $\{x \leq x_1\}$. This value is equal to $\mu(x_0)$. From all the above combinations of events, the one with the highest probability is $\{x \geq x_o$ or $x \leq -x_o\}$. Therefore, for the consistency principle to be satisfied,

$$\mu(x_0) = \mu(-x_0) \geq P(x \leq -x_0) + P(x \geq x_0) = 2P(x \geq x_0), \quad \text{for every } x_0 \geq 0 \tag{11}$$

where μ stands for value of the membership function and P stands for probability. Eq. (11) is a sufficient condition because, if Eq. (11) is satisfied for every x_o, then the possibility of x falling outside an interval $[x_1, x_o]$, which contains the most possible value of x, is always larger than the probability of the same event. Of course the possibility of x being inside the interval $[x_1, x_o]$ is always equal or higher than the probability of the same event, because this possibility is one. Inequality mentioned in Eq. (11) holds for:

$$-2.994 \leq x_0 \leq 2.994 \tag{12}$$

This range covers 99.7% of the values of a normally distributed variable. Note again, that in this study we are not interested in the tails of the distributions, because we are trying to measure probabilities on the order of ten percent.

To obtain the possibility distribution function of the response from those of the input parameters we use the *vertex method* [28] with 5 α-cuts at 0, 0.25, 0.50, 0.75 and 1. An α-cut is the range of values of a fuzzy variable whose membership function is equal or bigger than α. We use linear interpolation to obtain membership values other than those corresponding to the above α-cuts. A typical membership function of the response is presented in Figure 14.

In the fuzzy set optimization, we want to minimize the possibility that the acceleration will exceed the peak acceleration limit. If we consider one failure mode, this possibility is by definition [25],

$$\Pi(H_{peak} \geq H_{lim}) = \max\left[\mu_R(H_{peak})\right] \tag{13}$$

where μ_R stands for the membership function of the response and the maximum is for all values of H_{peak} that exceed H_{lim}. Furthermore, it makes sense to expect the acceleration limit to be to the right of the apex of the membership function of the response. If the opposite were true, it would mean that the structure would fail for the nominal values of the parameters. If the acceleration limit is to the right of the apex, then the possibility of failure is equal simply to the value of the membership function of the response at the failure limit. With this in mind, Eq. (8) can be rewritten as follows:

$$\Pi(H_{peak}^{(1)} \geq H_{lim}^{(1)} \text{ OR } H_{peak}^{(3)} \geq H_{lim}^{(3)}) = \max\left[\mu_R(H_{lim}^{(1)}), \mu_R(H_{lim}^{(3)})\right] \tag{14}$$

However, two peak acceleration values correspond to each membership value between 0 and 1. One lies to the left and one to the right of the apex. To guarantee that the optimizer will move in the correct direction, we modify the membership function in Figure 14 as shown in Figure 15, where we replace the part of the membership function to the left of the apex by linearly extrapolating, using points A and B. The new curve is not a valid membership function because its highest value exceeds 1, but if we use it, we will still obtain the correct optimum and we will eliminate the possibility of getting a false optimum design. We refer to the modified curve as a *pseudo-membership function.* Using this modification we do not need to calculate the possibility function of the response that lies to the left of the apex.

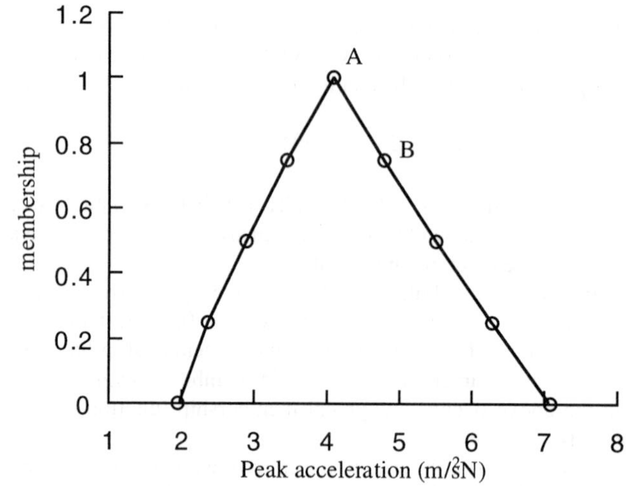

Figure 14 Typical membership function of the response for the laboratory truss.

In general, the vertex method requires the minimum and maximum values of the peak acceleration for each α-cut. This means that it requires solution of 2 optimization problems for each α-cut. Fortunately, in this study, the computational cost can be greatly reduced, because the maximum peak acceleration corresponds to one of the vertices of each α-cut. Specifically, parametric studies showed that the peak acceleration becomes maximum when the tip mass, tuning ratio (ratio of a damper natural frequency and the truss natural frequency) and loss factor take their extreme values (Figures 16 to 18). In addition, no minimization is needed because we do not need the part of the response to the left of the apex. Because there are 3 uncertain parameters, there are $2^3 = 8$ vertices in each α-cut. So, we need 4x8 = 32 function evaluations for the four α-cuts plus one function evaluation for the α=1 cut (apex). That makes a total of 33 function evaluations for each mode. This number is much smaller than the number of function evaluations required if optimizations were needed for each α-cut.

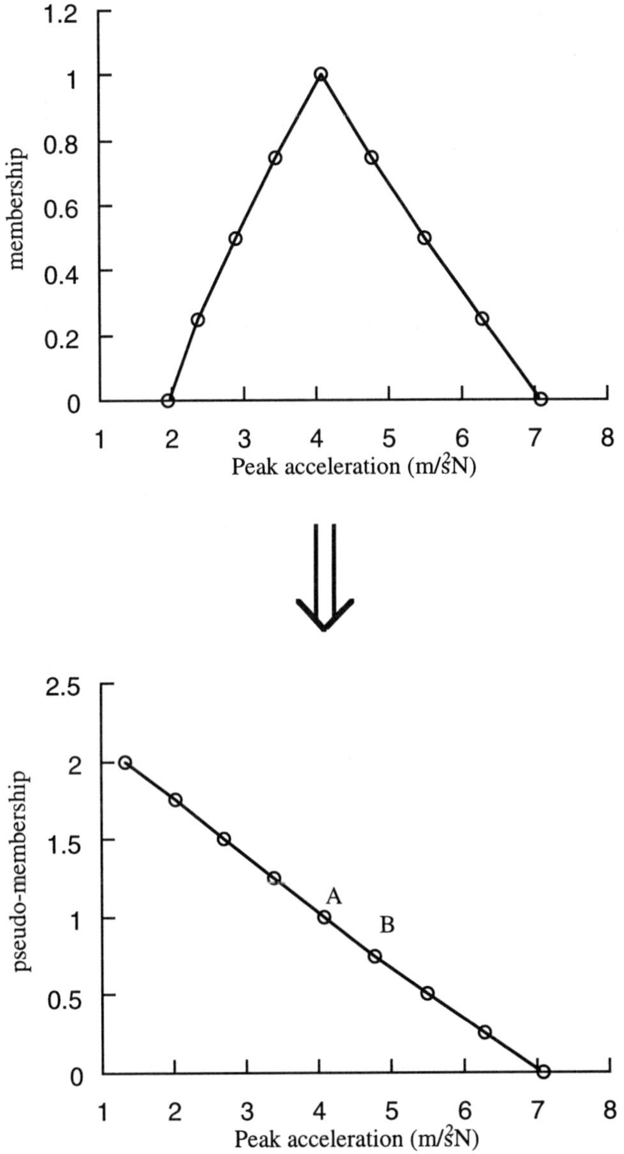

Figure 15 Modification of the membership function of the response.

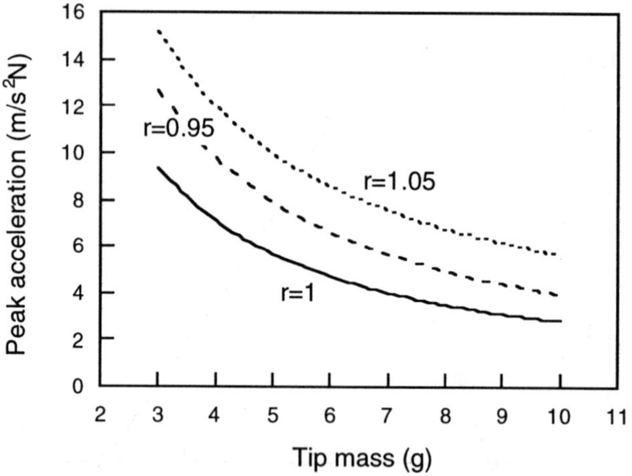

Figure 16 Effect of a type-3 damper tip mass on the peak acceleration of the third mode of the laboratory truss. The three curves correspond to different values of the tuning ratio (r). (From Ponslet [60]).

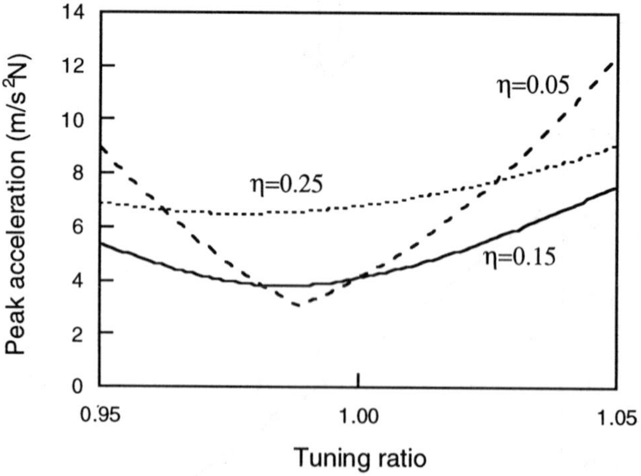

Figure 17 Effect of a type-3 damper tuning ratio on the peak acceleration of the third mode of the laboratory truss. The three curves correspond to different values of the loss factor (η). (From Ponslet [60]).

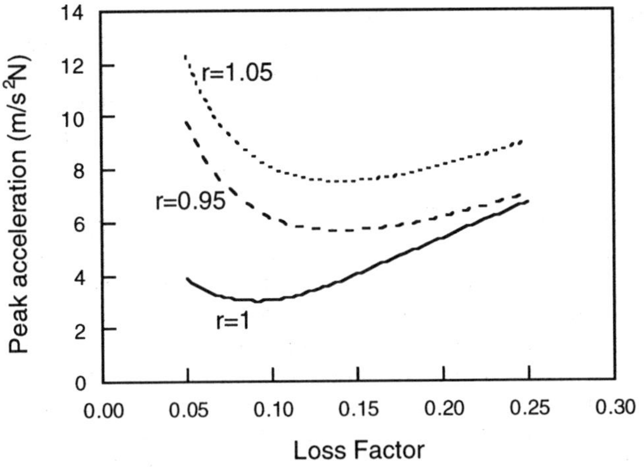

Figure 18 Effect of a type-3 damper loss factor on the peak acceleration of the third mode of the laboratory truss. The three curves correspond to different values of the tuning ratio (r). (From Ponslet [60]).

3.5. Selecting a Problem for Experimental Validation

Measuring probabilities of failure in the laboratory is time consuming because it requires measuring many realizations of a design. Moreover, measuring very small probabilities or very small differences in probabilities of failure requires a prohibitively large number of experiments. For this reason, we need a design problem that produces alternative designs whose probabilities of failure are both large but differ significantly.

The strategy for finding a problem that maximizes the contrast between fuzzy and probabilistic designs was to start with a design that, intuitively, is the fuzzy optimum and tune some problem parameters in a way that the probabilistic optimum design is considerably different than the fuzzy optimum. As it will become obvious from the following, the simplest way is to find a problem for which the fuzzy optimum has no added masses attached to the truss.

The following problem parameters were changed in order to find a problem that meets our criteria:

-- the mean values of the natural frequencies of the two types of dampers,

-- the scatter in the natural frequencies of each type of dampers,

-- the failure limits (the failure limits can be different for each mode).

We did not consider the types of probability distributions of the damper parameters in trying to find a problem that yields a considerable contrast between probabilistic and fuzzy designs.

In a problem with two (or more) failure modes, the fuzzy set approach tries to minimize the maximum of the individual failure possibilities (see Eq. (8)). This means that, unless a constraint prevents it, at the optimum, the possibilities of failure of modes 1 and 3 must be equal.

In this study, the only means of controlling the vibration is by adding tuning masses. Adding masses to a truss structure reduces its natural frequencies. Therefore, it improves the response of a structure equipped with tuned dampers only for those modes for which the corresponding tuned dampers are *undertuned* -- *i.e.*, their natural frequency is smaller than that of the corresponding structural mode. On the other hand, when the dampers are *overtuned* (*i.e.*, their natural frequency is larger than that of the corresponding structural mode), adding masses can only worsen the structural response of the respective mode.

On the basis of the above observations, we make one type of dampers undertuned and the other type overtuned. The failure limits for each mode are selected in a way that without any tuning masses on the structure, the possibilities of failure of modes 1 and 3 are equal. This is the fuzzy set optimum because any added mass will increase the possibility of one mode thus increasing the system failure possibility.

The probabilistic approach tries to minimize the system failure probability (Eq. (6)), which is a combination of both failure modes. Because of this, at the probabilistic optimum, the failure probabilities of the two individual failure modes are not necessarily equal. Instead, they depend on the relative cost of controlling each mode or the relative uncertainty in the amplitudes of the two modes. If the costs of controlling the two failure modes are the same but the uncertainty is different, the probabilistic optimization will provide a larger safety margin to the failure mode that has a larger uncertainty. If the uncertainty is similar but the costs of controlling the different failure modes are different, then the probabilistic optimization will provide a larger safety margin to the lower cost mode.

Added masses generally affect high frequency modes more than low frequency modes. Consequently, it is easier to control the natural frequency of the third mode of the truss with tuning masses than the first mode. It is also desirable to have less scatter in the natural frequency of the undertuned tuned absorbers so that adding tuning masses will improve this mode more than it will worsen the other. Taking into account all these facts, we select a design problem in which type-1 dampers are overtuned with a relatively large scatter and type-3 dampers are undertuned with a relatively small scatter.

In the selected problem, fuzzy optimization will not add any tuning masses. On the other hand, the probabilistic optimization will add tuning masses to reduce the failure probability of mode 3. Of course, at the same time the failure probability of mode 1 will increase, but at a slower rate than the rate of decrease of mode 3 because mode 1 is affected less by added masses and it has greater uncertainty.

Using the above procedure we identified the following design problem that gives large difference in the probabilities of failure of the probabilistic and fuzzy set optima. Type-1 dampers should have an average natural frequency of 110 Hz (overtuned) and a coefficient of variation of about 1.6%. Type-3 should have a mean natural frequency of about 182 Hz (undertuned) and a coefficient of variation of about 0.8%. The failure limits should be 5.00 m/(s^2N) for mode 1 and 6.75 m/(s^2N) for mode 3.

Table 1: Damper properties.

	Truss natural frequency (Hz)	Average damper natural frequency (Hz)	COV of damper natural frequency
Mode 1	100	110	1.6%
Mode 3	193	182	0.8%

3.6. *Approximate Solution Technique*

3.6.1. *Evaluation of Peak Acceleration*

As mentioned earlier (Paragraph 3.2.2), the design requirements limit the largest acceleration at specific nodes for a given excitation over prescribed frequency windows. These peak accelerations are evaluated in the frequency domain using the following numerical procedure: the frequency response function (FRF) from excitation force to response acceleration is computed using the mode superposition method, which is described below. The magnitude of the FRF is evaluated numerically with a relatively coarse step size within the frequency window of interest. The peak is first located approximately using a simple slope-reversal search (the slope is approximated using forward differences). The location of the peak and the corresponding value of the FRF are then refined by second order interpolation between the approximate peak and the 2 neighboring points in the FRF.

3.6.2. *2-mode Approximation*

The mode superposition method mentioned above requires the first few eigenvalues and eigenvectors of the truss with tuning masses and dampers. Obtaining exact eigenvectors and eigenvalues for a truss with two dampers requires solving a generalized complex eigenproblem with 38 degrees of freedom.

To reduce computational expense, we use an approximate 2-mode model that is valid when one damper is used for each structural mode. It uses a reduced basis made of one mode (either first or third, depending on the frequency of the peak) of the truss without damper plus one Ritz vector. The Ritz vector contains zeros everywhere except at the D.O.F. Associated with the damper. An analytical expression for the frequency response function from excitation force to response acceleration is easily derived for the resulting 2 D.O.F. model and is used instead of the numerical eigensolution and mode superposition method. This analysis has to be repeated with a different basis for each mode of interest. It is about 50 times faster than the full analysis. The associated error is about 10%. Detailed description of the full analysis and the 2-mode approximation can be found in [60].

All analyses in the probabilistic and fuzzy set based optimizations are performed using this 2-mode approximation and approximate mode shapes. The approximate mode shapes are obtained from the mode shapes of the original truss (without dampers or tuning masses), using a first order correction for the effect of the tuning masses [60].

A complete probabilistic analysis (1000 point Monte Carlo simulation) uses about 12 seconds of CPU time with the 2-mode approximation and about 500 seconds with the full analysis (using an IBM 3090 in vectorized mode). A complete fuzzy set analysis uses about 1 second of CPU time with the 2-mode approximation and about 45 seconds with the full analysis.

4. Analytical Comparison of Probabilistic and Fuzzy Set Based Optimization

In the previous section, we presented a probabilistic and a fuzzy set based approach for designing a damped truss. We also formulated the corresponding optimization problems. Finally, we identified a design problem for which the two alternate optimum designs have significantly different failure probabilities.

In this section, we first describe how we manufactured the sample of dampers used to compare the probabilistic and the fuzzy set based designs. Then, we explain how we calibrated the analytical models used to predict the response of the truss. Finally, we present the results of the probabilistic and fuzzy set optimizations, and compare their resulting probabilities and possibilities of failure. In the next section, we will compare the alternate designs experimentally based on the number of failures out of a number of realizations of each design.

The flow-chart of Figure 19 summarizes the approach for comparing the two methods both analytically and experimentally. Note that some of the results presented in this section are experimental. These results include the statistics of the dampers, presented in Subsection 4.1, and the calibration of analytical models, presented in Subsection 4.2.

4.1. Creating and Characterizing Dampers

We manufactured 29 dampers of each type and measured their properties. Because the dampers use a viscoelastic foam, their characteristics depend strongly on temperature. The effect of temperature changes was measured to be about 9% per ^{o}C on the loss factor, -0.9% per ^{o}C on the natural frequency, and 0.7% per ^{o}C on the identified tip masses. Note that, as mentioned in Paragraph 2.2.2, these three parameters are not directly measured but are estimated using regression. Therefore, parameters such as the damper tip mass can also be affected by temperature. To reduce this temperature effect, we use a temperature stabilization system that maintains an average temperature of 24.4^{o}C with a rapid oscillation of 0.8^{o}C. The period of that oscillation is about 15 minutes. All measurements are repeated 3 times and averaged to reduce the effect of the small temperature oscillation and other measurement errors. The average of these measurements is used to estimate the mean values, standard deviations and correlation coefficients of the parameters. These statistics are listed in Tables 2 and 3.

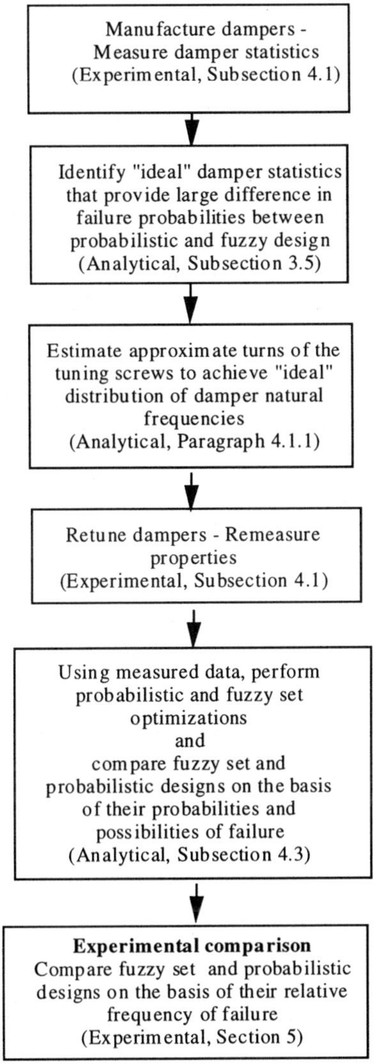

Figure 19 Flow-chart of approach for comparing probabilistic and fuzzy set based approaches. (Note: Each task is labeled as analytical or experimental. The subsection that describes each task is also specified).

Table 2: Type-1 dampers, statistics of parameters (sample of 29).

Parameter	m (g)	fn (Hz)	η	m_T (g)
Mean	6.942	105.1	0.1193	10.81
Standard dev.	0.108	3.195	0.00681	0
COV (%)	1.55	3.00	5.71	0
Correl. Coeff.				
m	1.000	0.804	0.729	
fn	0.804	1.000	0.491	
η	0.729	0.491	1.000	

Table 3: Type-3 dampers, statistics of parameters (sample of 29).

Parameter	m (g)	fn (Hz)	η	m_T (g)
Mean	7.613	198.1	0.1472	11.53
Standard dev.	0.070	2.175	0.00812	0
COV (%)	0.91	1.10	5.52	0
Correl. Coeff.				
m	1.000	0.218	0.692	
fn	0.218	1.000	0.055	
η	0.692	0.055	1.000	

The mean values and standard deviations of the dampers were different than those, determined in Subsection 3.5, that maximized the contrast between probabilistic and fuzzy set based designs. To reduce the uncertainty in the estimated failure probability from the tests, this study uses some ideas from *stratified sampling* [70] to create realizations of the dampers as explained in Paragraph 4.1.1. For the natural frequency, which most affects the performance of the dampers, we created a sample corresponding to the desired distribution determined in Subsection 3.5 in the following way. We sampled the probability distributions of the dampers' natural frequencies with a uniform step size in probability. Then, we estimated the number of turns of the screws needed to realize these desired distributions. The dampers were then adjusted and their properties were measured again.

After adjusting the tip screws, the dampers were measured again three times and their properties were averaged. The statistics of the dampers are listed in Tables 4 and 5 for damper types 1 and 3, respectively. The mean value and coefficient of variation of the natural frequencies of these dampers are close to the corresponding values found in Subsection 3.5 (Table 3.1). These statistics should lead to large differences in the failure probability of the fuzzy set based and probabilistic designs.

Table 4: Type-1 dampers after adjusting the tuning screws: statistics of parameters (sample of 29).

Parameter	*m (g)*	*fn (Hz)*	η	m_T *(g)*
Mean	7.033	110.4	0.1190	10.81
Standard dev.	0.131	1.752	0.00715	0
COV (%)	1.86	1.59	6.01	0
Correl. Coeff.				
m	1.000	0.682	0.648	
fn	0.682	1.000	0.247	
η	0.648	0.247	1.000	

Table 5: Type-3 dampers after adjusting tuning screws: statistics of parameters (sample of 29).

Parameter	*m (g)*	*fn (Hz)*	η	m_T *(g)*
Mean	7.366	181.7	0.1371	11.53
Standard dev.	0.093	1.231	0.00628	0
COV (%)	1.26	0.68	4.58	0
Correl. Coeff.				
m	1.000	0.333	0.760	
fn	0.333	1.000	0.079	
η	0.760	0.079	1.000	

Figure 20 shows normal probability plots of the measured distributions of tip masses, natural frequencies and loss factors of the 29 dampers of type-1, after adjusting the tuning screws. The same plots for type-3 dampers are given in Figure 21. The scale used on the probability axes of these plots is such that a perfectly normally distributed sample would lie on a straight line.

To quantify the consistency of each damper parameter distribution with the normal distribution, we performed the *chi-square test* [71] for all the damper parameters. This test determines if, based on the sample of a random variable, we should suspect the hypothesis that the random variable follows a particular probability distribution. This test requires that the data be classified in *k* mutually exclusive groups, where the observed frequency of occurrence for the ith group is f_i^o. Based on a theoretical distribution, the expected frequency of occurrence for the ith group, f_i^e, is recommended to be at least 5. Then we evaluate the quantity:

$$\chi^2 = \sum_{i=1}^{k} \frac{(f_i^o - f_i^e)^2}{f_i^e} \tag{15}$$

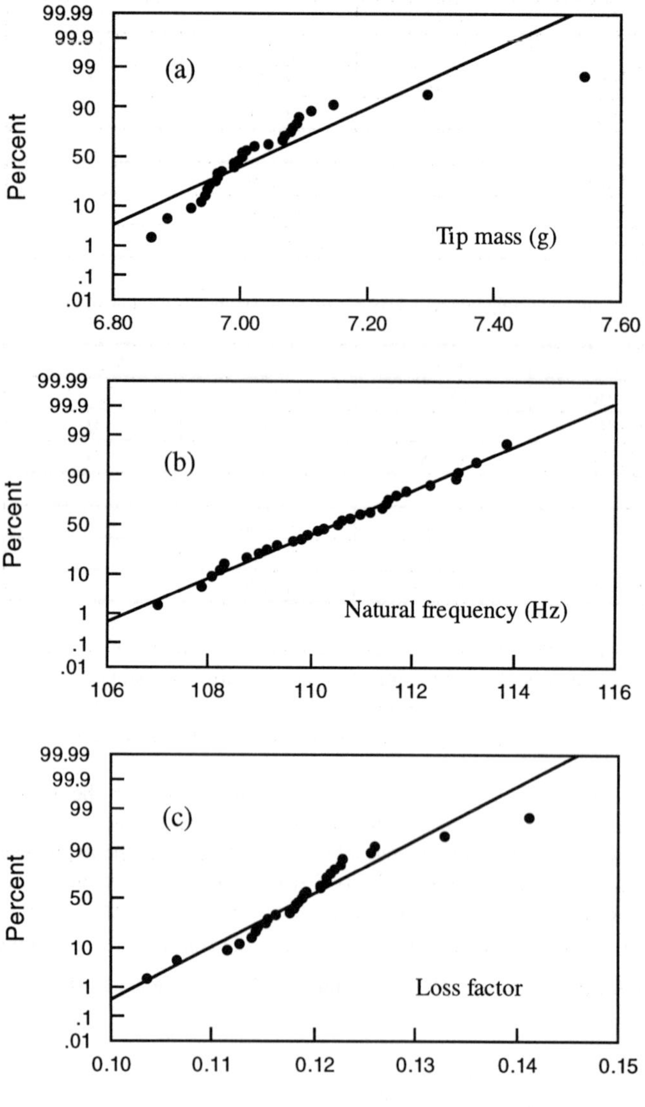

Figure 20 Type-1 dampers, distribution of parameters: (a) tip mass, (b) natural frequency, (c) loss factor.

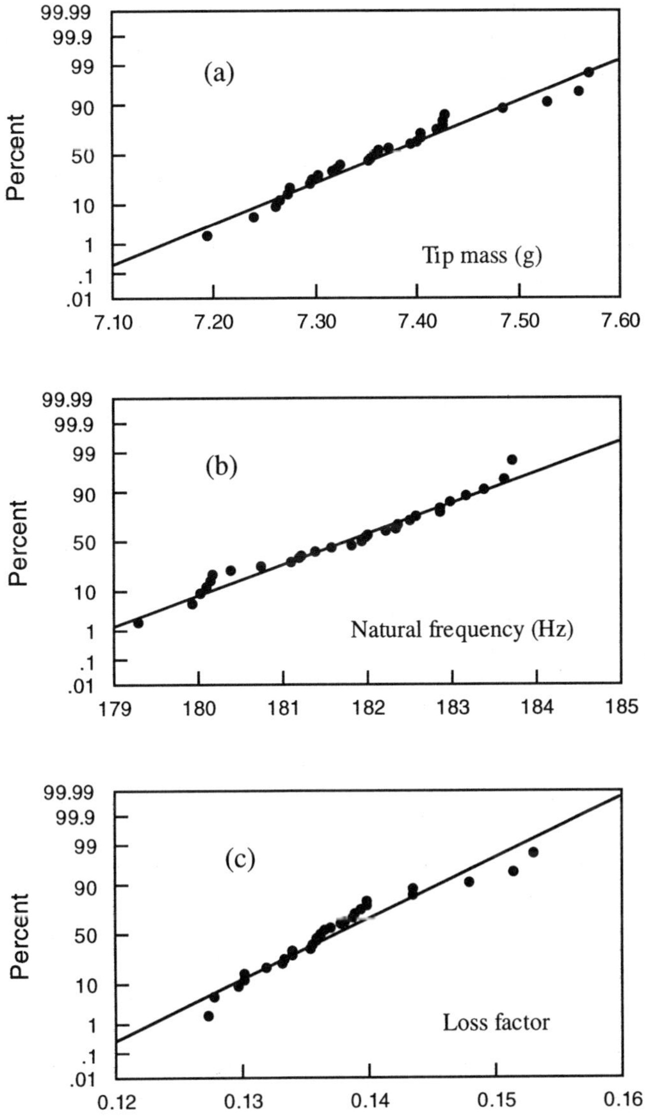

Figure 21 Type-3 dampers, distribution of parameters: (a) tip mass, (b) natural frequency, (c) loss factor.

This quantity follows approximately the chi-square distribution with *k-n-1* D.O.F. where n is the number of quantities obtained from the observed data. In this case, the mean and the standard deviation of the observed data were required to find the expected frequencies. Consequently, n=2. Then, we select a probability level α ($0<\alpha<1$) and, using tables of the chi- square distribution, compute the value χ^2_α, which will not be exceeded by χ^2 with a probability α:

$$\text{Prob}[\chi^2 < \chi^2_\alpha / \chi^2 \text{ is chi} - \text{square}] = \alpha, \text{ with } k - n - 1 \text{ d.o.f.} \tag{16}$$

where $\text{Prob}[\chi^2 < \chi^2_\alpha / \chi^2 \text{ is chi} - \text{square}]$ is the conditional probability of χ^2 being less than χ^2_α, given that χ^2 follows the chi-square distribution.

If even for a large value of α, say 95%, $\chi^2 > \chi^2_\alpha$ we should suspect the hypothesis that χ^2 follows the specified distribution, because the event $\chi^2 > \chi^2_\alpha$ is unlikely to occur. In this case the hypothesis that a sample is statistically consistent with a theoretical distribution is rejected. Otherwise, there is no reason to suspect that the hypothesis is wrong on the basis of the test results. The outcome of the test depends on the value of α, the higher the value of α the easier to pass the test. Typical values of α range between 0.9 and 0.99. In this study, we used a 95% probability level.

The results of the chi-square test for each damper parameter are listed in Tables 6 and 7 for damper types 1 and 3 respectively. We observe that in all cases $\chi^2 < \chi^2_\alpha$, with a good margin, except for the tip mass of type-1 dampers. Consequently, there is no reason to suspect the normalcy assumption for the remaining type-1 and type-3 damper parameters. Because the tip mass does not affect the response significantly, for simplicity, we assumed that its marginal probability distribution is normal as well.

Table 6: Type-1 dampers. Chi-square test results (α=0.95).

Parameter	Tip mass	Natural frequency	Loss factor
Number of groups (k)	5	5	5
Degrees of freedom (k-3)	2	2	2
χ^2	8.76	0.38	2.30
χ^2_α	5.99	5.99	5.99

Table 7: Type-3 dampers. Chi-square test results (α=0.95).

Parameter	Tip mass	Natural frequency	Loss factor
Number of groups (k)	5	5	5
Degrees of freedom (k-3)	2	2	2
χ^2	2.42	1.13	2.32
χ^2_α	5.99	5.99	5.99

4.1.1. *Creating a Sample of Dampers to Reduce Uncertainty in Experimentally Measured Failure Probability*

The classical technique to experimentally evaluate probabilities of failure is to pick a random sample of n realizations, test them, and count the number of failures. There are two sources of error associated with this technique. The first is a resolution error due to the limited size of the sample – we cannot distinguish between probabilities of failure that differ less than a minimum limit. The other type of error also depends on the sample size and is due to the randomness of the sample.

If a probability of failure P is estimated from a random sample of size n, the standard deviation of P is given by Eq. (10). There is very little interaction between the two modes of the structure - a type-1 damper has no measurable effect on the response of mode 3 and vice versa. As a result, we cannot create more than 29 independent combinations of type-1 and type-3 dampers, using the 29 type-1 and type-3 dampers available. If we used 29 random realizations of dampers, the standard deviation of P due to the randomness in the sample would obscure the difference in probabilities of failure. For example, if the failure probability of a design were P=30%, its standard deviation would be 8.5%. We need prohibitively large samples to reduce the uncertainty in P enough to allow meaningful measurements of the probability differences between the two approaches. For example, let us calculate the required number of samples we would need to compare the optimum fuzzy and probabilistic designs whose probabilities of failure were 38.2 and 25.9, respectively. Let us assume that the difference in probabilities of failure, $\Delta P = P_f - P_p$, is normally distributed. P_f is the probability of failure of the fuzzy set based design and P_p is the probability of failure of the probabilistic design. The mean and standard deviation of ΔP are:

$$\text{Mean: } \overline{\Delta P} = \overline{P_f} - \overline{P_p} \tag{17}$$

$$\text{Standard deviation: } \sigma_{\Delta P} = \sqrt{\sigma_{P_f}^2 + \sigma_{P_p}^2} \tag{18}$$

$\overline{P_f}$ and $\overline{P_p}$ are the mean values and σ_{P_f} and σ_{P_p} are the standard deviations of P_f and P_p, respectively. By substituting numerical values in Eq. (10), (17) and (18) we get $\overline{\Delta P} = 0.123$ and $\sigma_{\Delta P} = \dfrac{0.6542}{\sqrt{n}}$. To determine whether the probabilistic design is safer than the fuzzy set based design with enough confidence, it is reasonable to require that the mean of ΔP be equal to two standard deviations of ΔP (that corresponds to a probability of 97.72% confidence level that the probabilistic design is safer than the fuzzy set based design). Using Eq. (10), we find that n should be 113. It is too expensive to construct and test so many dampers.

Sampling uncertainty can be reduced if we use a non-random sample. Because we need to adjust the damper tuning screws to obtain the desired distributions of the natural frequencies of the dampers anyway, we can do it in such a way that the resulting uncertainty in the measured probabilities is reduced. This can be achieved by distributing the natural frequencies of the dampers uniformly along the probability axis of the probability distribution function of a normal sample. The process is illustrated in Figure 26. The horizontal axis represents the natural frequency of a damper and the vertical axis represents the cumulative probability distribution of the natural frequency. The solid line represents a normal distribution of the natural frequencies with the desired mean and standard deviation. We first distributed the natural frequencies of the 29 dampers of the sample uniformly along the probability axis. The first damper natural frequency has a probability of non-exceedance of 100/(29x2)%. The natural frequencies of the rest of the dampers correspond to probabilities of non exceedance obtained by incrementing the previous probability by 100/29%. Comparing those values with the measured natural frequencies of the 29 dampers, we obtained a desired change in natural frequency for each damper. Using measured sensitivity derivatives of the natural frequency with respect to the number of turns on the tuning screws, these changes in natural frequency were translated into adjustments of the tip screws.

We applied the above tailoring procedure to both types of dampers. The dampers were measured again after the adjustment. Each damper was measured three times and the parameters were averaged through the three measurements to reduce the effect of temperature. The resulting parameters after the averaging procedure were used to estimate the means, standard deviations, coefficients of variation and correlation coefficients between the parameters. These values were presented in Tables 4 and 5.

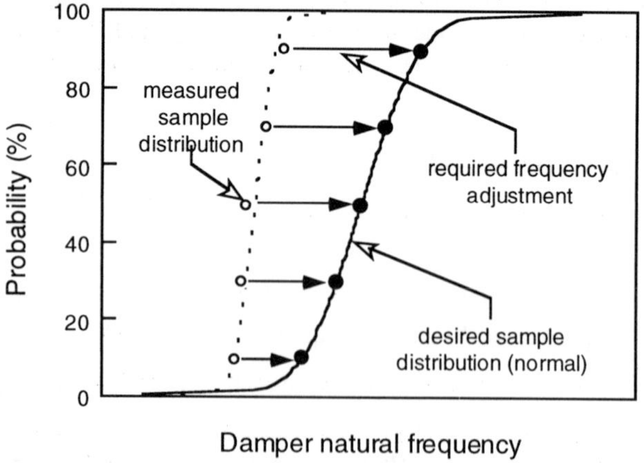

Figure 22 Creation of an "ideal" sample of dampers. (From Ponslet [60]).

4.2. *Validation and Calibration of Analytical Models*

4.2.1. *Structural Model Refinement*

As mentioned earlier in Paragraph 3.2.1, the truss is considered deterministic. The stiffnesses and loss factors of the support springs (see Figure 7) contained in the finite element model as well as the loss factor of the truss members (assumed to be the same for all members) have not been measured directly. Instead, they have been identified by fitting the analytical natural frequencies and damping ratios of the first 3 modes of the original truss (without tuning masses or dampers) to measured values. The resulting model predicts exactly the first and third natural frequencies and damping ratios.

4.2.2. *Calibration of Peak Acceleration*

We performed two series of tests on a few type-1 and type-3 dampers, chosen among the 29 available to cover the whole range of the parameter distributions. The parameters of each damper were measured 3 times and averaged to reduce experimental errors and temperature effects. The resulting sets of parameters were used to analytically predict the peak of the frequency response curve of the original truss (no tuning masses) with a damper at the location and in the direction corresponding to the largest amplitude in its target mode. Using the 2-mode approximation, the peak acceleration amplitudes were then measured. Again, the measurements were repeated 3 times and averaged.

These experimental results are plotted versus the corresponding analytical predictions in Figures. 23 and 24 for modes 1 and 3, respectively. The coefficient of correlation between analysis and experiment is equal to 98.4% for type-1 and 99.8% for type-3. This indicates that the analytical models and the measurements are precise.

Straight lines have been fitted to the data in each plot. These lines indicate a systematic mismatch between experimental and analytical results (the dotted lines in the figures represent an ideal one to one correspondence between analysis and experiment). The equations of these fitted lines, which are shown in Figures 23 and 24, were incorporated into the analysis to correct the mismatch. Note that this analytical-experimental comparison was made during the summer of 1994 [22], while the experimental comparison of probabilistic and fuzzy set optimum designs, described in Section 5, was done during the summer of 1995. Because of this difference, the regression equations might not be accurate.

4.3. *Optimization Results*

Using the data in Tables 4 and 5 for the statistics of the uncertain parameters we optimized the truss according to the probabilistic formulation in Eq. (6). Membership functions were also generated for the same uncertain parameters as described in Subsection 3.4. Using these membership functions we performed the fuzzy set based optimization according to the formulation in Eq. (7). The objective function has been modified using Eq. (14). During the optimizations, all analyses were performed using the 2-mode approximation (Paragraph 3.6.2). We estimated the probabilities of failure of both the probabilistic and fuzzy optimum designs using a Monte-Carlo simulation with a sample that contained 1000 random values of the natural frequency, loss factor and tip mass of each damper. The mean value, standard deviation and correlation coefficient of these damper parameters were the same as in Tables 4 and 5.

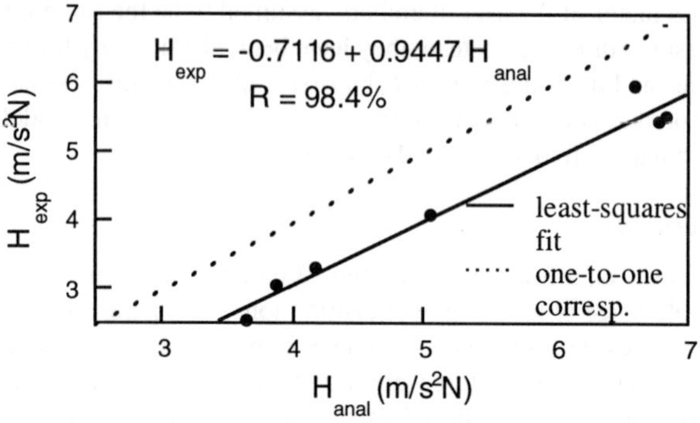

Figure 23 Mode 1, analytical-experimental correlation (2-mode approximation).

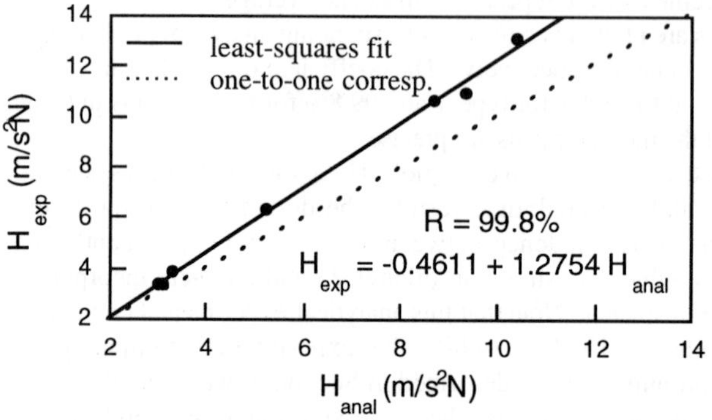

Figure 24 Mode 3, analytical-experimental correlation (2-mode approximation).

The failure limits were 5.00 m/(s^2N) for mode 1 and 6.75 m/(s^2N) for mode 3. These limits were found in Subsection 3.5 and correspond to a problem that is favorable to probabilistic optimization. Type-1 and type-3 dampers were attached to nodes 12 and 11, respectively. The designs are compared on the basis of their probabilities and possibilities of failure.

The fuzzy set based optimum design does not use any tuning masses, because such a design satisfies the optimality condition for the fuzzy set based optimum, which stipulates that the possibility of failure of the two modes are equal. The estimated probabilities and possibilities of failure of each mode and the system are listed in Table 8. Note that the possibilities of failure are only approximately equal, because this is a discrete optimization problem. Also listed in Table 8 are the standard deviations of the estimated failure probabilities, due to the finite sample size used in Monte Carlo simulation. The standard deviations were calculated using Eq. (10). The failure possibilities of the two modes are almost equal. This is because fuzzy set optimization tries to minimize the maximum of the failure possibilities of the two modes. The system possibility of failure is 0.816 and is equal to the maximum of the individual failure possibilities. Also, the probabilities of failure of each mode are almost equal at the optimum. The system probability of failure is 38.2%.

Table 8: Fuzzy set optimum, probabilities and possibilities of failure.

	Mode 1	Mode 3	System
Failure possibility	0.816	0.792	0.816
Failure probability, Pf (%)	20.8	23.2	38.2
Standard deviation in Pf (%)	1.3	1.3	1.5

The probabilistic optimum design uses 3 tuning masses, attached to node 7 of the truss. The probabilities and possibilities of failure of the probabilistic optimum are presented in Table 9. The system possibility of failure is 0.853, which is larger than that of the fuzzy optimum. On the other hand, the system probability of failure is 25.9%, which means that the probability of failure of the probabilistic design is 12.3% lower that the probability of failure of the fuzzy set based design. Note that each approach is safer by its own criterion.

Table 9: Probabilistic optimum, probabilities and possibilities of failure.

	Mode 1	Mode 3	System
Failure possibility	0.853	0.329	0.853
Failure probability, Pf (%)	25.6	0.6	25.9
Standard deviation in Pf (%)	1.4	0.2	1.4

Figures 25 and 26 compare the two alternate designs. Figure 25 shows the membership functions of the response of the two failure modes for each design. Only the part of the membership function lying to the right of the apex is shown. Figure 26 contains the histograms of the peak acceleration of the two failure modes for each design. The dashed vertical line in each plot in Figures25 and 26 represents the corresponding failure limit. Figures 25 and 26 provide insight into the way the probabilistic and fuzzy set approaches maximize safety. The fuzzy set approach equalizes the failure possibilities of the two modes (in this particular problem, this means that it also approximately equalizes the failure probabilities of the two modes). The probabilistic approach, on the other hand, uses the tuning masses at locations which influence more the mode that is easier to control . In this case, mode 3 is easier to control because it is more sensitive to added masses and has smaller scatter than mode 1. As a result, the probability of failure of mode 3 is drastically reduced (from 23.2% to 0.6%), while mode 1 degrades only slightly (from

20.8% to 25.6%) (Figure26). This results in a significant reduction in the system failure probability.

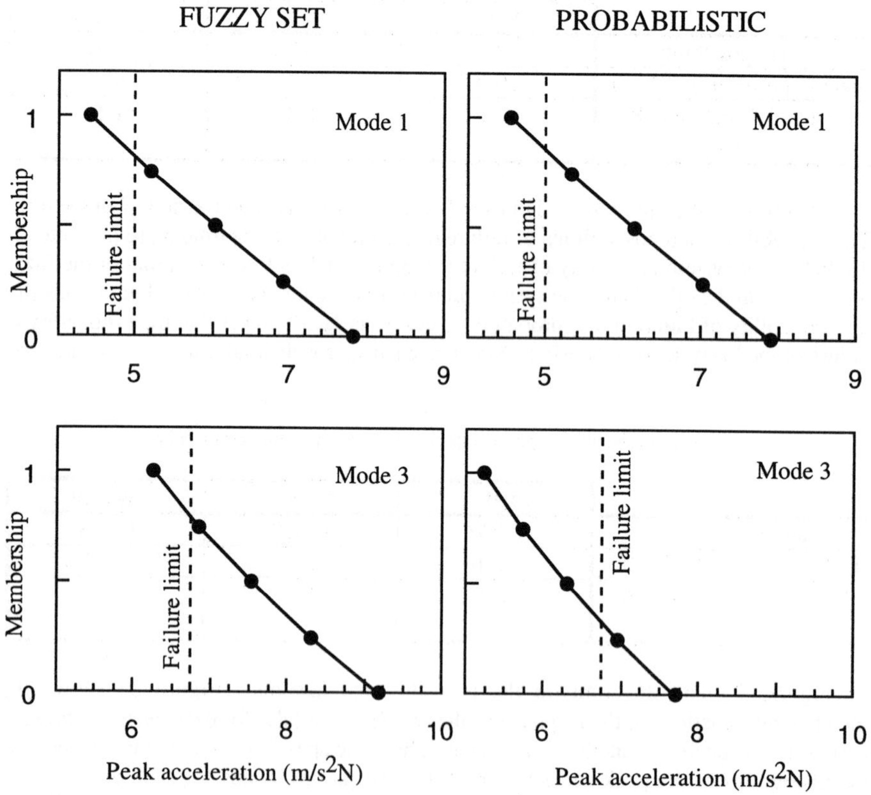

Figure 25 Fuzzy set and probabilistic optimum designs. Membership functions of the peak acceleration.

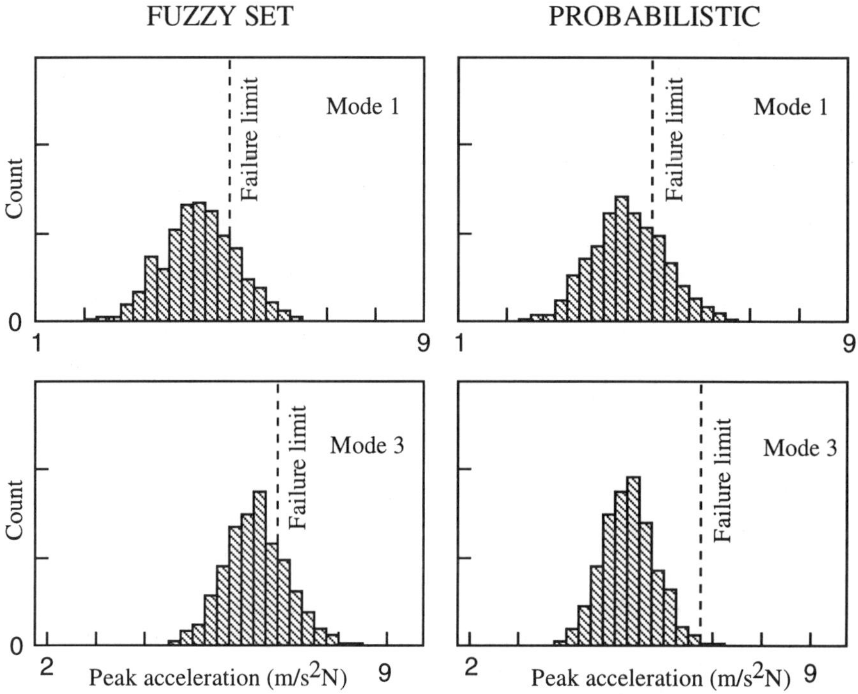

Figure 26 Fuzzy set and probabilistic optimum designs. Distributions of peak acceleration (Monte Carlo simulations with 1000 points).

4.4. Summary of Section 4

We designed alternate probabilistic and fuzzy set based optimum designs using same resources in both cases. The problem selected was favorable for the probabilistic optimization. The important difference between the two design methods is that the fuzzy set based approach does not consider what failure mode is easier to control - it simply tries to equalize the possibilities of failure of each failure mode. This is because, according to the most commonly used rule of fuzzy set calculus for the possibility of the union of events, this possibility is equal to the maximum of the possibilities of these events. On the other hand, the probabilistic approach tries to reduce the failure probability of the failure mode that is easier to control.

Of course, one can use a different definition of the possibility of the union of events, such as the sum of the possibilities of these events. In this case, there would still be problems for which the probabilistic optimum would have considerably lower failure probability than the fuzzy one. An example of such a problem is the case where the natural frequencies of type-1 and type-3 dampers are highly correlated. Because the above definition of the possibility of the union of events neglects correlation between modes, probabilistic design would still be better than the fuzzy design.

The above conclusions about the differences between the two approaches should be independent of the metric of failure used by the fuzzy set approach. For example, if instead of the membership function of the maximum allowable acceleration, we had used the area under the membership curve to the right of the maximum allowable acceleration as the metric of failure, our conclusions would still hold. Indeed, the fuzzy set approach would try to equalize the areas under the membership curves for the two failure modes, and still would not consider how easy it is to control each mode.

The most important conclusion of this section is that under ideal conditions, where we have sufficient information about random uncertainties, accurate models for predicting the response of a structure and failure is crisp, probabilistic methods can yield significantly better designs than fuzzy set based methods, because they take more information into account than fuzzy set based methods.

5. Experimental Comparison

In the previous section, we analytically found a pair of probabilistic and fuzzy set optimum designs. The probabilistic optimum design had a lower probability of failure than its fuzzy set counterpart. However, this comparison was based on <u>analytically predicted</u> probabilities of failure. Due to modeling errors and incomplete knowledge of the probability distributions, the predicted probabilities of failure can be significantly different from the actual ones [1]. Therefore, we need to verify <u>experimentally</u> that the probabilistic design is better despite modeling and other errors.

We constructed 29 nominally identical realizations of both the probabilistic and the fuzzy set optimum designs. The comparison of probabilistic and fuzzy optimum designs is based on the number of realizations of each alternate design that fail out of the total number of 29 realizations. Following that, we estimate the error in the experimentally measured failure probabilities. We also compare the experimental and analytical results and discuss some sources of error that may explain any differences between analytical and experimental results.

5.1. Experimental Results

We prepared 29 pairs of type 1 and type 3 dampers by pairing randomly the 29 dampers of each type (all dampers were used only once). The dampers in each pair are attached to the truss, thus creating 29 realizations of the fuzzy and probabilistic optimum designs. Each realization is tested 3 times in the laboratory and the measured peak acceleration amplitudes are averaged through the 3 measurements. This averaging is intended to reduce the effect of measurement errors and temperature changes. For each design, the probabilities of failure are estimated from these 29 values by counting the number of designs that vibrate more than the prescribed maximum levels. Note that the fuzzy set optimum design has no tuning masses on the truss. Also, the probabilistic optimum design has three tuning masses at node 7 of the truss.

With a sample of 29 measurements, the resolution of the probability measurement is about 3.45%. This resolution is sufficient to measure the differences in the failure probabilities between probabilistic and fuzzy set based designs, because the difference in the analytical predictions is 12.3% (see Tables 8 and 9).

The results of the experiment are listed in Table 10. The experimental probabilities of failure of the two designs are compared to predicted values from Monte Carlo simulations using 1000 points and the 2-mode approximation. In parentheses are the number of failures out of the 29 realizations. We observe that the experiments verify that the probabilistic design is superior over the fuzzy set optimum design, as was predicted by the analysis.

Table 10: Fuzzy set and probabilistic optimum designs: Comparison of experimental (sample of 29) and analytical probabilities of failure (Monte Carlo, sample of 1000, 2-mode approximation).

	Fuzzy set optimum		Probabilistic optimum	
	Analytical	Experimental	Analytical	Experimental
Mode 1 (%)	20.8	27.6 (8/29)	25.6	31.0 (9/29)
Mode 3 (%)	23.2	41.4 (12/29)	0.6	10.3 (3/29)
System (%)	38.2	62.1 (18/29)	25.9	41.4 (12/29)

Figure 27 shows the cumulative probability distributions of the peak acceleration amplitudes of the fuzzy set based design for modes 1 and 3, respectively. The same plots are shown in Figure 28 for the probabilistic design. The vertical line corresponds to the peak acceleration limits.

Table 11 lists the differences in the failure probabilities obtained analytically and experimentally for the two design approaches. The experimental probabilities of failure are higher than the analytical ones for both failure modes. For the fuzzy set based design, the analysis predicted almost equal probabilities of failure for the two failure modes. However, the experiment gave a much higher probability of failure for mode 3. In the probabilistic design, the increase in probability of failure between analysis and experiment was again larger for mode 3. However, the increase in the failure probability of mode 3 was smaller for the probabilistic design than for the fuzzy set based design.

Table 11: Differences between experimental and analytical failure probabilities.

	Design	
	Fuzzy set	Probabilistic
Mode 1 (%)	6.8	5.4
Mode 3 (%)	18.2	9.7

The above results can be explained based on the following observations. The probabilities of failure of the fuzzy set based design and the probabilistic design in mode 1 are close. However, in mode 3, the probabilistic design has much smaller failure probability than its fuzzy set counterpart, because mode 3 has a smaller scatter than mode 1, so it is easier to reduce the probability of failure of mode 3 by slightly reducing the mean acceleration amplitude. Because of the sharpness of the distribution (see Figs. 26, 27 and 28) the probability of failure of mode 3 is more sensitive than that of mode 1 to modeling errors and other unmodeled uncertainties. Moreover, the probability of failure of the fuzzy optimum is more sensitive to modeling errors, which results in a shift of the entire probability density function of the acceleration amplitudes to the right (see Figure 26). As a result, the discrepancy between analytical and experimental results is larger for the probability of failure of the fuzzy set based design for mode 3.

In this study, modeling and other errors during the experiment affected both approaches in the same direction because they led to underestimating the acceleration. As a result, the experimental distributions of the accelerations are shifted to the right relative to the analytical distributions (Figs. 27 and 28). Because of this systematic shift, the analytical predictions of the relative reliability of each method were verified experimentally. We do not know if this would be true if the errors affected each approach in a different direction. In Subsection 5.4, we will examine sources of error that can explain the difference between the analytical predictions and experimental results.

Although the experiment validated that the probabilistic optimum design is safer than the fuzzy set optimum design, it also showed that if there is little information or large modeling errors it might be better to use fuzzy design. To see this, let us assume that the maximum allowable failure probability was 0.35. Then, according to the results in Table 9, probabilistic design could be judged acceptable, whereas in reality it would be unacceptable (Table 10). Fuzzy set analysis, due to its conservatism, could have protected us from accepting such an unsafe design, because it gives a failure possibility of 0.816, which is too close to one. Although this is not as precise as failure probability, there is a clear implication that the obtained design is likely to fail. Therefore, if a designer used fuzzy set optimization and assessed the risk of failure using the possibility of failure, it is likely that he or she would conclude that the truss must be redesigned to increase reliability using more resources (*e.g.*, more dampers).

5.2 Estimation of Error in Measured Failure Probabilities

In this subsection, we will provide estimates for two types of error:
1) *Experimental error*, due to non-ideal conditions during the experiment (*i.e.*, temperature variation, small differences in the orientation of the damper on the truss, etc.).
2) *Resolution error,* due to finite number of experiments.

Note that, these two types of error are independent. They cannot be combined because we do not know the probability distribution of the resolution error.

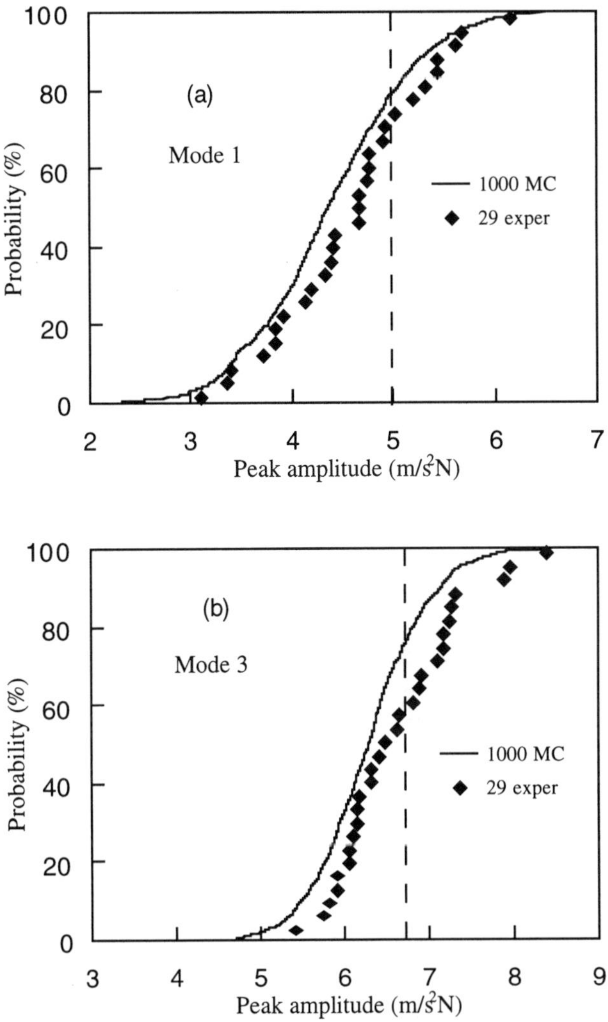

Figure 27 Fuzzy set based design, cumulative probability distributions from analysis (Monte Carlo simulation-1000 samples) and experiment (29 samples): (a) Mode 1, (b) Mode 3.

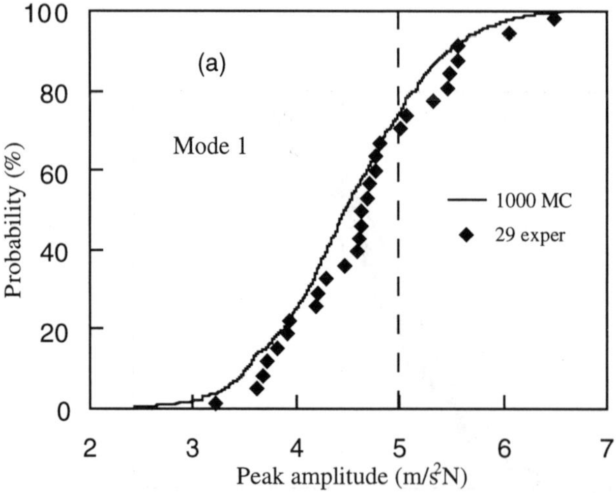

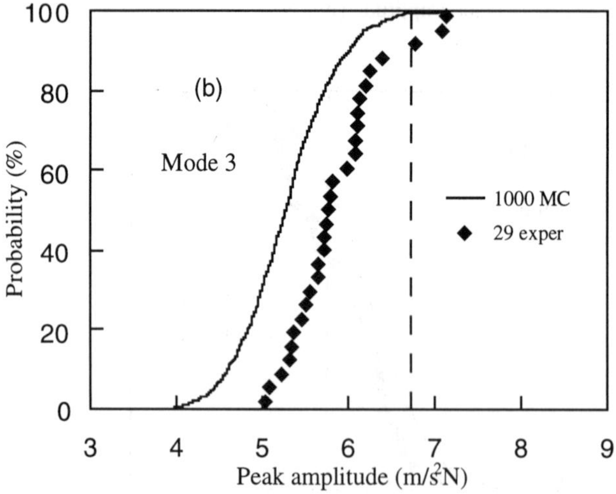

Figure 28 Probabilistic design, cumulative probability distributions from analysis (Monte Carlo simulation-1000 samples) and experiment (29 samples): (a) Mode 1, (b) Mode 3.

5.2.1 Estimation of Experimental Error

Experimental error implies lack of repeatability. To accurately estimate the experimental error we would have to determine the statistics of the measured failure probabilities. This could be impractical, because it requires repeating the whole experiment (*i.e.*, measuring the response of the truss three times for each of the 29 dampers) many times and calculating the statistics of the obtained failure probabilities. In this paragraph, we will provide an estimate of the error of the measured failure probabilities using Monte Carlo simulation and the existing experimental measurements.

We assume that the measured structural response, R_{m_i}, when damper i is attached to the structure can be expressed as:

$$R_{m_i} = R_i + e_i \tag{19}$$

where R_i is the response of the damper that we would measure if there were no experimental error and e_i is the experimental error (including the effect of temperature variation). We assume that the mean value of the error term is zero and that the error is normally distributed. R_i is constant for a given damper, because it depends only on the damper properties. The mean and the variance of R_{m_i} are:

$$\text{Mean: } \overline{R_{m_i}} = R_i + \overline{e_i} = R_i \tag{20}$$

$$\text{Variance: } \sigma^2_{Rm_i} = \sigma^2_{R_i} + \sigma^2_{e_i} = \sigma^2_{e_i} \tag{21}$$

We have three structural response measurements for each damper. The average of these three measurements provides an estimate of R_i while their variance provides an estimate of $\sigma^2_{e_i}$. Note, that these estimates are very crude because they were obtained using only three sample points. We observed that $\sigma^2_{e_i}$ was independent from R_i and that it varied significantly from damper to damper. Assuming that e_i is the same for all dampers, we obtained a very crude approximation of its variance, σ^2_e, by averaging the variances of the 29 measurements, *i.e.*,

$$\sigma^2_e = \frac{1}{29}\sum_{i=1}^{29} \sigma^2_{e_i} \tag{22}$$

We can simulate the experimental procedure for estimating the probability of failure using Monte Carlo simulations as follows. For each damper, we assume that the average of three measurements is equal to R_i. For each damper, we generate three sample values of the error, e_i, following the distribution $N(0, \sigma_e)$. We then add these values to R_i, which yields three sample values of the measured response of the truss when this damper is attached to the truss. We average these three responses as we do in the experiment. Then we compare the average value of the truss response for each damper to the corresponding failure limit and estimate the probability of failure by counting how many of the 29 realizations fail. We repeat the whole procedure 1000 times and we calculate the statistics of the failure probabilities. The standard deviations of the failure probabilities are listed in Table 12 for both designs.

Table 12: Standard deviation of failure probability (%) due to experimental error, obtained from Monte Carlo simulation with 1000 repetitions.

	Design	
	Fuzzy set	Probabilistic
Mode 1 (%)	4.2	4.1
Mode 3 (%)	4.1	2.6
System (%)	4.0	4.6

Assuming again (as we did in Paragraph 4.1.1) that the difference in failure probabilities of the alternate designs, ΔP, is normally distributed, from Eq. (18) we get that the standard deviation, $\sigma_{\Delta P}$, is 6.10%. The difference in the experimentally estimated failure probabilities of the two alternate designs, ΔP, is 20.7%, which is equal to 3.39 standard deviations of ΔP. This corresponds to a 99.37% probability that the actual failure probability of the fuzzy design is larger than that of the probabilistic design. These results provide satisfactory confidence in the experimental results.

5.2.2 Estimation of Resolution Error

As explained in Paragraph 4.1.1, the sample of dampers used in the experiments was not random but it was obtained by sampling the distributions of the natural frequencies of the dampers using a uniform step size in probability. Figure 29 shows plots of the peak acceleration amplitude versus the natural frequency of type-1 and type-3 dampers, respectively, for the range of damper frequencies used in this study. Two designs are examined: One with no tuning masses on the structure (fuzzy set optimum), and one with 3 tuning masses at node 7 of the truss (probabilistic optimum). The remaining two damper parameters are assumed constant and equal to their mean values. The relation between the peak acceleration and the natural frequency is monotonic in the range of natural frequencies examined in this study. Therefore, if the natural frequency is sampled with a uniform step size in probability, then the response (peak acceleration) is also sampled with a uniform step size in probability.

By neglecting the experimental error (discussed in Paragraph 5.2.1) and the error due to randomness of the sample, the only error left is a resolution error due to the limited size of the sample. This type of error is illustrated in Figure 30. This figure shows part of the cumulative probability distribution plot of the response. The probability of the peak acceleration exceeding a limit is 100% minus the cumulative probability of that limit. There are 20 realizations whose peak acceleration is less than that corresponding to point B.

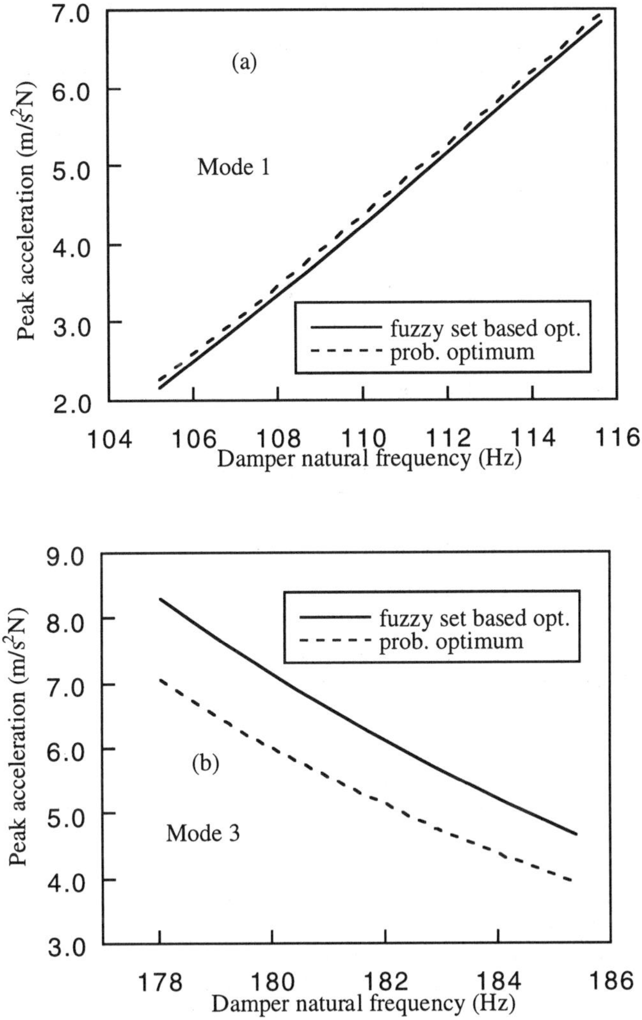

Figure 29 Peak acceleration vs. damper natural frequency: (a) Mode 1, (b) Mode 3.

Therefore, if the failure limit lies between points A and B of Figure 30, we will always measure 9 failures out of 29 samples (*i.e.*, there are 9 realizations whose peak acceleration is higher than the failure limit). This translates into a probability of failure of 31.03% (9/29). But, as can be seen from Figure 30, the actual probability of failure can be anywhere in the range from 29.31% to 32.76% or equivalently 31.03% plus or minus half the resolution. Since resolution is approximately 3.45%, for a probability of failure of p%, the actual probability of failure can be between (p-1.725)% and (p+1.725)%. Note that the above error analysis is approximate because it does not take into account that the tip mass and loss factor are not sampled with a uniform step size in probability. However, their effect on the response is much smaller than the effect of the natural frequency. Note that the resolution error is much smaller than the error due to randomness in the sample (see Paragraph 4.1.1).

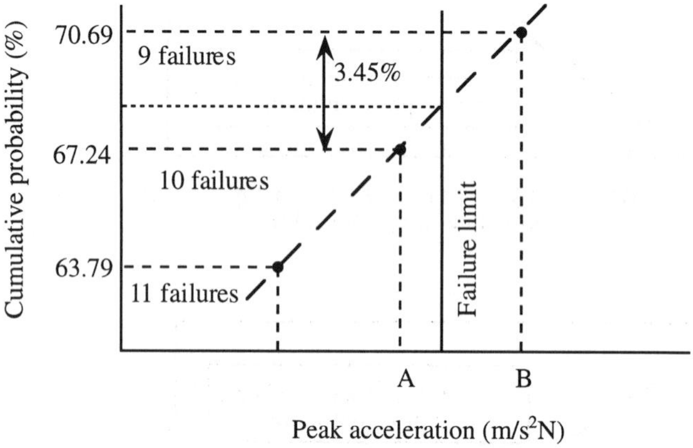

Figure 30 Resolution error of measured failure probabilities.

5.3 Error in Analytically Predicted Probabilities of Failure

In Subsection 5.1, we observed a considerable discrepancy between the analytical predictions and the experimental results. In this subsection, we will examine the effect of two sources of error on the analytically predicted failure probabilities:

a) Errors in calibration of peak acceleration.

b) Difference in the average temperature when we measured damper properties and when we measured the response of the truss.

Using the exact model instead of the calibrated 2-mode approximation we concluded that the modeling error due to the 2-mode approximation is negligible compared to the above two errors.

5.3.1 Calibration of Peak Acceleration

In Paragraph 4.2.2, we mentioned that we corrected the analytical predictions of the peak acceleration amplitudes using a linear regression equation. This correction, however, was determined using measurements taken about one year prior to the experiments described in the previous section. To assess the effect of this discrepancy we took new measurements on a number of dampers to determine a new correction equation and used this equation to correct the structural response predictions obtained from the analysis.

We followed the procedure described in Paragraph 4.2.2. The experimental results are plotted versus the corresponding analytical predictions in Figs. 31 and 32 for modes 1 and 3, respectively. The coefficient of correlation between analysis and experiment is equal to 97.3% for mode 1 and 98.2% for mode 3. The equations of the straight lincs fitted to the data of each plot are shown in Figs. 31 and 32. These equations replace the old ones (obtained in Paragraph 4.2.2) in the analysis. Note that the difference between the old and new regression equations used to calibrate the response is considerable for both modes 1 and 3.

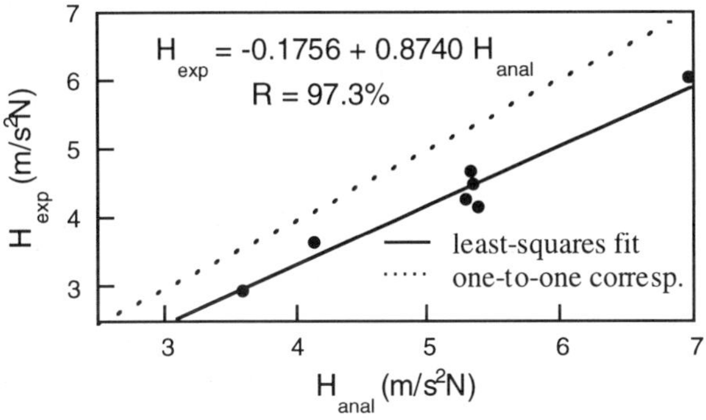

Figure 31 Mode 1, analytical-experimental correlation (2-mode approximation).

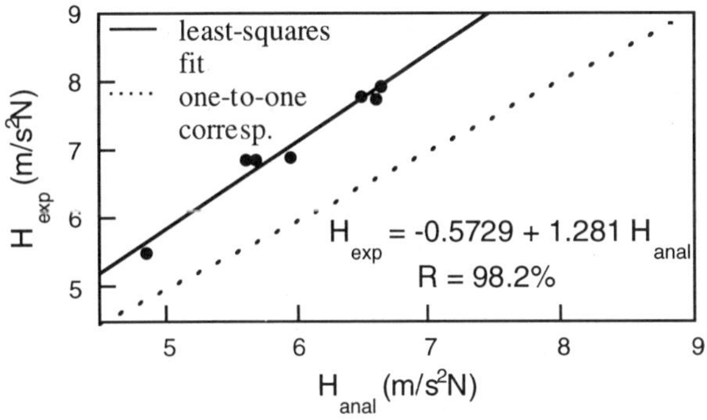

Figure 32: Mode 3, analytical-experimental correlation (2-mode approximation).

5.3.2 Difference in the Average Temperature When Damper Properties and Structural Response Were Measured

Because the dampers are made of viscoelastic material, their properties are very sensitive to temperature. The laboratory temperature during the structural response measurements was on the average approximately 0.7 ^{o}F higher than the temperature when the properties of the dampers were measured. Because these properties were used in the analysis, the temperature variation can be responsible for the discrepancy between analysis and experiment.

We estimated the effect of this temperature difference the following way. From tests we determined the sensitivity of each parameter with respect to temperature. Table 13 lists these sensitivities for each damper parameter. Using these sensitivities and linear extrapolation we corrected the mean values of the damper parameters as shown in Table 14. For simplicity, we assume that, because the temperature variation is small, the standard deviations and the correlation coefficients of the damper properties are not affected.

Table 13: Sensitivities of damper parameters with respect to temperature.

	Tip mass (kg/^{o}F)	Natural frequency (Hz/^{o}F)	Loss factor (1/^{o}F)
Type-1 dampers	2.88×10^{-5}	-0.41	5.20×10^{-3}
Type-3 dampers	2.95×10^{-5}	-0.98	6.73×10^{-3}

Table 14: Mean values of damper properties when corrected for temperature.

	Tip mass (g)	Natural frequency (Hz)	Loss factor
Type-1 dampers	7.050	110.15	0.12267
Type-3 dampers	7.390	181.02	0.14181

5.3.3. Updated Analytical Estimation of Failure Probabilities

Using corrected properties due to temperature and updated regression equation for correction of analytical predictions, we generated 1000 new samples of each type of damper using the mean values of the damper parameters in Table 14. Using these samples and the new analytical-experimental correction determined in Paragraph 5.3.1, we calculated the probabilities of failure for the fuzzy set and the probabilistic optimum designs using Monte Carlo simulation. The new probabilities of failure are listed in Table 15. The differences between these values and the experimental ones are presented in Table 16. Observe that the difference between the predicted and measured failure probability of mode 3 of the fuzzy set based design is reduced by about 60% compared to the values of Table 11. The difference in mode 3 of the probabilistic optimum design is also reduced by about 20%. Mode 1 was not affected significantly, because it has larger scatter, and therefore, the corresponding failure probability is not sensitive to small variations in the mean values of the damper parameters.

Table 15: Analytical probabilities of failure (Monte Carlo, 1000 samples, 2-mode approximation, new analytical-experimental correction, damper properties corrected for temperature).

	Fuzzy set based optimum	Probabilistic optimum
Mode 1 (%)	20.6	24.1
Mode 3 (%)	34.3	2.4
System (%)	48.6	25.8

Table 16: Differences between corrected analytical and experimental failure probabilities.

	Design	
	Fuzzy set based	Probabilistic
Mode 1 (%)	7.0	6.9
Mode 3 (%)	7.1	7.9

The differences between analytical and experimental failure probabilities are considerably larger than one standard deviation of the corresponding experimental error (Table 12). Thus, the experimental errors we accounted for are not the only factors responsible for the discrepancies. One should keep in mind that the standard deviations of the experimental error estimated in Paragraph 5.2.1 were approximate.

Figure 33 shows the cumulative probability distribution of the peak acceleration of the fuzzy set based design for modes 1 and 3, respectively. The same plots are shown in Figure 34 for the probabilistic design. The corrections brought the analytical curves of mode 3 closer to the experiment, without significantly affecting mode 1. This implies that the discrepancy in the average temperature during the experiments and the error in the calibration equation were responsible for a large portion of the discrepancy between analytical and experimental failure probabilities.

Note that when we took the measurements to determine the new correction equation (see Paragraph 5.3.1) the temperature was the same as when the damper properties were measured. Consequently, the correction equations do not account for any part of the error due to change in the average temperature when the damper properties were measured and when the structural response measurements were taken.

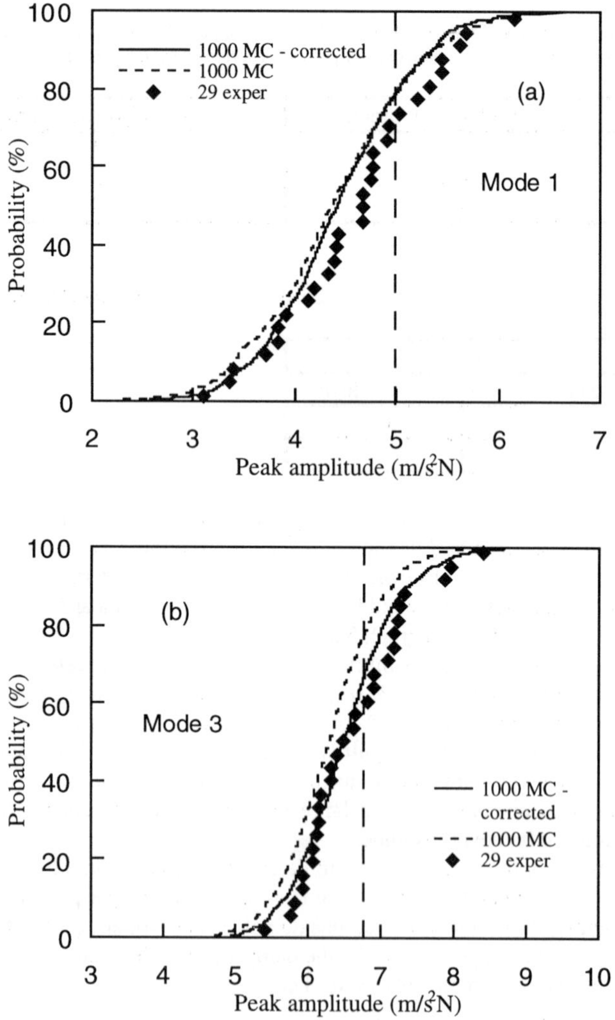

Figure 33 Fuzzy set based design, cumulative probability distributions from analysis (Monte Carlo simulation-1000 samples, original and corrected sample) and experiment (29 samples): (a) Mode 1, (b) Mode 3.

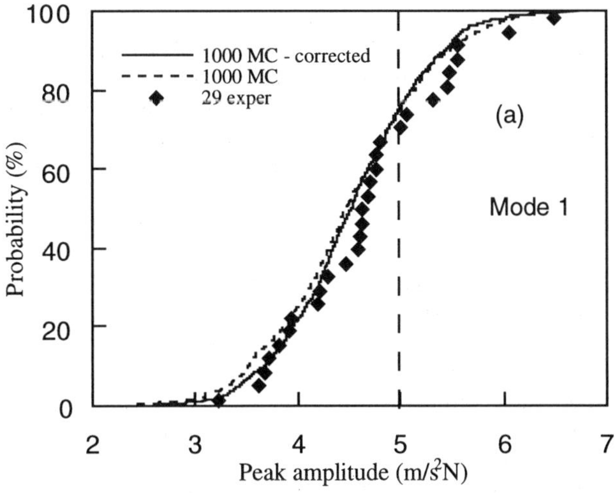

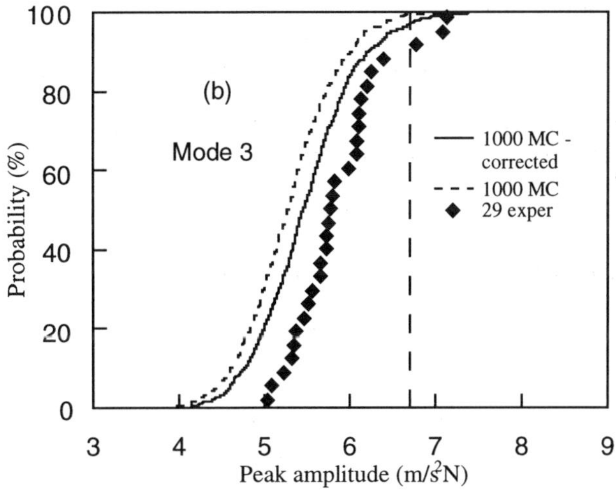

Figure 34 Probabilistic design, cumulative probability distributions from analysis (Monte Carlo simulation-1000 samples, original and corrected sample) and experiment (29 samples): (a) Mode 1, (b) Mode 3.

6. Conclusions and Future Work

The primary objectives of this work were to understand the differences in the way that probabilistic methods and fuzzy set based methods maximize reliability and experimentally test optimal designs obtained using each approach. We applied probabilistic and fuzzy set optimization to a practical design problem and compared the failure probabilities of the optimum designs. We used a modular structure and a non-destructive definition of failure. The same resources were available for both approaches. The research has provided us with the following conclusions and key observations:

- Probabilistic optimization tries to reduce the probabilities of failure of the modes that are easier to control, in order to minimize the system failure probability. This is because the system probability of failure is related to the sum of the probabilities of failure of the individual failure modes. This is the reason that, at the optimum, modes whose performance function has small scatter and is sensitive to the design variables casually have lower failure probabilities than the other modes.
- Fuzzy set optimization does not consider which failure mode is easier to control but simply tries to equalize the possibilities of failure of all failure modes. This is because, according to the most commonly used rule for calculating the possibility of failure of the union of failure events, this possibility is equal to the maximum of the possibilities of failure of the individual failure events.
- Analytical results showed that, when we have sufficient information about uncertainties, accurate models for predicting the response of a structure and crisp definition of failure, probabilistic methods can yield significantly better designs than fuzzy set based methods.
- For a problem selected to provide large contrast in the probabilities of failure of the two approaches, the higher reliability of the probabilistic approach, predicted by the analysis, was verified experimentally.
- There was considerable discrepancy between the experimentally measured rates of failure and the analytically predicted values. Two possible sources of such a discrepancy were examined: errors in the regression equation used to calibrate the analytically calculated peak acceleration and a temperature shift between the time when the damper properties were measured and the structural response was measured. Accounting for these two factors in the analysis brought the analytical results significantly closer to the experiment.

This comparison considered a problem that was favorable to probabilistic design. To complete the comparison the authors plan to consider a broad spectrum of problems in the near future. They expect that the third and fourth observations above are likely to change after considering these problems.

Although in the experimental comparison the improved reliability of the probabilistic design was confirmed for the problem studied, one should keep in mind that in this problem the statistical characteristics of the random parameters could be estimated from measurements with good accuracy. Also, modeling error had been minimized. However, designers rarely have the luxury of complete information about uncertainties and accurate models for predicting the response of a system. Therefore, further research is needed to determine which approach is better when there is limited statistical information and large modeling errors are present. These are the situations where fuzzy set based methods could be better because they are more conservative than probabilistic methods.

This study was also based on a crisp definition of failure. That means that there is an abrupt transition from a completely acceptable to a completely unacceptable design. However, we can also characterize as satisfactory to a certain degree or unsatisfactory to another degree. Fuzzy set based methods can handle this fuzziness in the definition of failure. Further research should investigate the effect of such imprecision in the definition of failure on the efficacy of the fuzzy set based method relative to other design methods.

7. Acknowledgements

The authors appreciated Professor Zimmermann's comments on the fairness of the comparison of probabilistic and fuzzy set based methods.

8. References

1. Ben-Haim, Y., and Elishakoff, I., *Convex Models of Uncertainty in Applied Mechanics*, (Elsevier Science Publishing Company, New York, NY, 1990).
2. Ben-Haim, Y., and Elishakoff, I., *Journal of Applied Mechanics, Transactions ASME,* **Vol. 58,** No. 2, (1991), pp. 354-361 .
3. Elishakoff, I., and Ben-Haim, Y., *Structural Safety*, **Vol. 8**, No. 1-4, (1990), p. 103-112 .
4. Elishakoff, I., Elisseeff, P., and Glegg, S., *AIAA Journal*, **Vol. 32**, No. 4, (1994), pp. 843-849 .
5. Elishakoff, I., Cai, G. Q., and Starnes, J. H. Jr., *International Journal of Non-Linear Mechanics*, **Vol. 29**, No. 1, (1994), pp. 71-82 .
6. Michael, W., and Siddall, J. N., *ASME Journal of Mechanical Design,* **Vol. 103**, No. 4, (1981), pp. 842-848 .
7. Balling, R. J., Free, J. C., and Parkinson, A. R., *ASME Journal of Mechanisms, Transmissions, and Automation in Design*, **Vol. 108**, No. 4, (1986), p. 438 .
8. Chen, J. S., Wang, C. S., and Zug, P., *ASME Journal of Mechanisms, Transmissions, and Automation in Design,* **Vol. 106**, No. 4, (1984), pp. 510-517 .
9. Bandler, J. W., *Journal of Optimization Theory and Applications*, **Vol. 14**, No. 1, (1974), p. 99-113 .
10. Bryne, D., and Taguchi, G., *Quality Progress,* (1987), pp. 19-26 .
11. Phadke, M., *Quality Engineering Using Robust Design* (Prentice Hall, Englewood Cliffs, NJ, 1989).
12. Otto, K. N., and Antonsson, E., *ASME Journal of Mechanical Design*, **Vol. 115**, No. 1, (1993), pp. 5-13 .
13. Charnes, A., and Cooper, W. W., Management Sciences, **Vol. 6**, (1958), pp. 73-79.
14. Frangopol, D. M., and Moses, F., *in Advances in Design Optimization*, Chapter 13, (Adeli, Chapman and Hall ed., New York, NY, 1994).

15. Moses, F., and Stevenson, J. D., *Journal of the Structural Division*, (1970), pp. 221-244 .
16. Khalessi, M. R., Lin, H.-Z., and Alvarez, M. S., *in Proceedings of the AIAA/ASME/ASCE/AHS/ASC 35th Structures, Structural Dynamics, and Materials Conference*, Hilton Head, SC, **Vol. 2**, (1994), pp. 714-723 .
17. Fox, E. P., and Safie, F., *in Proceedings of the AIAA/SAE/ASME/ASEE 28th Joint Propulsion Conference and Exhibit,* Nashville, TN, (1992).
18. Nikolaidis, E., and Burdisso, R., *Computers and Structures*, **Vol. 28**, No. 6, (1988), pp. 781-788 .
19. Reddy, M. V., Grandhi, R. V., and Hopkins, D. A., *in Proceedings of the AIAA/ASME/ ASCE/AHS/ASC 34h Structures, Structural Dynamics, and Materials Conference*, **Vol. 2,** (1993), pp. 990-999.
20. Hasselman, T. K., and Chrostowski, J. D., *in Proceedings of the AIAA/ASME/ASCE/AHS/ASC 31st Structures, Structural Dynamics, and Materials Conference,* Long Beach, CA, **Vol. 4**, (1990), pp. 1945-1951 .
21. Hasselman, T. K., and Chrostowski, J. D., *in Proceedings of the AIAA/ASME/ASCE/AHS/ASC 33rd Structures, Structural Dynamics, and Materials Conference*, Dallas, TX, **Vol. 3,** (1992), pp. 1272-1284 .
22. Ponslet, E., Maglaras, G., Haftka, R.T., Nikolaidis, E., Sensarma, P., and Cudney, H.H., *in Proceedings of the AIAA/USAF/NASA/OAI 5th Symposium on Multidisciplinary Analysis and Optimization*, Panama City, FL, part 1 (1994), pp. 554-559 .
23. Melchers, R. E., *Structural Reliability, Analysis and Prediction*, (Ellis Horwood, Chichester, UK, 1987), pp. 93-94,.
24. Zadeh, L. A., *Information and Control,* **Vol. 8**, No. 3, (1965), pp. 338-353 .
25. Zadeh, L. A., *Fuzzy Sets and Systems,* **Vol. 1**, (1978), pp. 3-28.
26. Zadeh, L. A., *Information Sciences*, **Vol. 8**, (1975), pp. 199-249 .
27. Kaufmann, A., and Gupta, M. M., *Introduction to Fuzzy Arithmetic*, (van Nostrand Reinhold Company, New York, 1985).
28. Dong, W. M., and Shah, H., *Fuzzy Sets and Systems*, **Vol. 24**, (1987), pp. 65-78.
29. Dubois, D., and Prade, H., *Fuzzy Sets and Systems: Theory and Applications*, (Academic Press, New York, 1980).
30. Zimmermann, H. J., *Fuzzy Set Theory and its Applications*, (Kluwer-Nijhoff Publishing, Boston, MA, 1985).
31. Wood, K. L., and Antonsson, E. K., *Mechanism and Machine Theory*, **Vol. 25**, No. 3, (1990), pp. 305-324 .
32. Wood, K. L., Antonsson, E. K., and Beck, J. L., *Research in Engineering Design*, **Vol. 1**, No. 3, (1990), pp. 187-203.
33. Thurston, D. L., and Carnahan, J. V., *Journal of Mechanical Design*, **Vol. 114**, No. 4, (1990), pp. 648-658 .
34. Buckley, J. J, *Risk Analysis*, **Vol. 3**, No. 3, (1983), pp. 157-168 .
35. Kubic, W. L., and Stein, F. P., *American Institute of Chemical Engineers Journal*, **Vol. 34**, No. 4, (1988), pp. 583-601.

36. Allen, J. K., Krishnamachari, R. S., Masetta, J., Pearce, D., Rigby, D., and Mistree, *Structural Optimization,* **Vol. 4**, No. 2, (1992), pp. 115-120.
37. Fang, J. H., and Chen, H. C., *In The American Association of Petroleum Geologists Bulletin,* **Vol. 74**, No. 8, (1990), pp. 1228-1233.
38. Brown, C. B., *Journal of Engineering Mechanics Division, ASCE,* **Vol. 105**, No. EM5, (1979), pp. 855-871 .
39. Ayyub, B. M., and Lai, K.-L., *Naval Engineers Journal,* **Vol. 104**, No. 3, (1992), May, pp. 21-35 .
40. Dong W., Chiang, W.-L., Shah, H. C., and Wong, F. S., *Civil Engineering Systems,* **Vol. 6**, No. 4, (1989), pp. 170-179 .
41. Chou, K. C., and Yuan, J., *Journal of Structural Engineering,* **Vol. 119**, No. 11, (1993), pp. 3276-3290 .
42. Laviolette, M., and Seaman, J. W., Jr., *IEEE Transactions on Fuzzy Systems,* **Vol. 2**, No. 1, (1994), pp. 4-15 .
43. Dubois, D., and Prade, H., *IEEE Transactions on Fuzzy Systems,* **Vol. 2**, No. 1, (1994), pp. 16-21.
44. Wilson, N., *Vagueness and Bayesian Probability, IEEE Transactions on Fuzzy Systems,* **Vol. 2**, No. 1, (1994), pp. 34-36 .
45. Klir, G. J., *IEEE Transactions on Fuzzy Systems,* **Vol. 2**, No. 1, (1994), pp. 27-31 .
46. Lindley, D. V., *IEEE Transactions on Fuzzy Systems,* **Vol. 2**, No. 1, (1994), p. 37.
47. Gaines, B. R., *Information and Control,* **Vol. 38**, No. 2, (1978), pp. 154-169 .
48. Natvig, B., *Fuzzy Sets and Systems,* **Vol. 10**, No. 1, (1983), pp. 31-36 .
49. Henkind, S. J., and Harrison, M. C., *IEEE Transactions on Systems, Man, and Cybernetics,* **Vol. 18**, No. 5, (1988), pp. 700-714 .
50. Bordley, R. F., *Fuzzy Sets and Systems,* **Vol. 33**, No. 3, (1989), pp. 347-354 .
51. Dubois, D., and Prade, H., *European Journal of Operational Research,* **Vol. 40**, No. 2, (1989), pp. 135-154 .
52. Dubois, D., and Prade, H., *in IEEE 2nd International Conference on Fuzzy Systems,* Piscataway, NJ, (1993), pp. 1059-1068.
53. Vadde, S., Allen J. K., and Mistree, F., *Computers and Structures,* **Vol. 52**, No. 4, (1994), pp. 645-658 .
54. Ang, A. H-S., and Tang, W. H., *Probability Concepts in Engineering Planning and Design, Design, Risk and Reliability,* **Vol. 2**, (New York, John Wiley and Sons, 1984).
55. Nikolaidis, E., and Kaplan, P., *Uncertainties in Stress Analyses on Marine Structures,* (Ship Structure Committee, Report-363, 1991).
56. Wood, K. L., Otto, K. N., and Antonsson, E. K., *Fuzzy Sets and Systems,* **Vol. 52**, No. 1, (1992), pp. 1-20 .
57. Chiang, W.-L, and Dong, W. M., *Probabilistic Engineering Mechanics,* **Vol. 2**, No. 2, (1987), pp. 82-91.

58. Hasselman, T. K., Chrostowski, J. D., and Ross, J. D., *in Proceedings of the AIAA/ASME/ASCE/AHS/ASC 35th Structures, Structural Dynamics, and Materials Conference*, Hilton Head, SC, **Vol. 1**, (1994), pp. 72-83.
59. Dong, W. M., Chiang, W.-L., and Wong, F. S., *Computers and Structures*, **Vol. 26**, No. 3, (1987), pp. 415-423.
60. Ponslet, E., *Analytical and Experimental Comparison of Probabilistic and Deterministic Optimization,* Ph.D. Dissertation, Dept. of Aerospace Engineering, Virginia Polytechnic Institute and State University, Blacksburg, VA, (1995).
61. Haftka, R. T., and Kao, P. J., *in ASME Winter Annual Meeting*, Dallas, TX, (1990).
62. Gangadharan, S. N., Nikolaidis, E., Lee, K., and Haftka, R. T., *in Proceedings of the AIAA/ASME/ASCE/AHS/ASC 34th Structures, Structural Dynamics, and Materials Conference,* La Jolla, CA, (1993), **Vol. 1**, pp. 534-543.
63. Van Wamelen, A., Haftka, R. T., and Johnson, E. R., *in Proceedings of the 8th Technical Conference on Composite Materials*, American Society of Composites, Cleveland, OH, (1993).
64. *MATLAB User's Guide*, (The MathWorks, Inc., Natick, MA, 1992).
65. Klir, G. J., and Yuan, B., *Fuzzy Sets and Fuzzy Logic*, (Prentice Hall, Upper Saddle River, 1995).
66. Ross, T. J., *Fuzzy Logic with Engineering Applications*, (McGraw Hill, New York, 1995)
67. Maglaras, G., and Nikolaidis E., *Structural Optimization*, **Vol. 2**, No. 3, (1990), pp. 163-172 .
68. Fox, E. P., *in Proceedings of the AIAA/ASME/ASCE/ AHS/ASC 34th Structures, Structural Dynamics, and Materials Conference*, La Jolla, CA, (1993), pp. 714-723.
69. Wood, K. L., and Antonsson, E. K., *Journal of Mechanisms, Transmission, and Automation in Design*, **Vol. 111**, No. 4, (1989), pp. 616-626.
70. McKay, M. D., Beckman, R. J., and Conover, W. J., *A, Technometrics,* **Vol. 21**, No. 2, (1979), pp. 239-245.
71. Walpole, R .E., and Myers, R. H., *Probability and Statistics for Engineers and Scientists*, (Macmillan Publishing Co., New York, NY, 1972).

Uncertainty Modeling in Vibration and Fuzzy Analysis of Structural Systems, pp. 319–345
edited by B. M. Ayyub, A. Guran and A. Haldar
Series on Stability, Vibration and Control of Systems Series B: Vol. 10

FUZZY LOGIC CONTROL SYSTEM DESIGN OF AN IMPACTING FLEXIBLE LINK

DAN BOGHIU and DAN B. MARGHITU
Department of Mechanical Engineering, Auburn University
Auburn, Alabama 36849, USA

ARDÉSHIR GURAN
American Structronics and Avionics
16661 Ventura Blvd
Encino, California 91436, USA

ABSTRACT

This chapter discusses the application of modern fuzzy logic control of mechanical structures. An overview of intelligent control with emphasis on fuzzy logic control design techniques is given. Then, a fuzzy logic controller was designed and employed to suppress the unwanted vibrations of a rotating elastic beam resulting after the impact with an external rigid body. The momentum balance method and an empirical coefficient of restitution was used in the collision of the two bodies. It is seen that the controller can be applied successfully to control the impact between two bodies.

1. Introduction

Intelligence is one of the main characteristics of an autonomous mechanical structures. Future autonomous mechanical structures are required to achieve pinpoint accuracy in the presence of nonlinearities, disturbances, uncertainties and component failures. The design challenges of precision control, vibration suppression, health monitoring and autonomous operation under an unknown dynamic model, severe disturbance, significant uncertainties and unexpected failures can be accommodated using an intelligent control technique. Among typical intelligent control techniques, one can mention the neural control, fuzzy control and fuzzy neural control. However, in this chapter we emphasis on fuzzy logic control techniques. We first present a review of intelligent control design, then we apply a fuzzy logic strategy to control the vibrations in an impacting flexible link.

Artificial neural networks (ANNs) have recently emerged as a powerful, intelligent design paradigm in pattern recognition, data association, and control. Their distinguished properties can be summarized as follows:

• Accommodating nonlinear and multiple degree-of-freedom systems. The Neural network is capable of modeling and controlling nonlinear, multiple degrees of freedom dynamical systems.

• Uncertainty handling capability. The learning mechanism of neural networks enables the neural network controller to handle nonparametric uncertainties and unmeasured disturbances.

• Massively parallel processing capability. Neural networks possess a massively parallel processing capability. This capability allows significant reduction of the computation time, facilitate the implementation of real-time controllers, and cut the hardware cost.

• Self-autonomy. The learning and generalization properties make the neural networks suitable for the implementation of an autonomous dynamical systems.

Fuzzy logic control (FLC) which is capable of simulating human thinking by incorporating the imprecision inherent in all physical systems, and which can convert the linguistic control strategy, based on expert knowledge, into an automatic control strategy. As a result, fuzzy logic control is particularly suitable for the control of dynamical systems in presence of nonlinearities and uncertainties.

In the design of a conventional controller (P, PI, PID or state feedback controller), what is modeled is the system or process to be controlled, whereas in the design of a fuzzy logic controller, the focus is the human operator's behavior. In the first case, the system is modeled analytically by a set of differential equations and their solution is used by the PID controller to adjust the systems control activities for each type of behavior required. In the fuzzy controller, these adjustment are handled by a fuzzy rule-based expert system, a logical model of thinking processes that a person might go through in the course of manipulating the system. This shift in the focus, from the process to the person involved, changes the entire approach to automatic control problems.

Compared with conventional controllers, fuzzy logic control offers the following unique advantages:

• Fuzzy logic control is able to make effective use of all available information. Information on a control system mainly comes from two sources: 1) sensors that provide numerical measurements of key variables, and 2) human experts who provide linguistic descriptions about the system and control instructions. Fuzzy logic control will be the best choice in situations where the most important information comes from human experts.

• Fuzzy logic control is a model-free approach. Fuzzy logic control does not require a well defined mathematical model of the system under control. As the mathematical model of the mechanical structure is imprecise and time-varying, the model-free feature of the fuzzy logic control is very appealing.

• Fuzzy logic control is able to handle nonlinear systems. Fuzzy logic control is essentially a nonlinear control technology. Since a nonlinear system can be approximated by fuzzy systems as asserted by the Universal Approximation Theorem, by carefully choosing the parameters of the fuzzy controllers, it is always possible to design a fuzzy controller that is capable of controlling the given nonlinear system.

Neural networks are data based and fuzzy logic is knowledge based. Thus, the two approaches can be made to complement and reinforce each other. The act of combining neural networks and fuzzy logic offers an opportunity in improving performance, robustness and accuracy of intelligent mechanical structures. A combined fuzzy neural network has a number of attractive features in control applications. Fuzzy neural networks naturally integrate existing expert knowledge and historical data into a unified framework. A fuzzy neural network can allow inaccurate knowledge to be used in the inference architecture and use the neural network training to refine the prior knowledge. This feature is particularly suitable to the high performance robot control system design since the robot dynamics are characterized by the nonlinear, time-varying, and uncertain behaviors.

The advantages of intelligent control approach are summarized as follows:

- The complexity and uncertainty issues are addressed via the distributed parallel processing, learning, and online reoptimization properties of neural networks.
- The nonlinear dynamics due to severe kinematic and inertial coupling as well as dynamics can be naturally incorporated into the design framework.
- The knowledge base and decision making logic furnished by fuzzy system leads to a human intelligence enhanced adaptive control scheme.

2. Fuzzy Logic Control

The development of the advanced robot technology involves various research activity within which the applications of modern control methodology and knowledge-base intelligent control approaches play a major role.

2.1. Background on FLC

During past several years, fuzzy set logic has been growing rapidly. Both theoretical results and applications have already demonstrated effective utilization of fuzzy logic control. Recently, some surveys and books[1,2,3,4] organized thousands of articles which had been published on this field. A number of very good books can also be found from references therein. Applications of this concept can be found, for instance, in artificial intelligence, computer science, control engineering, decision theory, pattern recognition, robotics, and aircraft control[5].

Theoretical advances have been made in many directions. In fact, it is extremely difficult for either a newcomer to the field to apply fuzzy set theorem to his problems to properly recognize the present "state of the art". Therefore, many applications use fuzzy set theory in an elementary level. Most theoretical publications are very specialized and assume that the readers have a background in fuzzy set theory. They are, thus, difficult to understand. The primary goal of this section is to bridge the gap between theories and applications, and to provide the design process of FLC.

Combining multivalued logic, probability theory, and knowledge base, FLC simulates human thinking by incorporating the imprecision inherent in all physical systems. In traditional logic and computing, sets of elements are "crisp"; whether or not an element belongs to a set is a matter of true and false. In contrast, fuzzy logic works by tuning the hard-edges world of binary control variables into soft grades with varying degrees of membership. For example, a temperature of 70°F can at the same time be both "warm" and "a little bit cool", a condition ignored by classical logic but a cornerstone of fuzzy theory.

Fuzzy systems make decisions based on inputs having the form of "linguistic variables" which will be explained later. The variables are tested with a few number of IF-THEN rules, which produce one or more responses depending on which rules are asserted. The response of each rule is weighted according to the confidence or degree of membership of its inputs, and the centroid of the responses may be calculated to generate the appropriate output. Fuzzy logic shares the IF-THEN structure of artificial intelligence (AI), but, unlike AI, it is not a symbolic process.

Another strength of fuzzy logic is that it deals with observed, rather than measured, variables of the systems. Generally, a fuzzy logic system measures one output parameter which, combined with its derived rate of change, serves as two of its input observed parameters to the controller. In system modeling, this means that users can indirectly process more variables than with a conventional control system. For traditional control modeling, the first step is to derive a mathematical model to describe the system. This requires a detailed understanding of all variables in the system which is not always easy or possible for a complicated system. In contrast, fuzzy modeling deals with the relationship of the output to the input and lumps many parameters together. This translates into the inclusion of high order variables, so that the resulting control system is of a high order and often provides a more accurate and stable response.

2.2. Design Procedure of FLC: A Case Study

Fuzzy logic control provides an effective means of capturing the approximate, inexact nature of the real world. Viewed in this perspective, the essential part of the FLC is a set of linguistic control rules related by the dual concepts of fuzzy implication and the compositional rule of inference. In essence, the FLC provides an algorithm which can convert the linguistic control strategy based on expert knowledge into an automatic control strategy.

Experience shows that FLC results are superior to those obtained by conventional control algorithms. In particular, this technique appears very useful for processes that are too complex to be analyzed by conventional quantitative techniques when available sources of information can only be interpreted qualitatively, inexactly, or uncertainly. Thus, fuzzy logic control may be viewed as an approach that combines conventional precise mathematical control and human-like decision making. However, one drawback of fuzzy control is the lack of rigorous stability

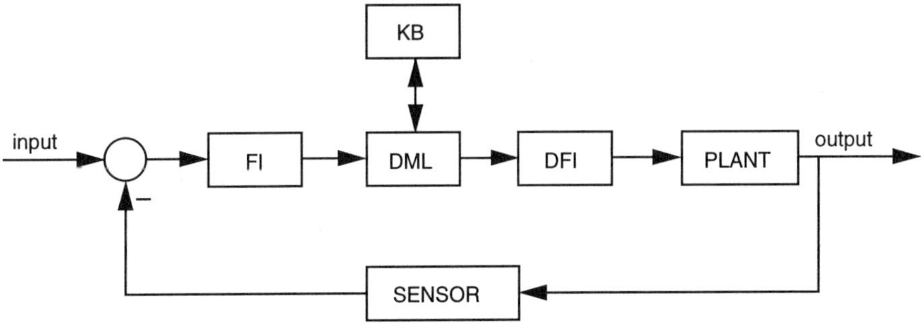

Fig. 1. Basic Fuzzy Logic Structure

and robustness analysis techniques. Most stability analysis methods for FLC are only approximates. As a results, there is no way to obtain a measure of robustness. Good performance is demonstrated through experience (as stated previously) instead of analytical means.

A basic FLC is shown in Figure 1 , and contains four principal components[6]:

- a *fuzzification* interface (FI), similar to an A/D converter in digital control. It has the following functions

  1. Measures the values of input variables.
  2. Performs a scale mapping that transfers the range of values of input variables into corresponding universes of discourse.
  3. Performs the function of fuzzification that converts input data into suitable linguistic values which may be viewed as labels of fuzzy sets.

- a *decision making logic* (DML), similar to a digital controller, is the kernel of a FLC. It has the capability of simulating human decision making based on fuzzy concepts. It can infer fuzzy control actions employing fuzzy implication and the rules of inference in fuzzy logic.

- a *defuzzification* interface (DFI), similar to a D/A converter in digital control. It performs the following functions:

  1. A scale mapping, which converts the range of value of output variables into corresponding universes of discourse,
  2. Defuzzification, which yields a nonfuzzy control action from an inferred fuzzy control action.

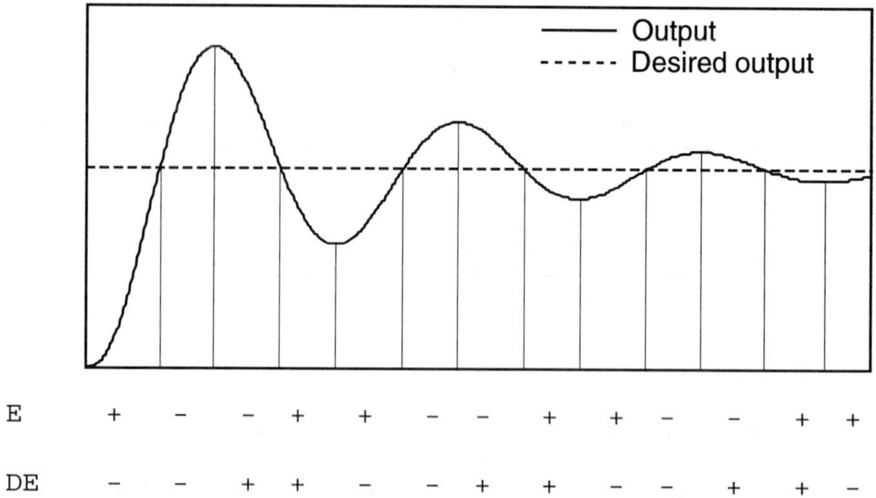

Fig. 2. Standard step response of a plant

- a *knowledge base* (KB), similar to digital control theorems. It comprises a knowledge of application domain and the attendant control goals. It consists of a data base and a linguistic (fuzzy) control rule base:

  1. The data base provides necessary definitions, which are used to define linguistic control rules and fuzzy data manipulation in FLC.
  2. The rule base characterizes the control goals and control policy of the domain experts by means of a set of linguistic control rules.

Using a standard step response of a plant (Figure 2), we try to demonstrate the utility of FLC. Figure 2 shows the system response of a plant. The inputs for FLC are error (E) between the desired position and the plant output, and its rate of change (i.e., the error derivative (DE)). The output of FLC is the control input (CI) to the plant. Thus, we are discussing a two-input one-output fuzzy logic control system. In the following, four principle components will be examined.

The Fuzzification Interface

The fuzzification can be defined as a mapping from an observed input space (controller inputs) to fuzzy sets in certain input universe of discourse. As in control applications, the observed data are usually crisps. This strategy interprets a crisp value x as a fuzzy set A with membership function $\mu_A(x)$ belonging to [0, 1]. The most common shapes of fuzzy sets are the triangles and trapezoids.

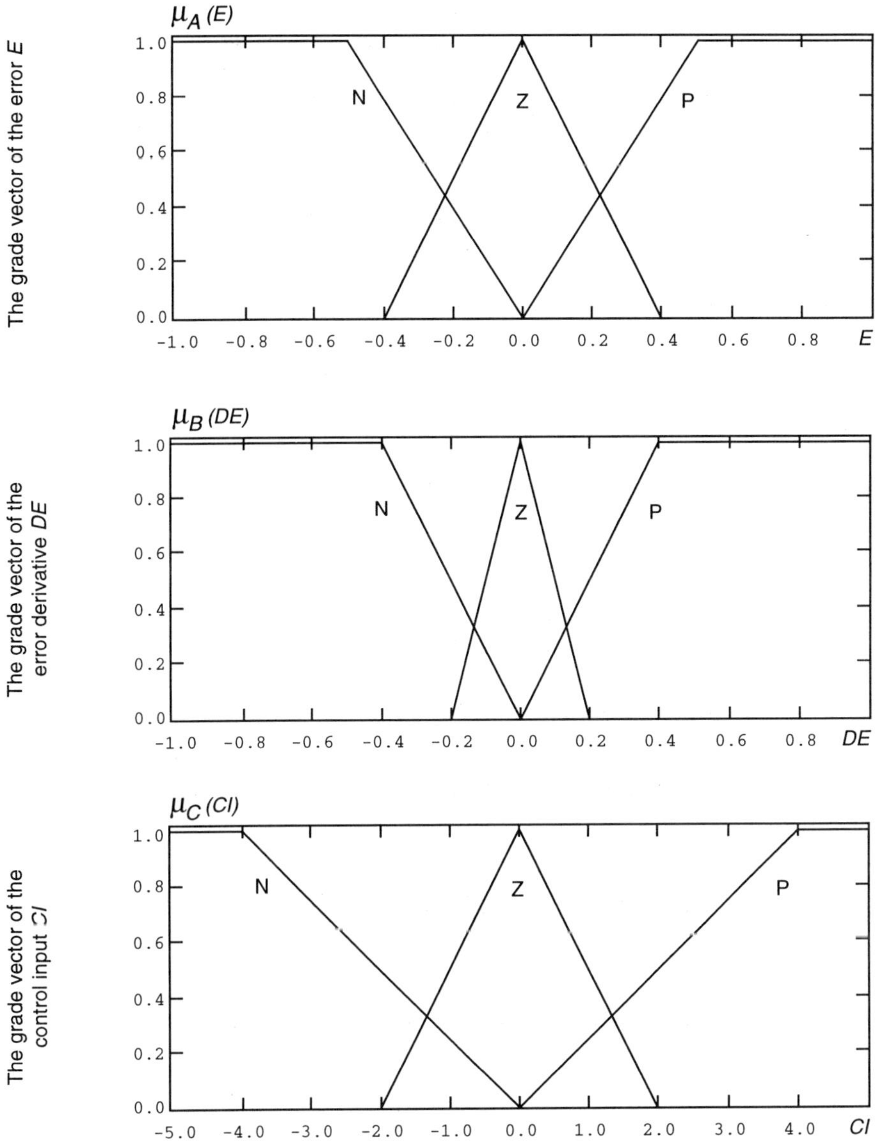

Fig. 3. Diagram representation of the member functions for error (E), error derivative (DE) and control input (CI)

For this case it is assumed that E and DE are all in the range of [-1, 1], and CI belongs to [-5, 5]. The fuzzy set representation of variables E, DE and CI are decomposed into three fuzzy sets (Figure 3), denoted by negative (N), zero (Z) and positive (P) fuzzy sets. Generally, the fuzzy set representation is symmetric with respect to vertical axis and the number of fuzzy sets is odd. In Figure 3, an error of 0.3 (E=0.3), belong to the zero (Z) and positive (P) fuzzy sets. The degree of truth to Z is $\mu_{AZ}(0.3) = 0.25$ and to P is $\mu_{AP}(0.3) = 0.6$.

From Figure 3 one can see that the fuzzy sets and membership functions of E, DE and CI can be described as follows:

- The membership functions of error E

$$\mu_{AN}(E) = \begin{cases} 1 & -1 \le E \le -0.5 \\ -2E & -0.5 < E < 0 \\ 0 & 0 \le E \le 1 \end{cases}$$

$$\mu_{AZ}(E) = \begin{cases} 0 & -1 \le E < -0.4 \\ 2.5(E+0.4) & -0.4 < E \le 0 \\ -2.5(E-0.4) & 0 < E < 0.4 \\ 0 & 0.4 \le E \le 1 \end{cases}$$

$$\mu_{AP}(E) = \begin{cases} 0 & -1 \le E \le 0 \\ 2E & 0 < E < 0.5 \\ 1 & 0.5 \le E \le 1 \end{cases}$$

- The membership functions of the error derivative DE

$$\mu_{BN}(DE) = \begin{cases} 1 & -1 \le DE \le -0.4 \\ -2.5DE & -0.4 < DE < 0 \\ 0 & 0 \le DE \le 1 \end{cases}$$

$$\mu_{BZ}(DE) = \begin{cases} 0 & -1 \le DE < -0.2 \\ 5(DE+0.2) & -0.2 < DE \le 0 \\ -5(DE-0.2) & 0 < DE < 0.2 \\ 0 & 0.2 \le DE \le 1 \end{cases}$$

$$\mu_{BP}(DE) = \begin{cases} 0 & -1 \le DE \le 0 \\ 2.5DE & 0 < DE < 0.4 \\ 1 & 0.4 \le DE \le 1 \end{cases}$$

- The membership functions of the control input CI

$$\mu_{CN}(CI) = \begin{cases} 1 & -5 \le CI \le -4 \\ -0.25CI & -4 < CI < 0 \\ 0 & 0 \le CI \le 5 \end{cases}$$

$$\mu_{CZ}(CI) = \begin{cases} 0 & -5 \leq CI < -2 \\ 0.5(CI+2) & -2 < CI \leq 0 \\ -0.5(CI-2) & 0 < CI < 2 \\ 0 & 2 \leq CI \leq 5 \end{cases}$$

$$\mu_{CP}(CI) = \begin{cases} 0 & -5 \leq CI \leq 0 \\ 0.25CI & 0 < CI < 4 \\ 1 & 4 \leq CI \leq 5 \end{cases}$$

Membership functions indicates the degree to which a value belongs to the class labeled by the linguistic description such as positive, zero and negative.

Thought the fuzzy sets of the error E and its derivative DE are defined between [-1, 1], E and DE are not restricted to these ranges, due to the extension of the fuzzy sets to $\pm\infty$. This means that, for instance, any error E larger to 1 will belong to the positive (P) fuzzy set.

The range of the control input CI is determined by the maximum value of the control force/moment generated by the actuator.

The Decision-Making Logic

The decision making logic is in fact the way in which the controller output is generated. It uses the input fuzzy sets, and the decision is taking according to the values of the inputs (error E and its derivative DE). A *consequent table* is required and the controller output is generated using "If-Then" rules. The rule has two parts, an *antecedent* and a *consequent*. The antecedent is the "If" part and the consequent is the "Then" part. The consequent table is a matrix of 3 columns (the number of grades of the E fuzzy-set) and 3 rows (the number of grades of DE fuzzy-set). The consequent table is shown in Table 1.

E	N	Z	P
$DE=N$	N	N	P
$DE=Z$	N	Z	P
$DE=P$	N	P	P

Table 1. The consequent table for case study

To see how the decision making logic is applied, suppose that E=0.3 and DE=-0.3. Using Figure 3, one can obtain that the fuzzy pairs of E are (0.3, 0.6) $\in P$ and (0.3, 0.25) $\in Z$, while fuzzy pair of DE is (-0.3, 0.75) $\in N$. From Table 1, there are two DML rules excited by this kind of inputs, *i.e.*

Rule 1: If E is P and DE is N, then CI is P.

Rule 2: If E is Z and DE is N, then CI is N.

Figures 4 and 5 illustrates the operations of linguistic fuzzy logic rules, where linguistic "and" is represented by *minimum* operator. That is

$$\mu_{CP}(CI) = min(\mu_{AP}(0.3), \mu_{BN}(-0.3)) = 0.6$$
$$\mu_{CN}(CI) = min(\mu_{AZ}(0.3), \mu_{BN}(-0.3)) = 0.25$$

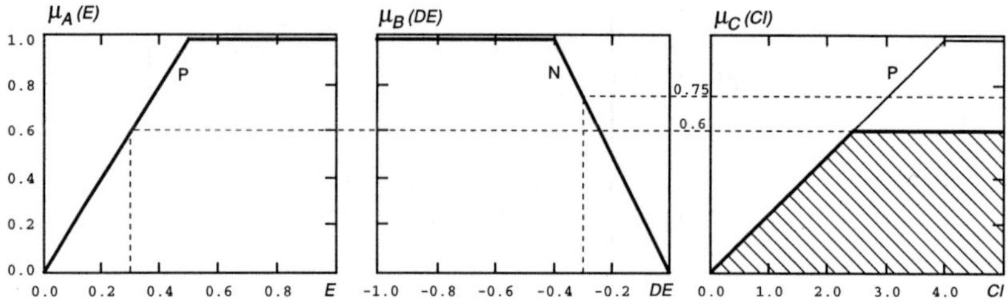

Fig. 4. Influence of Rule 1 on the fuzzy sets

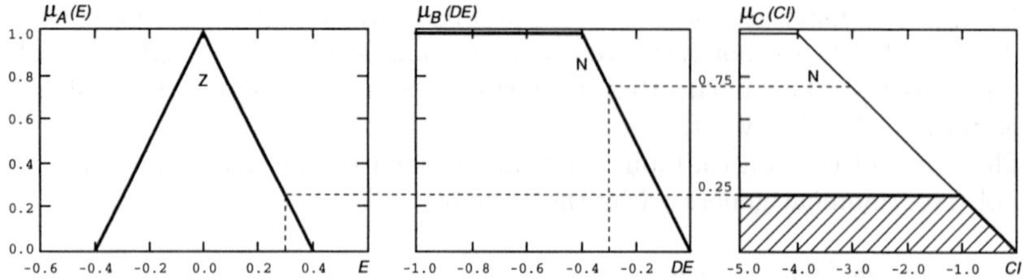

Fig. 5. Influence of Rule 2 on the fuzzy sets

The Defuzzification Interface

Defuzzification is the name given to the operation of obtaining a crisp number from a fuzzy set based on the grades of the fuzzy set.

At present, the commonly used strategies may be described as the maximum criterion, the mean of maximum. and the center of area.

1. *The maximum criterion method.* The maximum criterion produces the point at which the possibility distribution of the control action reaches the first maximum value. For this example, the control input $CI = \mathrm{C1} = 2.4$.

2. *The mean of maximum method* (MOM). The MOM strategy, provides a control action which represents the mean value of all local control actions whose membership function reach the maximum. More specifically, in the case of discrete universe, the control action may be given as

$$CI = \sum_{j=1}^{k} \frac{CI_j}{k} \tag{1}$$

where CI_j is the support value at which the membership function reaches the maximum value $\mu_C(CI_j)$, and k is the number of such support values. For this example, the control $CI = \mathrm{C2} = 0.7$.

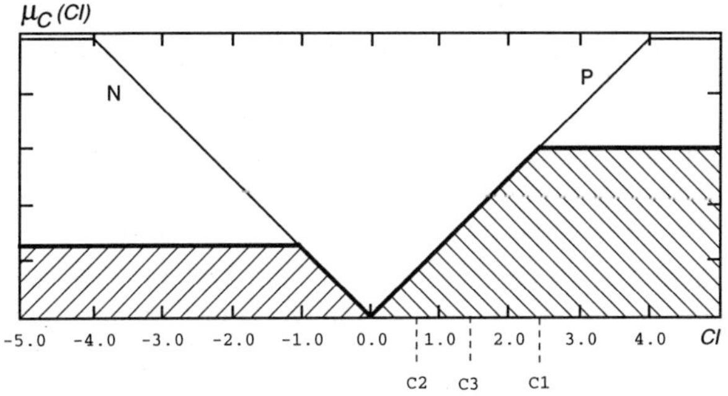

Fig. 6. Graphical representation of defuzzification strategies

3. *The center of area method* (COA). The widely used COA strategy generates the center of gravity of the possibility distribution of a control action. In the case of a discrete universe, this method yields

$$CI = \frac{\sum_{j=1}^{n} \mu_C(CI_j) CI_j}{\sum_{j=1}^{n} \mu_C(CI_j)} \tag{2}$$

where n is the number of term set of the control input. For this example, the control $CI = \text{C3} = 1.4$.

Figure 6 shows a graphical interpretation of these defuzzification strategies. From the results of applications[7], the MOM strategy yields a better transient performance while the COA strategy provides a better steady-state characteristics.

The Knowledge Base
The knowledge base of a FLC consists of two components, a data base and a fuzzy rule base. In what follows, we shall discuss the construction of the KB in FLC.

A. Data Base
The concepts associated with the data base are used to characterize fuzzy control rules and fuzzy data manipulation in an FLC. These concepts are subjectively defined and are based on experience and engineering judgment. In this sense, it is noted that the correct choice of the membership functions of a term set plays an essential role in the success of an application.

1. Discretization and Normalization of Universes of Discourse.

   (a) Discretization of a Universe of Discourse. Discretization of a universe of discourse is frequently referred to as quantization. In effect, quantization discretizes a universe into a certain number of segments. Each segment is labeled as a generic element, and forms a discrete universe. A fuzzy set is then defined by assigning the grade for membership value to each generic element of the new discrete universe. A look-up table based on discrete universes, which defines the output of a controller for all possible combinations of input signals, can be implemented by off-line processing in order to shorten the running time of the controller.

(b) Normalization of Universe of Discourse. The normalization of a universe requires a discretization of the universe of discourse into a finite number of segments, with each segment mapped into a suitable segment of normalized closed interval [-1,1].

2. Fuzzy Partition of Input and Output Spaces.

   A linguistic variable is associated with a term set, with each term in the term set defined on the same universe of discourse. A fuzzy partition then determines how many terms should exist in a term set. The primary fuzzy sets (linguistic terms) usually have meanings, such as NB: negative big; NM: negative medium; NS: negative small; ZE: zero; PS: positive small; PM: positive medium; and PB: positive big.

3. Completeness. Intuitively, a fuzzy control algorithm should always be able to infer a proper control action for every state of process. This property is call "completeness". The completeness of a FLC relates to its data base, rule base, or both.

4. Membership Function of a Primary Fuzzy Set. There are two methods used for defining fuzzy sets, depending on whether the universe of discourse is discrete or continuous: numerical or functional.

B. Rule Base

A fuzzy system is characterized by a set of linguistic statements based on expert knowledge. The expert knowledge is usually in the form of IF-THEN rules, which are easily implemented by fuzzy conditional statements in fuzzy logic.

1. Choice of Process State Variables and Control Variables of Fuzzy Control Rules. Fuzzy control rules are more conveniently formulated in linguistic rather than numerical terms. The proper choice of process state variables and control variables is essential to the characterization of the operation of a fuzzy system. In particular, the choice of linguistic variables and their membership functions has a strong influence on the linguistic structure of FLC. Typically, the linguistic variables in an FLC include the state, state error derivative, state error integral, etc.

2. Source and Derivation of Fuzzy Control Rules. There are four modes of derivation of fuzzy control rules. The derivations are based on expert experience and control engineering knowledge, operator's control actions, fuzzy model of a process, and learning. These four modes are not mutually exclusive, and it seems that a combination of them could be necessary to construct an effective method for the derivation of fuzzy control rules.

3. Justification of Fuzzy Control Rules. There are two principal approaches to the derivation of fuzzy control rules. The first is a heuristic method in which a collection of fuzzy control rules is formed by analyzing the behavior of a controlled process. The control rules are derived in such a way that the derivation from a desired state can be corrected and the control objective can be achieved. The other approach to generating the rule base of an FLC is analogous to the conventional controller designed by pole placement. The fuzzy control rules of an open-loop system and a desired closed-loop system are initially given. The purpose is to synthesize a linguistic control element based on the fuzzy models described above. The main idea is to invert the low order linguistic model of a certain open loop system. However, linguistic inversion mappings are usually incomplete or multivalued. This method is restricted to relatively low order systems but it provides an explicit solution for rule generation of the FLC, assuming that fuzzy models of the open and closed systems are available.

4. Types of Fuzzy Control Rules. Depending on their nature, two types of fuzzy control rules, state evaluation fuzzy control rules and object evaluation fuzzy control rules, are currently in use in the design of the FLC.

   (a) State Evaluation Fuzzy Control Rules. Most FLC's are state evaluation fuzzy control rules which, in the case of 3-input 1-output case, are characterized by the following rule:

   $$\text{Rule } i\text{: if } a \text{ is } A_i,\ b \text{ is } B_i, \text{ and } c \text{ is } C_i \text{ then } d \text{ is } D_i$$

   where a, b, c, and d are linguistic variables representing the process state variables and the control variable A_i, B_i, C_i, and D_i are the linguistic values (corresponding fuzzy sets) of the linguistic variables.

   (b) Object Evaluation Fuzzy Control Rules: This method predicts present and future control actions and evaluates control objectives. It is called object evaluation or predictive fuzzy control. A typical rule is described as:

   $$\text{Rule } i\text{: if } (\ c \text{ is } C_i \rightarrow (a \text{ is } A_i \text{ and } b \text{ is } B_i)) \text{ then } c \text{ is } C_i$$

   In linguistic terms, the rule is interpreted as: "if the performance index a is A_i and index b is B_i then a control command c is chosen to be C_i, then this rule is selected and the control command C_i is taken to be the output of the FLC".

5. Number of Fuzzy Rules: There is no general procedure in deciding the optimal number of fuzzy control rules. A number of factors are involved in the decision, for example, performance of the controller, efficiency of computation, human operator behavior, and the choice of linguistic variables.

6. Consistency of Fuzzy Control Rules: If the derivation of fuzzy control rules is based on the human operator experience, the rules may be subject to different performance criteria. In practice, it is important to check the consistency of fuzzy control rules in order to minimize the possibility of contradiction.

During the past several years, fuzzy logic has found numerous applications in field ranging from earthquake engineering to finance[8]. Fuzzy control has particularly emerged as one of the most active and fruitful areas for research in the application of fuzzy set theory. In many applications, the FLC-based systems have proved to be superior in performance to conventional systems. The first successful industrial application of the FLC was the cement kiln control system developed by the Danish cement plant manufacturer F. L. Smith in 1979. An ingenious application is Sugeno's fuzzy car, which has the capacity of learning from examples[9,10]. Other applications of FLC include the heat exchange process control[11], waste water process control[12], activated sludge process control[13,14], traffic junction control[15], cement kiln[16,17], aircraft flight control[5], tuning process control[18], robot control[19,20,21], model-car parking and tuning[9,10], automobile speed control[22], power system and nuclear reactor control[23], fuzzy memory devices[24], automatic container crane operation systems[25], fuzzy computers[26], structural control[3] and structronic systems[4].

In the following sections we illustrate yet another application of FLC, by using this theory as a tool to control elastic collision in presence of frictional impact. The renewed interest in this area is partly due to the discovery of serious shortcomings in the most widely used approach to solve rigid body collision problem in classical mechanics and partly due to practical need for a series of design tasks that if undertaken will suppress the unwanted vibrations resulting after impact in mechanical structures. We first formulate the impact problem, then we design a fuzzy logic controller to suppress or reduce the nonlinear vibrations of the system after collision. We will show that the fuzzy logic controller performs satisfactory for a wide range of the restitution coefficient.

3. The system model

The diagram of the system[27] is shown in Figure 7. The system consists of a slender flexible beam AB cantilevered onto a rigid massless base, with negligible dimensions. The base is attached to a rigid link OO_1 (of length L_o) which rotates with a constant angular velocity Ω in the horizontal plane. The flexible beam has length L, a constant flexural rigidity EI and a uniformly distributed mass per unit length $\rho = m/L$, where m is the total mass of the beam. The base can perform small rotational

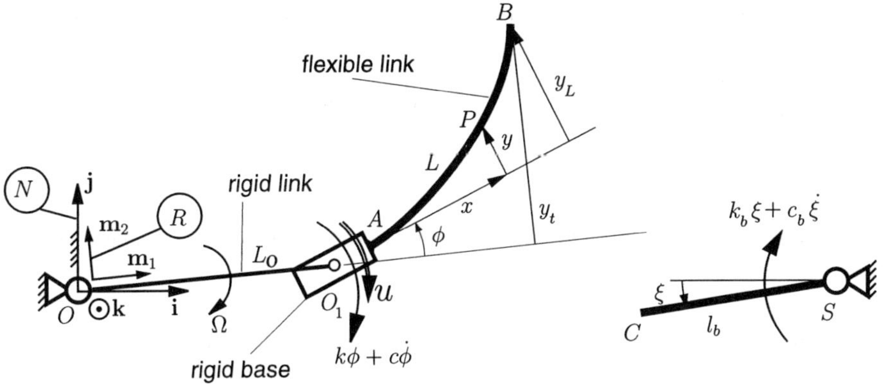

Fig. 7. Impacting system model

deflections $\phi(t)$ about the z axis passing through the point O_1. A spring (of constant k) and a damper (of constant c) are connected to the base and the rigid link in order to avoid large rotation of ϕ. The torque created by k and c is $k\phi + c\dot{\phi}$ (Figure 7).

A second rigid link, of length l_b and mass m_b can rotate around the fixed point S. The rigid link is connected to the ground through a spring (of constant k_b) and a damper (of constant c_b), avoiding large values of the angle ξ. The spring and the damper are designed such that, after each impact between the two links, the angle ξ goes rapidly to zero, before a new impact reoccurs.

Two reference frames are considered: a "fixed" reference frame (N), of unit vectors $\mathbf{i}$, $\mathbf{j}$ and $\mathbf{k}$, whose origin is at O, and a rotating reference frame (R), of unit vectors $\mathbf{m}_1$, $\mathbf{m}_2$ and $\mathbf{m}_3$, with the origin at O and attached to the rigid link OO_1. The unit vectors are related by the transformation

$$\begin{bmatrix} \mathbf{m}_1 \\ \mathbf{m}_2 \\ \mathbf{m}_3 \end{bmatrix} = \begin{bmatrix} \cos\Omega t & -\sin\Omega t & 0 \\ \sin\Omega t & \cos\Omega t & 0 \\ 0 & 0 & 1 \end{bmatrix} \begin{bmatrix} \mathbf{i} \\ \mathbf{j} \\ \mathbf{k} \end{bmatrix}. \tag{3}$$

Let x be the position of any point P on the elastic beam with respect to the end A of the base, and y be the elastic deflection. The position vector of the point P is

$$\mathbf{r}_P = (L_o + x\cos\phi - y\sin\phi)\mathbf{m}_1 + (y\cos\phi + x\sin\phi)\mathbf{m}_2. \tag{4}$$

The elastic deflection y of the beam is computed as

$$y(x,t) = \sum_{i=1}^{n} \Psi_i(x) q_i(t), \tag{5}$$

where $q_i(t)$ are the generalized elastic coordinates, and $n \in \mathcal{N}$ is the total number of vibration modes($\mathcal{N}$ is the set of natural numbers). The functions $\Psi_i(x)$ are chosen as the mode shapes of a cantilever beam and are defined by the expression

$$\Psi_i(x) = \cosh(z) - \cos(z) - \frac{\cosh(\lambda_i) + \cos(\lambda_i)}{\sinh(\lambda_i) + \sin(\lambda_i)}(\sinh(z) - \sin(z)), \tag{6}$$

where

$$z = \frac{x\lambda_i}{L}, \tag{7}$$

and λ_i ($i = 1,...,n$) are the first n^{th} consecutive roots of the transcendental equation

$$\cos(\lambda)\cosh(\lambda) = -1. \tag{8}$$

The additional degree of freedom of the flexible link is due to the rotation $\phi(t)$ of the base.
The velocity of the point P, in the fixed reference frame (N), is computed with the expression

$$\mathbf{v}_P = \frac{{}^R d\,\mathbf{r}_P}{dt} + \boldsymbol{\Omega} \times \mathbf{r}_P, \tag{9}$$

where the first term of the right hand side represents the derivative with respect to time in the moving reference frame (R), and $\boldsymbol{\Omega} = -\Omega\mathbf{k}$.
The total kinetic energy of the system is

$$K = \frac{\rho}{2}\int_0^L \mathbf{v}_P \cdot \mathbf{v}_P dx + \frac{m_b l_b^2 \dot{\xi}^2}{6}. \tag{10}$$

The total potential energy of the system is computed as

$$U = \frac{EI}{2}\int_0^L \left(\frac{\partial^2 y}{\partial x^2}\right)^2 dx + \frac{1}{2}k\phi^2 + \frac{1}{2}k_b\xi^2. \tag{11}$$

Using Lagrange's method, the nonlinear equations of motion are

$$\mathbf{M}\ddot{\mathbf{x}} + \mathbf{f}(\mathbf{x}) = \mathbf{d}u. \tag{12}$$

For the simulations presented here a single mode approximation was used. Simulations involving three, five and seven modes were performed and no perceptible difference was found with respect to the dynamic behavior. Thus one mode is considered adequate for accurately describing the elastic motion for the simulation reported in this paper[28,29]. With one mode approximation ($n = 1$) we have the following notations:

- $\mathbf{x}$ is the generalized coordinates vector, defined by $\mathbf{x} = [\phi,\ q_1,\ \xi]^T$;
- $\mathbf{M}$ is the mass matrix defined as

$$\mathbf{M} = \begin{bmatrix} mL^2/3 + q_1^2 G_1 & F_1 & 0 \\ F_1 & G_1 & 0 \\ 0 & 0 & m_b l_b^2/3 \end{bmatrix}; \tag{13}$$

The functions F_1 and G_1 are defined as

$$F_1 = \int_0^L x\rho\Psi_1(x)\,dx, \qquad G_1 = \int_0^L \rho\Psi_1^2(x)\,dx. \tag{14}$$

- $\mathbf{f}(\mathbf{x})$ is a nonlinear vector

$$\mathbf{f}(\mathbf{x}) = \begin{bmatrix} k\phi + 2(q_1\dot{\phi} - \Omega q_1)\dot{q}_1 G_1 + L_o\Omega^2(\sin\phi + q_1\cos\phi V_1) \\ -\Omega^2 q_1 G_1 + (2\Omega + \dot{\phi})q_1\dot{\phi}G_1 + EIq_1H_1 + L_o\Omega^2\sin\phi V_1 \\ k_b\xi + c_b\dot{\xi} \end{bmatrix}; \tag{15}$$

where the functions V_1 and H_1 are

$$V_1 = \int_0^L \rho\Psi_1(x)\,dx, \qquad H_1 = \int_0^L \left(\frac{\partial\,\Psi_1(x)}{\partial\,x}\right)^2 dx. \tag{16}$$

- $\mathbf{d} = [1,\ 0,\ 0]^T$ is the input vector;
- u is the control torque applied to the moving base, as shown in Figure 7.

The equations of impulsive motion[30] were determined using the basic assumption that the configuration of the bodies was held constant in the analysis of the collision process, with no significant change in mass and moments of inertia i.e., $\mathbf{x}(t) \approx$ constant. Furthermore, it was customarily assumed that each body exerts an impulsive force on the other at the common point of contact $B = C$. Let $\mathbf{F}_c$ be the impact force, which in this case has only a vertical component

$$\mathbf{F}_c = [0,\ F_c,\ 0]\,. \tag{17}$$

The friction during impact was neglected.

Let $\mathbf{v}$ be the vector of generalized speeds for the flexible link, defined as

$$\mathbf{v} = \left[\dot{\phi},\ \dot{q}_1,\ \dot{\xi}\right]^T = [v_j]_{j=1,2,3}\,. \tag{18}$$

An integrated form of the differential equations can be written as

$$\frac{d}{d\,t}\frac{\partial K}{\partial v_j} = Q_j, \tag{19}$$

where Q_j are the generalized impulsive forces during impact.

Equation (19) establishes a relationship between the time derivative of the generalized vector $\mathbf{v}$ and the contact force F_c, which leads to the matrix form

$$\mathbf{M}\mathbf{v} = \mathbf{D}(\mathbf{x})F_c, \tag{20}$$

where $\mathbf{D}(\mathbf{x})$ is a vector that depends on pre-impact positions. Taking into account that at the impact moment ξ can be expressed in terms of ϕ and q_1, only the first two components of the generalized vector $\mathbf{v}$ are independent. The vector $\mathbf{D}(\mathbf{x})$ is defined as

$$\mathbf{D}(\mathbf{x}) = \begin{bmatrix} L\cos(\phi+\theta) - q_1\Phi(L)\sin(\phi+\theta) \\ \Phi(L)\cos(\phi+\theta) \end{bmatrix}. \tag{21}$$

Let $\mathbf{v}_B^-$ and $\mathbf{v}_B^+$ be the velocities of the elastic beam tip before and after impact, respectively. Let $\mathbf{v}_C^-$ and $\mathbf{v}_C^+$ be the velocities of the impacted rigid end link before and after impact, respectively. With these notations, one can write

$$\mathbf{v}_C^+ - \mathbf{v}_B^+ = e(\mathbf{v}_B^- - \mathbf{v}_C^-), \tag{22}$$

where e is the coefficient of restitution. Solving the system of equations (20) and (22) the unknown velocities after impact ($\dot\phi^+$, $\dot q_1^+$, $\dot\xi^+$) are determined.

4. Fuzzy logic control

In this section a fuzzy logic controller is designed such that the elastic vibration of the beam are eliminated. The total deflection y_t of the beam tip can be computed with the expression (Figure 7)

$$y_t = L\sin\phi + y_L\cos\phi \tag{23}$$

where y_L is the elastic deflection of the beam tip, computed as

$$y_L(t) = y(L,t) = \Psi_1(L)\,q_1(t). \tag{24}$$

The inputs of the fuzzy logic controller are the error $E = -\phi$ and its time derivative $DE = -\dot\phi$. The output of the controller is the torque u applied to the base of the elastic link.

The Fuzzification Interface

The fuzzy sets associated with the error E, DE and u are shown in Figure 8. The error E fuzzy set is a combination of trapezoids and triangles. It has 7 grades, that correspond to *negative huge* (NH), *negative big* (NB), *negative small* (NS), *zero* (ZE), *positive small* (PS), *positive big* (PB) and *positive huge* (PH). The attention must be focused on the zero grade, since the steady state behavior is very sensitive to its choice. One can see that the support of the error E is between -1 and 1[rad]. However, the angle ϕ is not restricted to belong to this support, by extending the fuzzy set to $\pm\infty$. This means that, for instance, any error E larger to 1[rad] will belong to the positive huge (PH) set.

For the error time derivative DE, the support is between -2 and 2[rad/s], but DE can take any value, by extension. The fuzzy set associated is also a combination of trapezoids and triangles. It has only 3 grades, which correspond to *negative* (N), *zero* (Z) and *positive* (P). One can also note that the zero grade is narrow compared to the other grades.

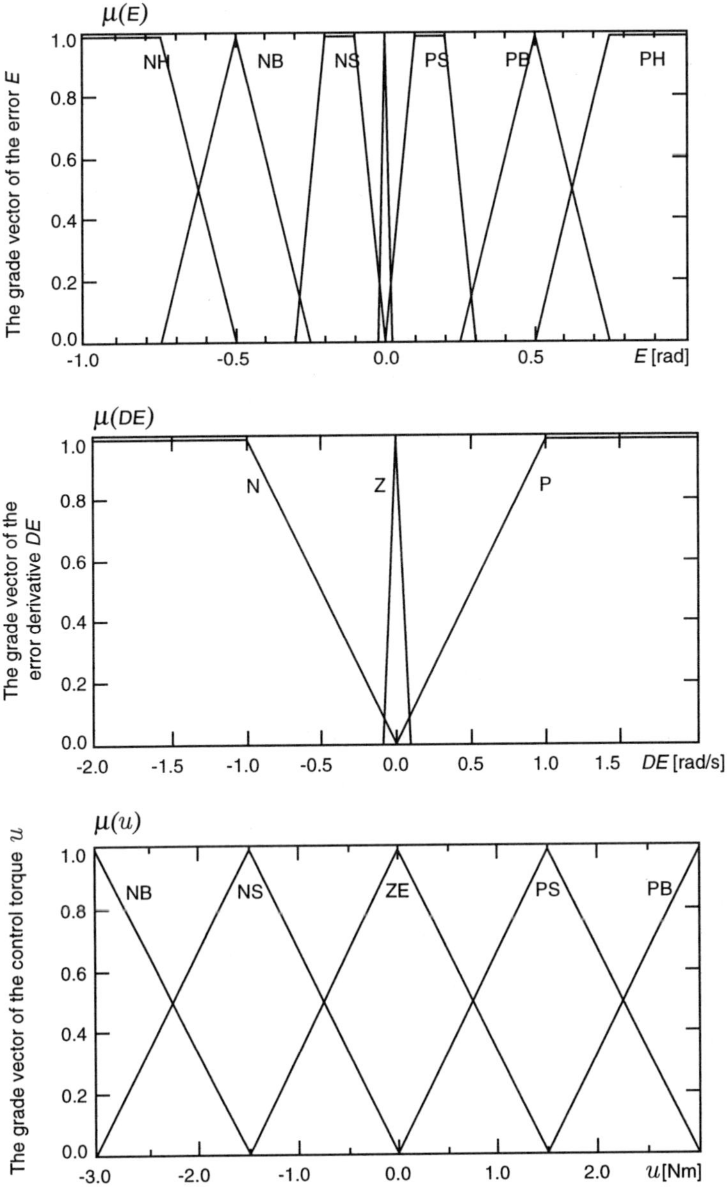

Fig. 8. Fuzzy sets of the impacting system

The fuzzy set of the controller output u consists of equidistant triangle grades. The control force has 5 components, that correspond negative big (NB), negative small (NS), zero (ZE), positive small (PS), positive big (PB) values of the controller output.

The Decision-Making Logic

The consequent table is a matrix of 7 columns (the number of the error E fuzzy-seta) and 3 rows (the number error time derivative DE fuzzy-sets). The consequent table is defined in Table 2.

E	NH	NB	NS	ZE	PS	PB	PH
$DE = N$	NB	NS	NS	ZE	ZE	PS	PB
$DE = Z$	NS	NS	ZE	ZE	ZE	PS	PS
$DE = P$	PB	NS	ZE	ZE	PS	PS	PB

Table 2. The consequent table for fuzzy logic controller design

The Defuzzification Interface

The center of area method is used to generate the numerical value of the control torque u from its fuzzy set.

5. Simulations and results

5.1. Numerical Data for Simulations

The following numerical values were used for simulations:

- for the elastic beam (aluminum) and its massless base: distance between the massless base and rotation center L_o=0.02[m], length of beam L=0.5[m], radius of beam 0.005[m], mass/unit length ρ=0.231[kg/m], angular velocity of rotation Ω=5[rad/s], spring constant k=2[Nm/rad] and damping constant c=2[Nms/rad]. One can also obtain V_1=0.0834, F_1=0.0303, G_1=0.1066 and H_1=98.8989.
- for the impacted rigid link and its support: beam mass m_b=0.3[kg], spring constant k_b= 16[Nm/rad] and damping constant c_b=5.6[Nms/rad].

5.2. Results Analysis

A hybrid computer program (written in C) was developed to simulate the impact of the elastic beam with the rigid link. The MATLAB fuzzy toolbox[31] was used to generate the control torque $u(t)$. The integration of the nonlinear equations of motion was performed using FORTRAN IMSL library[32]. The equations of motion were integrated every 10[μs]. The control torque was generated every 1[ms].

The initial conditions were $\phi = 0.05$[rad], q_1=0[m], ξ=0[rad] and zero velocities at t=0[s]. First, we consider a kinematic coefficient of restitution e=0.5 for the impact of the two bodies.

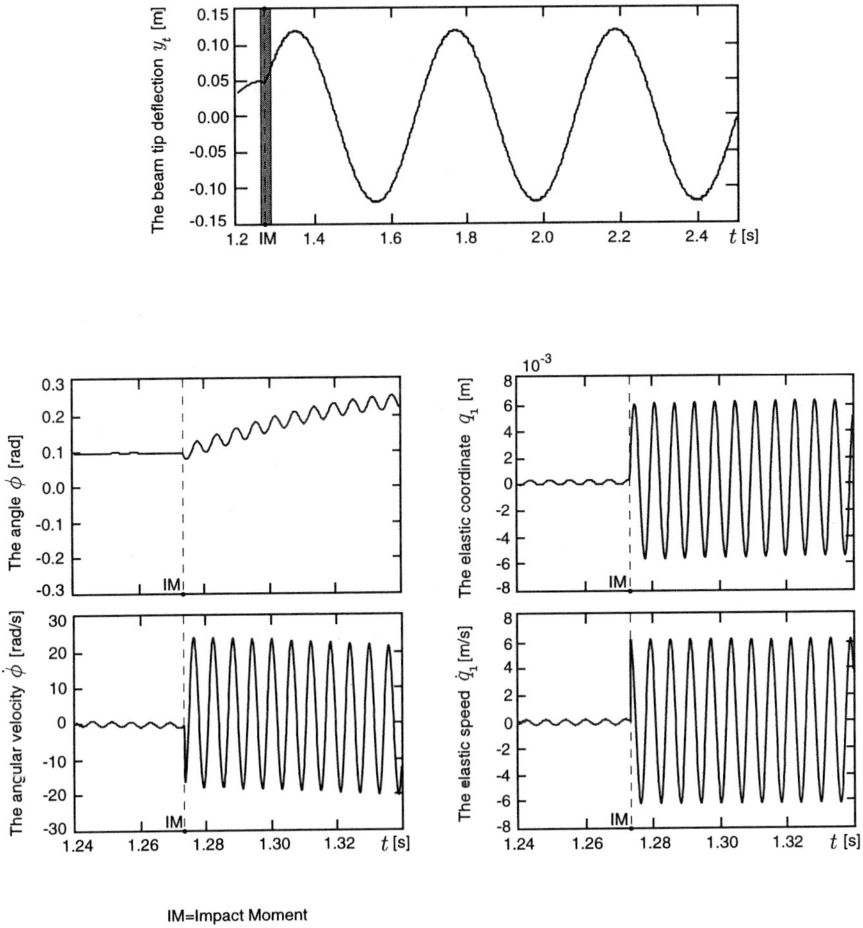

Fig. 9. Uncontrolled case for e=0.5

The uncontrolled system is shown in Figure 9. The top picture represents the total deflection of the beam tip y_t. The fist impact occurs at the impact moment (IM) t=1.273[s]. The next impact reoccurs at t=2.5[s] (not shown in this figure). The bottom four pictures represent a zoom on the interval 1.24 − 1.34[s], and the behavior of the angle ϕ, angular velocity $\dot{\phi}$, elastic coordinate q_1 and its time derivative $\dot{q}_1$ during the impact is shown. One can see that $\dot{\phi}$ has a jump from 0 to -18[rad/s], while $\dot{q}_1$ has a jump from 0 to 6[m/s]. Because the system is uncontrolled, the elastic coordinate q_1, as well the angle ϕ are oscillatory, making the total deflection y_t be oscillatory.

Figure 10 shows the behavior of the controlled system, for the same initial conditions at t=0[s]. The top picture represents the total deflection y_t of the elastic beam, and bellow the applied control torque is shown. The angle ϕ, the angular velocity $\dot{\phi}$, the elastic coordinate q_1 and its time derivative $\dot{q}_1$ are shown at the bottom of Figure 10. One can see that the elastic coordinate q_1 decreases rapidly to small values, as well as the angular velocity $\dot{\phi}$ and $\dot{q}_1$. Since q_1 becomes very small in a short time after impact, the angle ϕ has a similar behavior with the total deflection of the beam tip y_t. Due to the control torque applied, at $t = 1.8$[s] y_t is practically zero. The control torque u has sharp transitions in the impact moment and after that. It was found that these sharp changes are due to the narrow zero fuzzy sets of both the error E and its time derivative DE. When the zero sets were enlarged, the control torque became smoother, but the performances of the system were reduced, i.e., a longer stabilization time was required.

Several simulations were performed for different values of the restitution coefficient, from 0.1 to 0.9, simulating the impact of an elastic beam with rigid links of different materials. Figure 11 depicts the total deflection of the beam tip y_t obtained with three different values of the restitution coefficient. When friction is absent at the colliding end, different definitions of e yield exactly the same results[30].

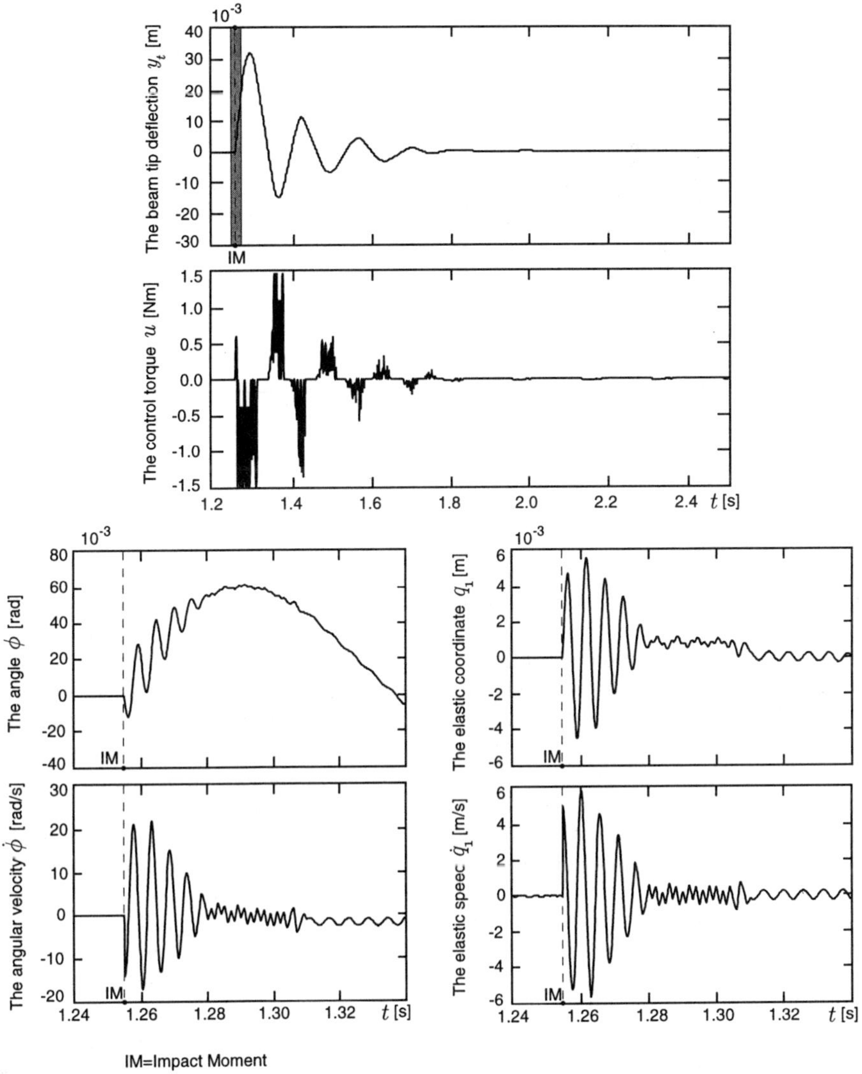

Fig. 10. Controlled case for e=0.5

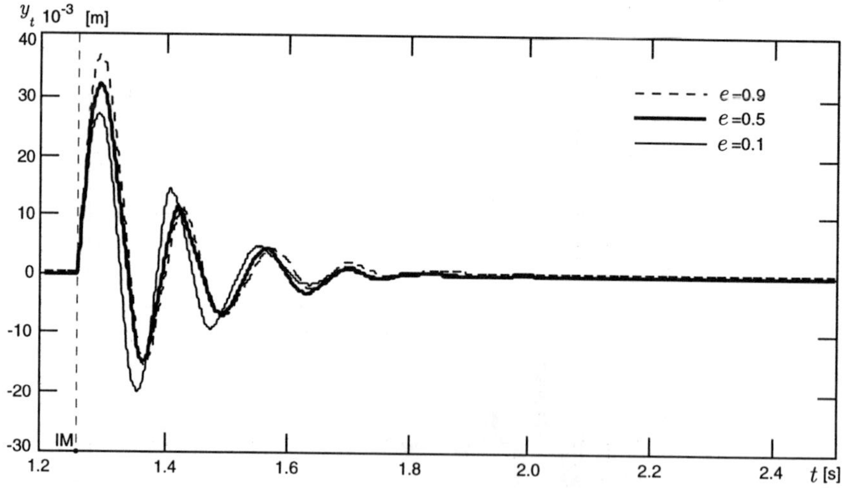

Fig. 11. The controlled beam tip deflection y_t for different values of the coefficient of restitution

6. Conclusions

In this chapter an integrated fuzzy logic control design technique is developed. The detailed design schemes, algorithms and procedures are presented. The method is used to suppress unwanted vibrations of an impacting rotating beam with a rigid link. The controller was designed using fuzzy logic strategy. The design of such a controller do not require a prior knowledge of the equations of motion. Several simulations were performed showing the behavior of the uncontrolled and controlled system. For the uncontrolled case, after impact, oscillations appear that do not vanish before the occurrence of a new impact. Large values of the total beam tip deflection were noticed. For the controlled case, the elastic vibrations practically vanish after a short period of time after impact. In less than 0.5[s] after impact, the total beam tip deflection, as well as the rigid angle ϕ of the base, vanish and remain zero until a new impact occurs. The behavior of the controller was tested for different values of restitution coefficient. It was found that for all cases the controlled system behaves in a similar way, the deflection of the elastic beam tip vanishing in a short period of time after each impact.

Due to its flexibility, the fuzzy logic controller can be successfully applied to a more general form of this problem, e.g. to control the impact between two or multiple bodies.

7. References

1. M. Sugeno, Industrial Applications of Fuzzy Logic Control, Amsterdam, North Holland, 1985.

2. H.-J. Zimmermann, Fuzzy Set Theory and Its Applications, Boston, Kluwer Academic Publishers, 1991.

3. B. Ayyub, A. Guran, and, A. Haldar, Uncertainty Modeling in Vibration, Control and Fuzzy Analysis of Structural Systems, World Scientific, Singapore, 1997.

4. A. Guran, H. S. Tzou, and G. Anderson, Structronics: Active Structures, Devices and Systems, World Scientific, New York, 1996.

5. P.M. Larkin, A fuzzy logic controller for aircraft flight control. In M. Sugano, editor, *Industrial Applications of Fuzzy Control*, Amsterdam, North Holland, 1985.

6. C. F. Lin, Advanced Control Systems Design, Prentice Hall, Englewood Cliffs, New Jersey, 1994.

7. C.C. Lee, Fuzzy logic in control systems: fuzzy logic controller-part I and II, *IEEE Trans. Syst. Man Cyber.*, 20, 1990.

8. P. M. Larsen, Applications of fuzzy set theory, *IEEE Trans. Sys. Man. Mach. Cybern.*, 15:175-189, 1985.

9. M. Sugeno and M. Murakami, An experimental study on fuzzy parking control using a model car, In M. Sugeno, editor, *Industrial Applications of Fuzzy Control*, Amsterdam: North Holland, 1985.

10. M. Sugeno and M. Murakami, Fuzzy control of a model car, *Fuzzy Set System*, 16:103-113, 1985.

11. J. J. Ostergaad, A fuzzy logic control of a heat exchange process. In M. M. Gupta, G. N. Saridis, and B. R. Gains, editors, *Fuzzy Automation and Decision Process*, Amsterdam: North Holland, 1977.

12. W. J. M. Kickert and H. R. Van Nauta Lemke, Application of a fuzzy controller in warm water plant, *Automatica*, 12:301-308, 1976.

13. O. Itoh, K. Gotoh, T. Nakayama, and S. Takamizawa, Application of fuzzy control to activated sludge process. In *Proc. 2nd IFSA Congress*, pages 282-285, 1987.

14. R. M. Tong, M. B. Beck, and A. Latten, Fuzzy control of the activated sludge waste water treatment process, *Automatica*, 16:695-701, 1990.

15. C. P. Pappis and E. H. Mandani, Fuzzy logic controller for traffic junction, *IEEE Trans. Sys. Man Cyber.*, 10:707-717, 1977.

16. P. M. Larsen, Industrial applications of fuzzy logic control, *Int. J. Man. Mach. Studies*, 12:3-10,1980.

17. I. G. Umbers and P. J. King, An analysis of human decision making in cement kiln control and implication for automation, *Int. J. Man. Mach. Studies*, 12:11-23, 1980.

18. Y. Sakai, A fuzzy controller in tuning process automation. In M. Sugeno, editor, *Industrial Application of Fuzzy Control*, Amsterdam: North Holland, 1985.

19. K. Ciliz, J. Fei, K. Usluel, and C. Isik, Practical aspects of the knowledge based control of a mobile robot motion, *Proceedings of 30th Midwest Symposium on Circuit and Systems*, Syracuse, New York, 1987.

20. C. Isik, Identification and fuzzy rule-based control of a mobile robot motion, *Proc. IEEE Int. Symp. Intelligent Control*, Philadelphia, Pennsylvania, 1987.

21. R. Tanscheit and E. M. Scharf, Experiments with the use of a rule-based self-organizing controller for robotics applications, *Fuzzy Set Systems*, 26:195-224, 1988.

22. Y. Kasai and Y. Morimoto, Electronically controlled continuously variable transmission, *Proc. Int. Congress on Transportation Electronics*, 1988.

23. J. A. Bernard, Use of rule-base system for process control, *IEEE Control Systems Magazine*, 8(5):3-13, 1988.

24. M. Togai and H. Watanabe, Expert system on a chip: an engine for real-time approximate reasoning, *IEEE Expert System Magazine*, 1:55-62, 1986.

25. Y. Yasunobu and T. Hasegawa, Predictive fuzzy control and its application for automatic container crane operation system, In *Proc. 2nd IFSA Congress*, Tokyo, Japan, 1987.

26. T. Yamakawa, A simple fuzzy computer hardware system employing min and max operations-a challenge to 6th generation computer, *Proc. 2nd IFSA Congress*, Tokyo, Japan, 1987.

27. D. Boghiu and D.B. Marghitu, The control of an impacting flexible link using fuzzy logic strategy, accepted for publication in the *Journal of Vibration and Control*, 1996.

28. A. Yigit, R.A. Scott and A.G. Ulsoy, Flexural motion of radially rotating beam attached to a rigid body, *Journal of Sound and Vibration*, 121(2):201-210, 1988.

29. D.B. Marghitu and Y. Hurmuzlu, Nonlinear Dynamics of an Elastic Rod with Frictional Impact, *International Journal of Nonlinear Dynamics*, 10:187-201, 1996.

30. D.B. Marghitu and Y. Hurmuzlu, Three-dimensional rigid-body collisions with multiple contact points, *Journal of Applied Mechanics*, 62:725-732, 1995.

31. M. Beale and H. Demuth, Fuzzy Systems Toolbox - For Use With MATLAB. PWS Publishing Company, Boston, Massachusetts, 1994.

32. Visual Numerics Inc., IMSL Math/Library. FORTRAN subroutines for mathematical applications, 1994.

Author Index

N

O

P

Q

R

X

Y

Z

Subject Index

N

O

P

R

S

T

U

V

W

Z

About the Editors

Bilal M. Ayyub is a professor of civil engineering at the University of Maryland at College Park. He completed his BSCE degree in 1980 from the University of Kuwait, and completed both the MSCE (1981) and PhD (1983) in civil engineering at the Georgia Institute of Technology. Dr. Ayyub has an extensive background in risk analysis, simulation, and marine and navigation structures. He has completed various research projects that were funded by the NSF, USCG, USACE, USN, DOD, and ASME. He is the co-author of two textbooks, the co-editor of two books, and the co-author of about 250 publications. Dr. Ayyub is the double recipient of the ASNE Jimmie Hamilton Award for the best paper in the *Naval Engineers Journal* in 1985 and 1992. Also, he was co-recipient of the ASCE Outstanding Research Oriented Paper in the *Journal of Water Resources Planning and Management* for 1987, the ASCE Edmund Friedman Award in 1989, and the K.S. Fu Award from NAFIPS in 1995.

Ardéshir Guran is a faculty member in the Electrical Engineering - Systems Department at the University of Southern california, Los Angeles. He received his B.Sc. in Structural Engineering, and M. Eng. in Civil Engineering from McGill University, Canada; and M.S. in Mathematics and Ph.D. in Mechanical Engineering from the University of Toronto, Canada. He taught at Rensselaer Polytechnic Institute, the Catholic University of America, and was an Invited Professor at Technical University of Hamburg, Germany, Technical University of Vienna, Austria, University of Bordeaux, France, Delft University of Technology, the Netherlands, and Virginia Polytechnic Institute, USA. Dr. Guran is the editor-in-chief of World Scientific Series on Stability, Vibration and Control of Systems and served as Associate Editor of IASTED Journal of Modeling and Simulation. He published over 200 articles in the areas of nonlinear dynamics, structural stability, structronics, acoustics, wave propagation, gyroscopic systems and structural control. He is the co-founder, Vice-President and Director of R&D at American Structronics and Avionics Corporation in Encino, California.

Achintya Haldar is a professor of Civil Engineering and Engineering Mechanics at the University of Arizona, Tucson, Arizona. He received his M.S. and Ph.D. in Civil Engineering from the University of Illinois, Urbana, Illinois. He previously taught at Illinois Institute of Technology and Georgia Institute of Technology. An ASCE fellow, Dr. Haldar is now the chairman of the Technical Administrative Committee on Structural Safety and Reliability. He was an associate editor of the ASCE's Journal of Structural Engineering. He authored or co-authored numerous technical articles on stochastic computational analysis of civil engineering structures, fatigue/fracture reliability/maintainability of infrastructure, nonlinear random vibration and earthquake engineering. He received many awards for his research, including the first Presidential Young Investigator Award and ASCE's Huber Civil Engineering Research prize. In addition, Georgia Tech gave him its highest teaching award.